THE ECOLOGY OF PLACE

THE ECOLOGY OF PLACE

CONTRIBUTIONS OF PLACE-BASED RESEARCH TO ECOLOGICAL UNDERSTANDING

Edited by Ian Billick and Mary V. Price

THE UNIVERSITY OF CHICAGO PRESS   Chicago and London

IAN BILLICK is executive director of the Rocky Mountain Biological Laboratory in Crested Butte, Colorado. MARY V. PRICE is professor emerita of biology at the University of California, Riverside.

The University of Chicago Press, Chicago 60637
The University of Chicago Press, Ltd., London
© 2010 by The University of Chicago
All rights reserved. Published 2010
Printed in the United States of America

18 17 16 15 14 13 12 11 10   1 2 3 4 5

ISBN-13: 978-0-226-05042-3 (cloth)
ISBN-13: 978-0-226-05043-0 (paper)
ISBN-10: 0-226-05042-4 (cloth)
ISBN-10: 0-226-05043-2 (paper)

Library of Congress Cataloging-in-Publication Data
The ecology of place : contributions of place-based research to ecological understanding / edited by Ian Billick and Mary V. Price.
     p. cm.
    "Papers of the scientific symposium to celebrate the 75th birthday of the Rocky Mountain Biological Laboratory"—CIP data.
    Includes bibliographical references and index.
    ISBN-13: 978-0-226-05042-3 (cloth : alk. paper)
    ISBN-13: 978-0-226-05043-0 (pbk. : alk. paper)
    ISBN-10: 0-226-05042-4 (cloth : alk. paper)
    ISBN-10: 0-226-05043-2 (pbk. : alk. paper)   1. Ecology—Longitudinal studies—Congresses. 2. Ecology—Research—Congresses.   I. Billick, Ian, 1966– II. Price, Mary V. (Mary Vaughan), 1949– III. Rocky Mountain Biological Laboratory.
    QH541.2.E252 2010
    577.072—dc22

                                                                        2010015030

To the places that draw us outside
and the people who seek to understand them

CONTENTS

Figure 0.1. View of the Rocky Mountain Biological Laboratory from the side of nearby 12,500-foot-high Gothic Mountain, in the Elk Mountains of west-central Colorado. Photograph by Nickolas Waser, 2003.

PREFACE AND ACKNOWLEDGMENTS

Ian Billick and Mary V. Price

This volume was conceived during the organization of a scientific symposium to celebrate the 75th birthday of the Rocky Mountain Biological Laboratory (RMBL), an independent field station that was founded in 1928. RMBL occupies Gothic, Colorado, an abandoned mining town named for the mountain whose cliffs loom nearby. Generations of students and research scientists have studied and worked in this place, and more than 1,000 publications have emerged from research done there. As we contemplated the scientific contributions of the symposium speakers, we were struck by the fact that their work was known well beyond the small cadre of people who study montane systems or their particular taxonomic group. We wondered: Why do empirical studies of particular organisms in particular places interest people who work with other systems?

Ecological systems present such a diversity of organisms, environments, and ecological relationships, and involve processes that are so stochastic and contingent, that ecologists continually debate whether widely applicable generalizations exist in ecology and, if so, how best to identify them. Empirical field studies contribute to the search for generalization in two obvious ways. They constitute the empirical core data in a "bottom-up" search for broad empirical patterns that guide the development of explanatory general theory. They also constitute the empirical core data in "top-down" tests of theory. Given this role of empirical studies, it would seem that the fastest route to progress would be to repeat measurements or tests over as many systems as possible. Paradoxically, many ecologists—including participants in the RMBL birthday symposium—have adopted a third approach: they study a single system in depth, often over many years.

We have dubbed this third approach "the ecology of place" by analogy to the "sense of place" ascribed to Aldo Leopold, a pioneer of American ecology and founder of the fields of conservation biology, restoration ecology, and wildlife management. Leopold's career led him to appreciate the complex interconnections of soil, water, and biota (including humans) that define what he called the "biotic mechanism" (ecological processes) of "the land" (the ecosystem). Leopold argued that not only the motivation for conserving the health of the land but also the knowledge about how to do so stems from the intuitive understanding that arises from deep personal involvement in the natural history of a particular place. Might such natural-history under-standing of single systems contribute to general scientific understanding in ecology as well as to conservation know-how? And if so, how?

In recent years there has been much discussion about the adequacy of current ecological knowledge, but rather little discussion about how that knowledge has been gained. There is no better time than now, we thought, to initiate discussion of how alternative approaches contribute to ecological understanding, and how the infrastructure of science can facilitate those approaches. Given Ian's position as director of a field station, Mary's belief in the value of in-depth case studies, and the fact that pattern-based and hypothetico-deductive approaches had already received attention in the lit-erature, we decided to focus primarily on the "ecology of place" approach. To that end, we invited accomplished field ecologists and ecology-watchers (most from outside the RMBL community) to reflect on the role of "place" in their research and in the science of ecology more broadly. We were over-whelmed by the enthusiastic response we got to our invitations; ecologists love to talk about what their systems have taught them!

We hope the ideas presented here will stimulate broad discussion about how we ecologists do, can, and should approach our science, and about the legitimate role that natural history—the aspect of our work that so many of us find the most satisfying—can play in it. We invite you to explore with us the apparent paradox that general understanding can emerge from a deep understanding of the specific.

This project has had a long gestation, in part because it explores unusual territory, and it could not have been carried out without the input and sup-port of many people. We thank the chapter contributors for their enthusi-asm, patience, and willingness to adapt their remarks to a book theme that itself evolved along with the chapters. They, Dan Blumstein, Diane Camp-bell, Philippe Cohen, Marco Encalada, Robert McIntosh, Curt Meine, Frank Price, Molly Smith, Rosie Trevalyn, and Dirk Van Vuren kindly reviewed individual chapters and helped us all to see connections among ideas that we had missed. Several anonymous reviewers read the entire book, making invaluable comments and suggestions for improving coherence, which is often a problem with edited volumes. Scott Armbruster, Ed Ford, Ed John-son, Becky Irwin, Jim Reichman, and Jennie Reithel initiated chapters that they were unable, for various reasons, to finish. We thank them for their efforts and regret the loss of their chapters. Christie Henry at the University of Chicago Press kept us going for all these years with her gentle encouragement. She and the University of Chicago Press editorial staff have been indispensable in finding errors and polishing prose and artwork. Finally, we thank Nick Waser for enthusiastically taking on the role of "shadow editor" and for his enduring support throughout.

1 *The Ecology of Place*

Mary V. Price and Ian Billick

The Challenge of Diversity, Complexity, and Contingency

Any attempt to understand the ecological and evolutionary sciences must take into account the remarkable biological diversity of planet earth. Some 1.7 million individual species have been described to date. Although this may seem like a large number, the true species diversity is much larger. From the rates at which undescribed species are discovered, biologists estimate that between 4 and 100 million species share our planet (Penniski 2003). This number, however, does not begin to describe the true extent of biological diversity. Each species is genetically and phenotypically heterogeneous, with numerous recognizably distinct subgroups—variously called subspecies, races, or varieties. Even within seemingly homogenous subspecific units, local populations and individuals differ from one another. Simply cataloging this diversity is difficult enough. Ecologists and evolutionary biologists have taken on the additional challenges of understanding the interactions between all these organisms and the environments they inhabit, and the roles those interactions play in determining distribution, abundance, and evolutionary change.

Ecological interactions are complex as well as diverse, and their outcomes are contingent upon context. The phenotypic characters that vary among species, subspecies, and individuals influence how organisms respond to the many abiotic features of their environments that can influence their well-being, such as light intensity, temperature, moisture, substrate texture or chemistry, and fluid properties of air or water. Organisms also are embedded in a web of interactions with other species in a local community. This makes ecological systems particularly complex because species interactions can be mediated through indirect as well as direct linkages in the community, and through feedbacks with the physical environment. Hence, even the pairwise interaction between two species is not constant, but instead is contingent on their abundances, which are dynamic, as well as on biological and physical environmental contexts that vary spatially and temporally on multiple scales (Holzapfel and Mahall 1999, Thompson 2001, Sanford et al. 2003).

Diversity and contingency present enormous challenges to understanding the ecological systems of the planet because there are simply too few ecologists to study them all. By browsing websites in the global list of scientific ecological societies provided by the British Ecological Society (2007), we estimate that there are some 25,000 members of ecological societies worldwide. Even if that figure underestimates the number of field-oriented research scientists by an order of magnitude (which we doubt), this army is far too small to tackle each of the planet's ecosystems, much less to understand their dynamics and how they are interconnected. To be sure, we may be able to mobilize many more of the planet's six billion people in the quest for ecological understanding, as is discussed in chapters 12 and 18. This strategy has merit, particularly for knowing how to manage local systems, but it will not completely solve the problem.

We need, in addition, some way of efficiently extracting "portable ecological knowledge" (Cooper 2003) from a small sample of systems. This latter problem is one that every ecologist and evolutionary biologist confronts: How can we develop a general understanding of ecological systems when we know that the results of a field study typically depend on a host of factors unique to the focal system as well as on the particular location and time of the study? Are there generalizations—broadly applicable statements—to be had? If so, what is their nature, and how best can we identify them given all that diversity and contingency? The reverse inferential direction is equally problematic: Can ecological or evolutionary generalizations be applied in useful ways to individual systems?

Stung by a perception that theirs is a slow-moving field populated by "incompetent scientists working on impossible problems" (to quote an anonymous reviewer; also see Peters 1991, Shrader-Frechette and McCoy 1993), ecologists have debated these questions, often vigorously, ever since ecology emerged as a "self-conscious" discipline in the early part of the twentieth century (McIntosh 1985, Cooper 2003, Kingsland 2005). Some, dodging the "competence" and "tractability" possibilities, have attributed the apparent lack of progress to weaknesses of method: there is too much natural history; there is too little natural history; we need more experiments; we should be looking for patterns; we should be studying mechanisms; we should be integrating; we should be using strong inference (sensu Platt 1964) and rigorous hypothesis-testing; we do too much hypothesis-testing and not enough model-fitting; there is too much theory; there is too little theory, or theory of the wrong sort (e.g., May 1981, Pielou 1981, Colwell 1984, Salt 1984, Kingsland 1985, Bartholomew 1986, Peters 1991, Allen and Hoekstra 1992, Schrader-Frechette and McCoy 1993, Pickett et al. 1994, Brown 1995, Weiner 1995, Hilborn and Mangel 1997, Resetarits and Bernardo 1998, Belovsky et al. 2004, Simberloff 2004). Meanwhile, historians and philosophers of science have started to take an interest in the history and conceptual structure of ecology and related fields. Biographies of influential ecologists (Tobey 1981, Meine 1988, Crowcroft 1991, Slobodkin and Slack 1999, Newton 2006, Goodstein 2007) and treatments of the historical development of major concepts, traditions, schools, controversies, and subfields (Jackson 1981, Mayr 1982, Kingsland 1985, 2005, McIntosh 1985 and 1987, Crowcroft 1991, Golley 1993) have appeared at an increasing rate. Philosophers of science—and philosophically inclined scientists—are turning their attention to the issues of contingency, historicity, uniqueness, and complexity that characterize ecology and evolutionary biology, and have begun to explore their implications for such things as the structure of scientific concepts and generalizations, the nature of scientific inquiry, the relative importance of explanation vs. prediction, or the relationships between theoretical and empirical traditions in these fields (e.g., Ayala and Dobzhansky 1974; Hull 1974; Medawar 1974; Mayr 1982; Schoener 1986; Gasper 1991; Pickett et al. 1994; Cooper 2003; Mitchell 2003, 2009). Such philosophical analysis suggests that the process of scientific inquiry in ecology and evolutionary biology and the way in which ecological knowledge is used to solve applied problems may differ in fundamental ways from the heretofore dominant paradigms that have been drawn from the physical sciences (Shrader-Frechette and McCoy 1993, Cooper 2003).

Meeting the Challenge

Lack of perfect consensus concerning how their science should be done has hardly deterred ecologists and evolutionary biologists from going about their business of searching for generalizations. What basic strategies do they use?

An obvious approach involves comparison. Comparison underpins the "macroecological" search for broad patterns (Brown 1995), such as the species-area relationship of island biogeography, the decrease in species diversity with latitude, or the scaling of physiological functions with body size. Comparison is the foundation for assessments of the overall frequency of occurrence, the strength of various ecological processes such as density-dependent population growth, or the relative contributions of competition, predation, disturbance, and climate to the species composition of communities. Comparison also is fundamental to detecting broad temporal trends, such as changes in biodiversity or climate, and to deriving estimates of biogeochemical fluxes on large spatial scales. Statements about frequencies of occurrence and estimates of variables or trends represent one sort of generalization. They arise primarily from replication of observations, experiments, or measurements across taxa, space, and time.

Another approach involves the development and testing of general theory—broadly applicable, testable propositions. Theory is diverse. It can involve the quantitative description of predictive empirical patterns uncovered by the comparative approach, or the elaboration of logical consequences of empirically or deductively generated premises. It can be analytical, deriving consequences of premises through well-defined mathematical algorithms, or computational, deriving consequences through brute-force computer simulation of mechanistic processes. It can involve an entirely verbal and qualitative statement of relationships among entities. Successful theory—that which provides valid prediction or explanation of empirical phenomena—is the stuff of generalization.

Not all ecologists adopt a broad macroecological type of comparative approach or a broad theoretical perspective. Many focus instead on forging a deep understanding of particular organisms and places. This focus on the particular is expected of "applied" ecologists, whose principal goal is to understand how to manage or restore a single system. However, it is also common among "basic" ecologists, whose goal is to generalize, as is indicated by how they publish. They choose general ecological and evolutionary journals as opposed to journals that specialize on particular taxonomic groups or environments, they describe essential features of their systems in terms understandable to colleagues unfamiliar with them, they set their studies in general conceptual frameworks, and they discuss how their data lead to conclusions about those frameworks. Although the information generated by focused studies can contribute to generalizations by providing data points in comparative studies or tests of general theory, the primary goal of the focused studies of "basic" ecologists is instead to evaluate how well theoretical conceptual frameworks account for the properties of individual systems. The generalization in this case involves a statement about the applicability of alternative conceptual frameworks to a particular system with particular properties.

We use the term "ecology of place" for this third research approach, because that approach pursues general understanding through the sort of detailed understanding of a particular place—the "sense of place"—that has come to be associated with ecologist and conservationist Aldo Leopold.[1] In our usage, "place" does not equal "point in space," nor does it mean "field site." Spatial ecology (e.g., Kareiva 1994, Tilman 1994) focuses on how the spatial location of a point affects its properties via ecological exchanges with neighboring points. "Field site" is the location for a study, often chosen for reasons other than its intrinsic ecological properties (see Krebs, chapter 13).

What distinguishes these spatial notions from our notion of "place" is that in the latter, the particular ecological features of the point in space are central. We reserve "place" to represent all of those idiosyncratic ecological features—including spatial location and time period—that define the ecological context of a field study; and "ecology of place" or "place-based research" for research that assigns the idiosyncrasies of place, time, and taxon a central and creative role in its design and interpretation, rather than as a problem to be circumvented through replication or statistical control.

Elements of the Ecology of Place

Place-based approaches in ecology and evolutionary biology share many elements with modern "case study" research designs used in the social sciences (e.g., Platt 1992, Yin 1994; see box 1.1), perhaps because these fields are confronted with similar challenges. Both ecology and the social sciences study complex entities whose current state is the product of individual properties, history, and setting.

As described by Yin (1994) and Platt (1992), case studies in the social sciences focus in depth on phenomena in their real-world context; rely on non-random selection of one or a few cases whose unique characteristics facilitate meeting research goals; use multiple types and sources of information about each case to answer the research question; evaluate theory by whether it can be successfully modified to fit rich patterns in data associated with one or a few cases rather than on a simple falsification criterion; use a logic of analytical rather than statistical generalization—that is, they predict how a phenomenon is manifested in particular cases, rather than summarize frequency of occurrence across cases; and they require multiple-investigator teams to carry out a multifaceted research design.

Place-based ecological studies are conducted almost exclusively with intact systems in a natural field context, although they may be supplemented with focused study of isolated components of the system in laboratory or field microcosms. They often are long-term, involving many years of sampling to collect information about system dynamics, or a sustained effort to evaluate theory in multiple cycles of theory development and evaluation. Shifts in the direction of research are common. Study systems are generally chosen to follow up on prior observations, or because they represent a particular class of systems or are particularly tractable for answering a question of interest. Typically, ecology of place takes advantage of diverse types of information—natural history anecdote, knowledge of a site's history, observation of the system and its environmental context, data on responses to experimental treatments or natural perturbation, and comparison with

BOX 1.1. Comparison of case-study and ecology of place approaches

Social sciences case study	Ecology of place
Focuses in-depth on phenomenon in a real-world context	Involves sustained field studies of intact systems, often conducted over many years
Selects cases to be studied non-randomly	Chooses study systems to be representative, to be tractable for questions asked, or to follow up prior observations
Uses diverse types and sources of information	Integrates information derived from natural history anecdote, historical knowledge, observation, and/or experiment
Evaluates theory by its fit to case-specific patterns in diverse types of data rather than by simple Popperian falsification	Evaluates general theory in a repeated cycle of theory modification, evaluation, and further theory development
Uses analytical rather than statistical generalization	Generalizes by predicting how a process or phenomenon is manifested in systems with particular properties, rather than by summarizing its frequency of occurrence
Deploys multiple-investigator teams	Often includes exploration of various research threads with the same system by a "loose consortium" of investigators

similar systems. These data are used to modify general theory so that it yields expectations about a particular system, and they are also used to evaluate those expectations. The unique product of place-based research is an evolving model that accounts for observed properties and behavior of the system, and that also allows predictions about the applicability of the general theory to new systems with similar or different properties. Place-based ecological research is carried out by loose, often interdisciplinary consortia of collaborators who bring diverse perspectives and skills and follow different threads of investigation. It also benefits from informal exchange among independent investigators who work at the same site.

Purpose and Overview of the Book

The primary purpose of this book is to initiate discussion of a research approach—what we call the ecology of place—that is implicitly used by many ecologists but has not been formally recognized. Several published volumes have discussed alternative approaches in ecology and evolution, such as the comparative method (Harvey and Pagel 1991), macroecology (Brown 1995, Gaston and Blackburn 2000), or experimental ecology (Resetarits and Bernardo 1998). Discussion of case study approaches in ecology has primarily considered how ecological knowledge from diverse sources can be deployed in a scientifically defensible way to solve specific environmental problems—that is, in what respect case-specific knowledge is portable (National Research Council 1986, Shrader-Frechette and McCoy 1993). To our knowledge, this book is the first to consider how place-based research contributes to basic scientific understanding. We believe a logical starting point is to provide examples of how practitioners of ecology of place approach their science, in the hope of uncovering features that contribute to its success.

Accordingly, we challenged contributors to this volume, many of whom have adopted an approach that incorporates elements of ecology of place at one time or another in their careers, to contemplate three primary issues. First, how does one go about doing place-based research? Second, how does ecology of place extract general insights from the idiosyncrasies of place, and how can these insights be applied to previously unstudied systems? Third, how does the infrastructure of science affect the ability of scientists to do place-based research and gain general understanding from it? The authors have responded enthusiastically with essays that are so rich in ideas that we as editors have been hard pressed to organize them into thematic groups. We settled on the following book structure.

In this introductory chapter we have outlined what the ecology of place entails; subsequent chapters in all parts elaborate on this idea with concrete examples and discuss the value and implications of the approach for the science of ecology. The two chapters in part 1, "The Imprint of Place on Ecology and Ecologists," take a historical perspective. They consider how the attributes of individual places molded the development of the discipline of ecology and the thinking of Aldo Leopold, who pioneered aspects of the ecology of place approach. The five chapters in part 2, "The Idiosyncrasy of Place: Challenges and Opportunities," consider how adopting a place-based perspective allows us to extract insights from the spatial and temporal heterogeneity that is one manifestation of the idiosyncrasy of ecological systems. Part 3, "Building an Understanding of Place," illustrates how local knowledge—the "sense of place"—is captured in a conceptual model of a local system. Its chapters describe the process by which such models develop, and they point out how the structures and philosophies of research teams contribute to progress in model development. Part 4, "The Interaction between Local and General Understanding," explores the interaction between knowledge of specific systems and more global knowledge. Part 5, "Building Capacity for Place-Based Research," considers how investment in infrastructure and education can facilitate such research. In a concluding chapter we summarize what major insights have emerged from the case studies presented in this book because they were place-based. We also suggest programmatic steps for taking greater advantage of the power of place-based approaches to advance ecological understanding.

Each of the book's parts is headed up by introductory remarks that develop themes that run through the book and highlight how individual chapters relate to those themes. We make no claim that we have identified, much less mined, all the productive themes in this body of essays. You will surely find many other nuggets as you read the various accounts of how successful ecologists and evolutionary biologists have made sense of the diversity and intricate complexity of this beautiful blue-green planet.

References

Allen, T. F. H., and T. W. Hoekstra. 1992. *Toward a Unified Ecology*. Columbia University Press, New York.

Ayala, F. J., and T. Dobzhansky, eds. 1974. *Studies in the Philosophy of Biology: Reduction and Related Problems*. University of California Press, Berkeley.

Bartholomew, G. A. 1986. The role of natural history in contemporary biology. *BioScience* 36:324–29.

Belovsky, G. E., D. B. Botkin, T. A. Crowl, K. W. Cummins, J. F. Franklin, M. L. Hunter Jr., A. Joern, D. B. Lindenmayer, J. A. MacMahon, C. R. Margules, and J. M. Scott. 2004. Ten suggestions to strengthen the science of ecology. *BioScience* 54:345–51.

Berryman, A. A. 2003. On principles, laws, and theory in population ecology. *Oikos* 103:695–701.

British Ecological Society. 2007. http://www.britishecologicalsociety.org/articles/beyondthebes/othersocieties/.

Brown, J. H. 1995. *Macroecology*. University of Chicago Press, Chicago.

Colwell, R. K. 1984. What's new? Community ecology discovers biology. Pages 387–96 *in* P. W. Price, C. N. Slobodchikoff, and W. S. Gaud, eds., *A New Ecology*. John Wiley & Sons, New York.

Cooper, G. J. 2003. *The Science of the Struggle for Existence: On the Foundations of Ecology*. Cambridge University Press, Cambridge.

Crowcroft, P. 1991. *Elton's Ecologists*. University of Chicago Press, Chicago.

Dayton, P. K., and E. Sala. 2001. Natural History: The sense of wonder, creativity and progress in ecology. *Scientia Marina* 65:199–206.

Gasper, P. 1991. The Philosophy of Biology: Introductory Essay. Pages 545–52 *in* R. Boyd, P. Gasper, and J. D. Trout, eds., *The Philosophy of Science*. The MIT Press, Cambridge, MA.

Gaston, K. J., and T. M. Blackburn. 2000. *Pattern and Process in Macroecology*. Blackwell Science, Oxford.

Golley, F. B. 1993. *A History of the Ecosystem Concept in Ecology: More than the Sum of the Parts*. Yale University Press, New Haven, CT.

Goodstein, J. R. 2007. *The Volterra Chronicles: The Life and Times of an Extraordinary Mathematician 1860–1940*. American Mathematical Society / London Mathematical Society, Providence, RI.

Harvey, P. H., and M. D. Pagel. 1991. *The Comparative Method in Evolutionary Biology*. Oxford University Press, Oxford.

Hilborn, R., and M. Mangel. 1997. *The Ecological Detective: Confronting Models with Data*. Princeton University Press, Princeton, NJ.

Holzapfel, C., and B. E. Mahall. 1999. Bidirectional facilitation and interference between shrubs and annuals in the Mojave Desert. *Ecology* 80:1747–61.

Hull, D. L. 1974. *Philosophy of Biology*. Prentice-Hall, Inc., Englewood Cliffs, NJ.

Jackson, J. B. C. 1981. Interspecific competition and species distributions: The ghosts of theories and data past. *American Zoologist* 21:889–901.

Kingsland, S. E. 1985. *Modeling Nature: Episodes in the History of Population Ecology*. University of Chicago Press, Chicago.

———. 2005. *The Evolution of American Ecology, 1890–2000*. The Johns Hopkins University Press, Baltimore.

Kareiva, P. 1994. Space: The final frontier for ecological theory. *Ecology* 75:1.

Lawton, J. H. 1999. Are there general laws in ecology? *Oikos* 84:177–92.

May, R. M. 1981. The role of theory in ecology. *American Zoologist* 21:903–10.

Mayr, E. 1982. *The Growth of Biological Thought: Diversity, Evolution, and Inheritance*. Harvard University Press, Cambridge, MA.

McIntosh, R. P. 1985. *The Background of Ecology: Concept and Theory*. Cambridge University Press, Cambridge.

———. 1987. Pluralism in ecology. *Annual Review of Ecology and Systematics* 18:321–41.

Medawar, P. 1974. A geometric model of reduction and emergence. Pages 57–63 *in* F. J. Ayala, and T. Dobzhansky, eds., *Studies in the Philosophy of Biology: Reduction and Related Problems*. University of California Press, Berkeley.

Meine, C. 1988. *Aldo Leopold: His Life and Work*. The University of Wisconsin Press, Madison.

Mitchell, S. D. 2003. *Biological Complexity and Integrative Pluralism*. Cambridge University Press, Cambridge.

———. 2009. *Unsimple Truths: Science, Complexity, and Policy*. University of Chicago Press, Chicago.

National Research Council, Committee on the Applications of Ecological Theory to Environmental Problems (G. H. Orians, chair). 1986. *Ecological Knowledge and Environmental Problem Solving: Concepts and Case Studies*. National Academies Press, Washington, DC.

Newton, J. L. 2006. *Aldo Leopold's Odyssey*. Island Press, Washington, DC.

Penniski, E. 2003. Modernizing the tree of life. *Science* 300:1692–97.

Peters, R. H. 1991. *A Critique for Ecology*. Cambridge University Press, Cambridge.

Pickett, S. T. A., J. Kolasa, and C. G. Jones. 1994. *Ecological Understanding: The Nature of Theory and the Theory of Nature*. Academic Press, San Diego.

Pielou, E. C. 1981. The usefulness of ecological models: A stock-taking. *The Quarterly Review of Biology* 56:17–31.

Platt, J. R. 1964. Strong inference. *Science* 146:347–53.

Platt, J. 1992. "Case study" in American methodological thought. *Current Sociology* 40:17–48.

Resetarits, W. J., Jr., and J. Bernardo, eds. 1998. *Experimental Ecology: Issues and Perspectives.* Oxford University Press, NY.

Salt, G. W., ed. 1984. *Ecology and Evolutionary Biology: A Roundtable on Research.* University of Chicago Press, Chicago.

Sanford, E., M., S. Roth, G. C. Johns, J. P. Wares, and G. N. Somero. 2003. Local selection and latitudinal variation in a marine predator-prey interaction. *Science* 300:1135–37.

Schoener, T. W. 1986. Mechanistic approaches to community ecology: A new reductionism? *American Zoologist* 26:81–106.

Schrader-Frechette, K. S., and E. D. McCoy. 1993. *Method in Ecology: Strategies for Conservation.* Cambridge University Press, Cambridge.

Simberloff, D. 2004. Community ecology: Is it time to move on? *The American Naturalist* 163:787–99.

Slobodkin, L. B., and N. G. Slack. 1999. George Evelyn Hutchinson: 20th-century ecologist. *Endeavour* 23:24–30.

Stegner, W. 1992. The Sense of Place. Pages 199–206 *in* W. Stegner, *Where the Bluebird Sings to the Lemonade Springs.* Random House, New York.

Thompson, J. N. 2001. The geographic dynamics of coevolution. Pages 331–343 *in* C. W. Fox, D. A. Roff, and D. J. Fairbairn, eds., *Evolutionary Ecology: Concepts and Case Studies.* Oxford University Press, NY.

Tilman, D. 1994. Competition and biodiversity in spatially structured habitats. *Ecology* 75:2–16.

Tobey, R. C. 1981. *Saving the Prairies: The Life Cycle of the Founding School of American Plant Ecology, 1895–1955.* The University of California Press, Berkeley.

Turchin, P. 2001. Does population ecology have general laws? *Oikos* 94:17–26.

Weiner, J. 1995. On the practice of ecology. *Journal of Ecology* 83:153–58.

Yin, R. K. 1994. *Case Study Research: Design and Methods.* Second edition. Sage Publications, Inc., Thousand Oaks, CA.

Note

1. The phrase "sense of place" is used variously to refer to personal rootedness in places (e.g., Stegner 1992), the capacity of places to draw out human emotional attachments or, even though Aldo Leopold did not use the term to our knowledge, to the intimate scientific understanding of place that he developed and advocated.

Part 1

THE IMPRINT OF PLACE ON ECOLOGY AND ECOLOGISTS

Mary V. Price and Ian Billick

The two chapters in this first section examine, from a historical perspective, the role of "place" in ecological thought. Kingsland (chapter 2) considers how places have "worked in the minds of ecologists" to shape the conceptual framework of ecology and to serve as loci for its emergence as a professional discipline. Waller and Flader (chapter 3) trace how places worked in the mind of one ecologist—Aldo Leopold—to shape his pioneering vision of "the Land."

Kingsland argues that individual places have been central to the science of ecology in two fundamental ways. They have played a creative, not just an evaluative, role in the generation and evolution of ideas. In addition, she suggests that the establishment of place-focused research facilities—otherwise known as field stations—was central to the emergence of ecology as a formal discipline.

Where do ecological ideas come from? Kingsland suggests that most do not spring deductively from the minds of ecologists, like Athena from the head of Zeus. Instead, they emerge when ecologists absorb the essential spirit of individual places—their genius loci. The origin of major ideas in ecology can be thus be traced to encounters between ecologists and unique places. Examples include the genesis of ideas about succession in the dynamic landscapes of North America, or about the ecosystem in "watery places" like lakes that have distinct boundaries.

Knowledge of those places could only develop once biologists ventured outside their laboratories to study organisms in their natural context, instead of examining the morphology or physiology of specimens extracted from the field. Ironically, as Charles Elton (1927, p. 3) noted, ". . . the discoveries of Darwin, himself a magnificent field naturalist, had the remarkable effect of sending the whole zoological world flocking indoors, where they remained hard at work for fifty years or more, and whence they are now beginning to put forth cautious heads again into the open air." It was only after enough species had been described and classified, Elton suggested, that fieldwork could begin to disentangle their complex relationships in nature.

This movement into the field around the turn of the twentieth century, Kingsland notes, spawned "outdoor laboratories" such as the Desert Laboratory in Tucson, Arizona, or the Scripps Institute in San Diego, where field-oriented scientists of all stripes could bring diverse perspectives to bear on fundamental biological problems. Field stations further served as "condensation nuclei" for the exchange and crystallization of ideas that defined the emerging discipline of ecology and established a framework for training the students who would carry the field forward.

Field stations were not the only outdoor laboratories formed during this period. In response to excesses of timber harvest, agricultural expansion, and overgrazing in the late 1800s, the United States government established extensive national forests, wildlife refuges, and rangelands as well as agencies to manage the natural resources they contained. The need to understand these ecological systems in order to manage them well was another force that drew biologists into the field and shaped the science of ecology.

Waller and Flader (chapter 3) trace the impact of place on the evolution of the thinking of one such applied ecologist, Aldo Leopold. We have included a chapter on Leopold even though one reviewer advised us not to: "And please, try not to have a chapter on Aldo Leopold—a really minor figure so far as the science of ecology is concerned, and a red herring for this undertaking." While we agree with the reviewer's conclusion that Leopold's research findings are rarely cited in modern "mainstream" ecology, we would argue that this is not the best measure of his professional influence. After all, Leopold served as president of the Ecological Society of America, wrote the first textbook on wildlife management, and headed the first department devoted to that subject at a North American university. He also founded the fields of conservation and restoration ecology. His text on game management (Leopold 1933) reflects the best thinking of the time, citing the likes of Adams, Allee, Elton, Errington, Grinnell, and Murie, to name just a few, and anticipating such modern concepts as trophic cascades and nonlethal predator effects (Ripple and Beschta 2004; Estes, chapter 8; Peckarsky et al., chapter 9). Although citations to Leopold's research disappeared from ecology textbooks soon after Andrewartha and Birch (1954), his influence can be traced in a number of modern initiatives in applied ecology—systematic environmental monitoring, use of fire as a management tool, multiple-factor thinking, integration of humans into natural ecosystems, and private-public partnerships as a conservation tool, to name a few (Meine 2004; Waller and Flader, chapter 3; Curtin, chapter 12). A chapter on how Leopold came by his understanding of place is warranted here because his methods anticipate the place-based approaches illustrated throughout this book.

Waller and Flader's essay traces the roots of Leopold's sense of place to several key sources. Leopold's lifelong passion for wild places and wild creatures began with childhood ramblings and family excursions. During his extensive time spent outdoors as a hunter and forester, he honed a habit of systematic observation, recording meticulously what he saw where and when in a small notebook that he carried with him always, and fleshing out the notes upon his return to office or home. Leopold drew upon this personal, wide-ranging, natural history database on everything from soil erosion to animal behavior to make inferences about the workings of ecological systems and their responses to human activities. Motivated by a Progressivist-era belief in the importance of science-based management of natural resources, Leopold was not content just to theorize; he continually reevaluated his inferences, both by experiment and by assessing the fit of new observations to his conceptual model.

A hallmark of Leopold's approach was comparison. Comparing observations of a single place made at different times allowed him to see the dynamic nature of ecological systems. Comparison of places with different histories of land use or natural disturbance—what Diamond (1983) called "natural experiments"—allowed him to appreciate how the past had molded the features of landscapes. Leopold was also willing to experiment, because observation of how a system responded to manipulation allowed him to test experimentally (albeit with minimal replication) ideas developed through other forms of comparison.

Leopold's approach to understanding ecosystems was largely driven by his realization that one cannot be a narrow specialist and "see the land" well; the interconnections among components of the ecosystem require a more generalized and holistic perspective. Leopold went into the field with his eyes and mind open to matters seemingly unrelated to the job at hand. As an example, Waller and Flader point to his interest in understanding the increased erosion and vegetation shifts he saw in the forests of the southwestern United States. This interest led him to propose a novel conceptual model of interactions among fire, grazing, climate, vegetation, and watershed health—a task that was not part of his job as forest supervisor.

It is interesting to contemplate how many elements of the ecology of place (Price and Billick, chapter 1, box 1.1) are manifest in Leopold's work. Certainly he focused in-depth on phenomena in their real-world context. As an applied ecologist, he aimed to understand particular systems well enough to manage them. He deployed multiple types of information to develop case-specific understanding that he evaluated against diverse threads of new information about a given system, such as wildlife behavioral and population responses to habitat manipulation. It is also interesting to contemplate how Leopold's legacy of using comparison, knowledge of history, long-term observation of single systems, and experimentation to develop an understanding of a place is reflected in the narratives of this book, as well as in current initiatives for promoting place-based approaches to ecological understanding (see part 5).

References

Andrewartha, H. G., and L. C. Birch. 1954. *The Distribution and Abundance of Animals.* University of Chicago Press, Chicago.

Diamond, J. M. 1983. Laboratory, field, and natural experiments. *Nature* 304:586–87.

Elton, C. 1927. *Animal Ecology.* Macmillan, New York.

Leopold, A. 1933. *Game Management.* Charles Scribner's Sons, New York.

Meine, C. 2004. *Correction Lines.* Island Press, Washington, DC.

Ripple, W. J., and R. L. Beschta. 2004. Linking wolves and plants: Aldo Leopold on trophic cascades. *BioScience* 55:613–21.

2 The Role of Place in the History of Ecology

Sharon E. Kingsland

Figure 2.0. Daniel MacDougal, director of the Desert Botanical Laboratory, with Western writer Mary Austin (detail), ca. 1920. Courtesy Arizona Historical Society/Tucson (AHS 11477).

Abstract

The historical role of place-based research can be examined first by asking what role certain landscapes have had in the development or testing of theory, and second by exploring the early history of field stations and their relationship to the emergence of ecology as a discipline. The first theme is examined through several examples: Darwin's discussion of island populations in relation to his evolutionary theory; the application of the ecosystem concept and its relation to the study of lakes and coral atolls; and the development of the theory of ecological succession. The second theme is explored by reviewing how European and American biologists promoted field biology and the establishment of research stations around the turn of the twentieth century. The American experience, which drew on European models, is discussed in greater depth in order to explore the connection between the creation of research stations and the emergence of ecology as a distinct discipline. Several American biologists argued for the importance of field research and sought stable funding for field stations and marine laboratories, with varying degrees of success. The Carnegie Institution of Washington also created two research stations, its Desert Botanical Laboratory and its Station for Experimental Evolution, in 1903 and 1904 respectively. Such field stations aided the development of ecology as a new discipline both through their research activity and through their training of students.

Introduction

What role has place-based study had in the development of modern biology in general and the discipline of ecology in particular? First, we might reflect on how evolutionary biologists and ecologists have arrived at general theories, definitions, and concepts by studying certain kinds of landscapes. Second, we could consider what goals motivated the establishment of biological field stations at the time that biology was becoming a profession, and how their existence and organization influenced the growth of biology and the emergence of ecology as a new discipline. We do not have a detailed historical study of field stations, but at least in the history of early-twentieth-century biology there are a few studies from which to distill themes and observations that may be relevant to understanding current science as well as past science.

This essay explores the above themes with particular attention to American biology. Although ecology as a subject matter was defined first by Europeans, the idea that ecology was not only a new approach to natural history, but really a new discipline with its own methods and principles, was accepted more quickly in the United States than in other countries. Part of the reason lies in the extensive and varied landforms that Americans studied, but another important reason for the success of ecology was the establishment of field stations promoting long-term studies of distinctive regions. The American focus of this essay is meant to illustrate the connection between discipline formation and a certain type of place-based research. The larger point is that ecology was not destined to be defined as a separate discipline, but emerged as a discipline for particular historical reasons.

The Power of Place: Grand Facts and Theories

To begin with one of the greatest ecological thinkers, Charles Darwin, let us consider the change in his attitude toward island archipelagoes as he matured as a scientist. Of the places Darwin visited on the *Beagle* voyage, the Galápagos Islands have assumed special status as an inspirational location, but Darwin was not an evolutionist at the time of his visit there and did not consider the possibility that species could change until long after he had left. Although he noticed that mockingbirds collected on the different islands were different from each other, he at first paid no attention to those differences. When a knowledgeable resident of Charles Island pointed out that different species of tortoise inhabited different islands, Darwin did not grasp the significance of that observation. But on the final leg of the voyage home, his mind turned again to the differences in the animals found on the islands and very hesitantly he began to harbor "vague doubts" about the permanence of species (Browne 1995, p. 339). Back in England, taxonomist John Gould provided him with the expert opinion that the Galápagos finches were a new group confined to those islands, and that it seemed very likely that each island had its own species of finches. Darwin had thought the finches were only varieties, and their status as separate species took him by surprise (Browne 1995). Darwin's vague doubts about species permanence soon grew into a certainty that species could change.

If the initial Galápagos experience did little more than trigger a few doubts, by the time *The Origin of Species* was published over two decades later Darwin had greater appreciation of how certain locations could be critical testing grounds for evolutionary theories. In his discussion of the inhabitants of oceanic islands in the *Origin*, he pointed out that such places were important in generating the "grand facts" that could sustain or challenge a theory (Darwin 1964, chapter 12). He was impressed both by the number of endemic species found on islands and by the absence of certain classes of animals. These distributions seemed to be connected to the migratory ability of animals. Islands located very far from continents did not have mammals, and many oceanic islands lacked frogs, toads, and newts despite having a climate and terrain that could support these animals. On the other hand, islands had high proportions of endemic species, such as land birds. Most importantly, Darwin noted that inhabitants of islands had strong affinity to species on the nearest mainland, even though the species themselves were different. Twenty-five of the twenty-six species of Galápagos land birds were thought to be endemic, yet those species bore the "unmistakeable stamp of the American continent." In comparison, the Cape Verde Islands had a geology and climate similar to that of the Galápagos archipelago, but their species composition was strikingly different, with the Cape Verde species resembling African forms. The lack of correlation between climate, geology, and the species found on islands was striking—and unexpected if one believed that species were specially created and adapted to live in certain landscapes. This pattern of distribution was for Darwin a "grand fact" that could be explained not by special creation, but only by migration of species from nearby continents and their subsequent modification.

Island groups like the Galápagos also offered opportunities for comparative studies that could help test or refine theories of evolution. In fact, the distribution of species on the Galápagos Islands could have posed significant problems for Darwin's theory. Different islands had different, although related, species, but the islands were located close to each other and one might expect species to be constantly moving from island to island, keeping the populations mixed. How could islands situated close to each other, with similar climate and geology, come to have different, albeit closely related, species? Possibly deep-sea channels prevented the migration of some species, yet such barriers would not prevent all types of migration. Darwin argued that the expectation that the islands should all have the same species was an incorrect inference stemming from the erroneous belief that the physical conditions of a country were most important for its inhabitants. He suggested that it was the biological environment that mattered most. The biological differences among the islands would mean that the struggle for existence and the course of evolutionary divergence would operate differently in each location even though their physical landscapes and climate were similar. Once a species had diverged while populations were separated on the different islands, an immigrant wandering from one island to another would find itself at a competitive disadvantage with that island's native form. Closely related species would compete strongly and the one occupying its native ground would have the crucial advantage. Although Darwin did not say so explicitly, his discussion implied that certain kinds of landscapes, such as island groups, would be excellent places to study how selection operates by closely observing how populations behave, interact, and respond to environmental changes (see Grant and Grant, chapter 6).

In addition to these advantages for the study of evolution, ecologists appreciate islands because they represent simplified systems with well-defined boundaries. They can be used to describe the basic processes of ecological systems, although it is recognized that extrapolating from simple systems to more diverse continental systems may be difficult or impossible (Vitousek et al. 1995, p. 1). Not only islands but also small lakes and other watery systems, such as coral atolls, have been important in the development of ecological concepts, in part because of their well-defined boundaries. After Arthur G. Tansley proposed the term "ecosystem" in 1935 (Tansley 1935), the concept became embedded in ecology largely as a result of work done on landscapes that lent themselves to a holistic perspective. Raymond Lindeman (1942), in his classic study of Cedar Bog Lake in Minnesota, suggested that the choice of the lake as the subject of ecological analysis led to the recognition that distinctions between the biotic community and dead organisms or inorganic nutrients were artificial. In terrestrial ecology, where scientists studied plant and animal communities, such distinctions between the living community and the "environment" had been commonplace. In lakes, where the close connection between the living community and the non-living environment was particularly obvious, a systems viewpoint made more sense. Lindeman concluded that the "ecosystem" should be the fundamental ecological unit, not the biotic community.

Similarly, Eugene and Howard Thomas Odum promoted the ecosystem concept as fundamental to ecology from the 1950s on, and their theoretical perspective found support in their studies of salt marshes, coral atolls, and river systems (Hagen 1992, Kingsland 2005). Coral atolls were particularly suited to the ecosystem viewpoint because they were in a relatively constant environment and were fairly stable and isolated. For Eugene Odum one could think of the system as analogous to an organism, and measure its "metabolism." For Tom Odum the ecosystem was more like a machine governed by feedback mechanisms. The successful application of the ecosystem concept depended on finding natural systems that lent themselves to these analogies. When the Odums promoted the ecosystem concept in the 1950s, many ecologists rejected it as unrealistic because they were studying landscapes that did not have easily identifiable boundaries (Evans 1956).

Ramón Margalef, a Spanish ecologist, conjectured in the 1960s that even general attitudes toward theory development could reflect the kinds of landscapes studied (Margalef 1968). Margalef was exploring the problem of whether a theoretical superstructure could be built for ecology by drawing on new ideas generated in cybernetics, a postwar field concerned with problems of control and communication. He believed that American audiences might be receptive to his ideas because they had developed the concept of ecological succession and made it the central element of the discipline of ecology. Margalef's proposal that cybernetics could be applied to ecology depended on accepting the centrality of this concept of succession as a basic ecological process that defined the ecological community. Margalef believed that the American perception of ecological succession stemmed from the nature of the landscapes that early American ecologists studied, and he proposed that the Russian landscape similarly suggested a "dynamic approach in ecology and the theory of the climax" (Margalef 1968, p. 26). Thus the concept of ecological succession, which could be traced back to nineteenth-century European studies, was developed mainly in the American context, chiefly by Henry Chandler Cowles at the University of Chicago and by Frederic Edward Clements, whose work focused on Midwestern landscapes.

Margalef was drawing attention to the power of certain kinds of landscapes to generate theoretical insights about nature, insights that were fundamental to the definition of ecology as a subject. Not all landscapes, he argued, were suitable for the generation of theory, and differences in landscapes produced great differences in the kind of ecological work done in those places. Ecologists, he wrote, reflected "the properties of the ecosystems in which they have grown and matured," and schools of ecology were shaped by the "spirits of place" that belonged to the total landscapes. Desert ecologists would be inclined to focus on the effect of weather on communities. In Mediterranean and Alpine countries, on the other hand, the vegetation had a mosaic-like quality and was influenced by thousands of years of human occupation of the land. There the dominant ecological style was the plant sociology school of Josias Braun-Blanquet, whom Margalef characterized as the exponent of a "clear and careful bookkeeping and filing system." Scandinavia with its poor flora produced "ecologists who count every shoot and sprout" (Margalef 1968, pp. 26–27; see also Söderqvist 1986).

So it was not in Scandinavia but rather in the American Midwest where the concept of ecological succession was developed and made the central element of the new science of ecology. The originator of this important idea was Henry Cowles, who created an influential program in ecology at the University of Chicago. Frederic Clements later developed the concept and fully theorized it, but he credited Cowles as the original inspiration. No single idea has been more important for the definition of American ecology, yet the understanding of succession has been the subject of continuous debate and sometimes heated controversy. The reason lies in part in the relationship between the kinds of landscapes studied and the way succession was conceived.

Cowles had an unusual environment on his doorstep: the sand dunes in Indiana, to the east and south of Chicago and along the shore of Lake Michigan, which were well suited to the development of a theory of ecological succession (Engel 1983, Cittadino 1993). Cowles had been trained in the physiographic school of geology, which emphasized the dynamic changes occurring continually in the physical landscape. Physiographers studied landscapes in the making, and Cowles defined ecology along similar lines as the study of the ever-changing panorama of vegetation. The main subject for the ecologist was therefore the succession of plant societies and the laws that governed their development (Cowles 1899). The dunes exemplified these dynamic processes, as the topography changed constantly. Moreover, Cowles could see sequential changes in vegetation, from grasses growing on sands by the shore to shrubs further inland, and finally to coniferous and deciduous forests. Cowles inferred that in walking inland from the lake, the vegetational changes paralleled those that occurred over time, and one could read the history of the dunes by traversing the land over a mile or two inland from the shore. From this landscape Cowles formulated a general theory of ecological succession and vegetational climax, which he extended to other vegetation in the Chicago region.

Cowles never published a large synthetic work, but his early articles were original and enormously influential. Ecologist Charles Christopher Adams described Cowles's first analysis of succession, published in 1899, as a "startling revelation of a new kind of approach in biology" (Adams and Fuller 1940, p. 40). Cowles's observations of this exceptional landscape generated the approach that helped to define ecology as a subject, provided the first comprehensive study of succession in America, and became the basis for the training of many students at Chicago.

But successional sequences that appeared obvious to some observers were not as obvious to people who studied different kinds of landscapes. Henry Allan Gleason recalled that many Americans quickly took up the study of successional series and had no trouble recognizing such series in the field, whereas European plant ecologists were reluctant to admit the existence of succession or even persistently denied it (Kingsland 2005, p. 153). Tansley noted in 1926 how characteristic the study of succession was of the American school, and how this emphasis reflected the enormous changes occurring in the country (Tansley 1926). The United States being a "new" country in which ecological changes were happening quickly and dramatically, the successional point of view was seen there as particularly fruitful.

Even within the United States the perception of succession reflected the kind of landscape studied. Frederic Clements focused on the more stable grasslands and conifer forests of the western prairie, and he had a concept of succession that was more deterministic than Cowles's, although whether this difference was a result of the landscape studied or of his own rigidity of mind is unclear (Tobey 1981). The difference is illustrated by Clements's insistence that the plant community was a type of "complex organism" that developed along a fixed path in a way that was analogous to the development of an individual organism. Gleason studied the prairies and forests of Illinois and Michigan and grew skeptical that one could determine the sequence of changes that made up a successional series. He eventually became one of Clements's most vocal critics, advancing an "individualistic" thesis of vegetation which emphasized the role of chance in the formation of plant associations (Gleason 1926, McIntosh 1975, Nicolson 1990). Thus, taking a concept developed in one type of environment and importing it to another type of landscape could trigger modifications in the theory—in Gleason's case, even to the point of casting doubt on the reality of the plant association as a vegetational unit.

A similar process of transfer and modification, but with different outcome, is illustrated in the work of William Skinner Cooper, a student of Cowles. Cooper's doctoral thesis was a study of the balsam-birch-white spruce forest of Isle Royale in Lake Superior, but Cooper noted the inherent difficulty of studying succession in a climax community that had apparently reached a state of equilibrium with no successional changes occurring (Cooper 1913). He accepted the validity of the concept of succession, but was forced to admit that the investigation of plant succession could only be made indirectly in this kind of landscape. He could not observe succession occurring, but had to infer the process of succession by comparing vegetation at different stages in different locations.

Where Gleason was led by this kind of doubt to criticize the way ecologists thought of the plant association, Cooper instead searched for places where succession could be observed. Cowles had thought of ecology as the study of process, of change, and therefore Cooper believed that ecology was best studied in places where changes were occurring. He found the ideal location soon after his Isle Royale studies, in Glacier Bay, Alaska. In Cooper's studies of Glacier Bay, which began in 1916 and continued for many years, the need for longer-term study was more obvious than it had been on Isle Royale, for in Alaska change was occurring at a measurable pace as the glacier retreated. In Glacier Bay the processes of succession could be studied directly as vegetation invaded newly exposed regions, and therefore "the major part of a complete development cycle may be encompassed within the limits of a single lifetime" (Cooper 1923, p. 93). In offering an exceptional example of rapid vegetational change, Glacier Bay was ideal for the study of succession as a process and a confirmation that ecology was a science devoted to the study of dynamic processes (Rumore 2009).

One result was that Cooper's attitude toward certain ecological concepts—such as the concepts of successional development, progression and regression—shifted. He grew skeptical of ecological conceptions that implied a rigid view of succession, the most notorious being Clements's notion of the plant community as "complex organism." Such rigidity of mind had no place in ecology, Cooper thought, and he envisioned succession quite differently: not as the development of a "complex organism" but as a "universal process of change which is embodied in the great vegetational stream" (Cooper 1926, p. 410). Thus the power of place worked in the minds of ecologists not only to suggest research projects, hypotheses, and strategies for critiquing prevailing theories, but also to conjure up metaphors and analogies that served to organize thoughts about pattern and process and provided a way of speaking about nature.

Ecologists have long recognized that certain kinds of places may yield observations that serve as triggers for the formulation of theory or conversely can be used to challenge the central premises of theories—for example, the idea that succession leads to a stable climax community. Such challenges go beyond Popperian exercises that attempt to provide "yes or no" answers to specific hypotheses. Rather, they may cast doubt on the validity of basic assumptions about the organization of nature, and in doing so they may also demand fundamental rethinking of the objectives of ecology as a science. Studies of succession in different environments, for instance, rendered problematic the entire idea, derived from Clements, that nature developed along a predetermined path. The most interesting question was not whether nature was or was not like an organism, but something more fundamental: If plant communities did not develop "naturally" in a predictable way, then how should ecologists judge the relative health or deterioration of systems? What standards could be used to determine how humans should intervene to manage, preserve, or restore natural systems? It is a bold step to recognize when a place-based study demands more basic questioning of ecological viewpoints and assumptions, for such questioning can also undermine what has been considered the main purpose of ecological study, in this case predicting (and potentially controlling) the process of succession.

Whether one is formulating or challenging theory, a persistent question has been how much research, and how prolonged a study, is required before one dares to advance a new argument. Lindeman's now-classic article on the trophic structure of a lake ecosystem, for instance, was not recommended for publication because his observations were based on the study of just one lake. Reviewers wanted more data, in effect a career's worth of further study (Cook 1977). Lindeman's mentor, G. Evelyn Hutchinson, had to intervene to get the article into print, and by the time it was published its author had died. The "ecology of place" can be taken to mean the long-term study needed to appreciate the detailed working of a particular system, but equally it can be understood to mean the ability to realize when a case study presents the opportunity for a novel approach that may orient ecology along new lines. In modern times an appropriate question might be whether the funding practices and general conservatism of science may inhibit theory development by limiting the freedom to pose new questions or think in novel ways when new places are investigated. This question has implications for how place-based research centers are organized, how they operate and with what objectives, and how they are used to train students—problems that are discussed in the chapters of part 5.

The Role of Field Stations in Establishing Ecology As a Discipline

Just as ecologists and evolutionary biologists have recognized the importance of particular kinds of places for analyzing problems, advancing theories, or characterizing the nature of ecological systems, they have also recognized and championed the need to establish institutions to foster research in the field and train students in biology. Eugenius Warming, a Danish botanist and one of the pioneers of ecology, extolled the value of travel in tropical South America as a means of raising questions about the striking diversity of tropical flora as compared to that found in northern climates (Warming 1899). Andreas F. W. Schimper, German botanist and author of a seminal ecological treatise, translated into English as *Plant-Geography upon a Physiological Basis* (Schimper 1903), similarly emphasized how the development of ecological plant geography depended on travel outside of Europe, whether to tropical forests, the Sahara, or the arctic tundra. In these environments new questions about adaptation and the relationship between vegetation and climate struck the careful observer. Schimper promoted ecology as a modern science based on "accurately observed facts" and "critically conducted experiments," thus raising it above the dilettantism and "anthropomorphic trifling" that threatened to discredit the study of biological adaptation (Schimper 1903, volume 1, pp.vi–vii). His work was widely viewed as the start of a new era of superior ecological analysis (Cittadino 1990).

Among the "ecology boosters" in the early twentieth century the common refrain was a call for more field stations. Schimper recognized that his ability to advance ecological science owed much to the establishment of laboratories outside Europe, notably the Dutch Royal Botanic Garden at Buitenzorg (now Bogor), on the island of Java. Under the leadership of Melchior Treub between 1880 and 1909, the botanic garden added laboratory facilities, revived the garden's scientific journal, and swiftly became a leading institution for long-term studies of tropical vegetation, drawing many foreign botanists (Cittadino 1990). Treub and other botanists also took advantage of an exceptional nearby "natural experiment" and tracked the repopulation of the island of Krakatoa after the volcanic eruption of 1883. British and European colonial networks, such as the British network of colonial botanical gardens centered on Kew Gardens, were seen as a great service to tropical science. American scientists, hoping to catch up to European initiatives, argued for a Pan-American institution for the study of tropical America with the idea of facilitating practical research in agriculture and forestry, as well as general studies in ecology, systematics, biogeography, and other subjects (Olsson-Seffer 1911).

Schimper had hoped that a laboratory similar to the one at Buitenzorg might be established in an arctic environment. Picking up the cue, botanist Morten P. Porsild of the University of Copenhagen established a research station on Disko Island in Greenland in 1906 (now the Danish Polar Center of the University of Copenhagen) with funding both from private sources and from the Danish government. As an American observer remarked, this location was ideal for arctic research, being rich in flora and the northernmost point where all the different plant formations of Greenland were represented (Olsson-Seffer 1905). The proposal, endorsed by Danish institutions as well as scientists across northern Europe, also drew the approval of Americans, who did not yet have a comparable permanent station in Alaska.

The appeal for field stations was not only driven by a desire for long-term and detailed study, signaling the move away from the dilettantism to which Schimper referred, but also by the idea that ecology was becoming a more experimental science and that these experiments had to be performed in the field. Cowles noted that the most valuable ecological experiments "must be performed not in the laboratories, but out of doors" (Cowles 1904, p. 882). He had in mind the careful transplantation experiments of French botanist Gaston Bonnier, who set up experimental gardens near Paris and in the French Alps and Pyrenees to study the effect of environment on plants. As Tansley argued in 1904, given that ecology was moving from being a more descriptive science to being more experimental, one important means of progress would be "the establishment of laboratories in regions exhibiting specialised types of vegetation" (Tansley 1904, p. 200). He hoped that British ecologists might emulate a recent American model, the Desert Botanical Laboratory established by the Carnegie Institution of Washington near Tucson, Arizona, in 1903. I turn now to the American context in order to explore the relationship between institution-building and disciplinary development in more detail. As Tansley implied, the Americans, although indebted to European ideas and initiatives, were beginning to move rapidly ahead in developing ecology along with other branches of biology.

Field Stations and Ecology: The American Case

The development of ecology as a discipline depended crucially on the ability of scientists to persuade governments, universities, or private donors to establish and maintain field stations. As Cowles's comments quoted above suggest, American biologists saw field stations as providing an antidote to laboratory science at a time when laboratory research was dominating older forms of natural history. Laboratory research focused on problems in physiology, morphology, embryology, and cell biology but also extended to the analysis of animal behavior and problems in evolutionary biology. Rather than thinking of laboratory science as replacing natural history, leading American biologists believed it was better to take a broad approach that included field research as a balance to the narrowing tendencies of laboratory work. The analysis of certain problems, such as adaptation, simply required the study of organisms in their natural environments. Another motive stemmed from the desire to support agriculture and fisheries with science; this was the same period that saw the establishment of agricultural experiment stations in all the states and territories. Yet another motive was to bring American research up to the standards of European science by emulating the European model of establishing botanic gardens and marine and freshwater stations for research. The creation of new kinds of research centers and field stations was important for the professional development of biology, but biologists also argued that it aided the economic development of the country.

At the Johns Hopkins University, America's first graduate university, William Keith Brooks exemplified this broad vision of biological study. The university built a biological laboratory in 1883 for work in physiology and morphology, but Brooks persuaded the university trustees to fund the Chesapeake Zoological Laboratory (CZL) in 1878 for marine research (Benson 1988). Brooks was inspired by Alexander Agassiz's marine laboratory, part of his summer estate in Newport, Rhode Island, which opened in 1877 and operated until his death in 1910. His father, Louis Agassiz, had run a summer school for teachers at Penikese Island, Massachusetts, and initially Alexander also invited teachers to the Newport laboratory. However, he was not interested in organizing a school and it soon turned into a purely research laboratory, training many of the most influential men in American biology (Winsor 1991).

By the 1880s Brooks made summer research at the CZL a requirement for Hopkins graduate students in morphology, among whom were several of the next generation's leaders in biology (including limnologist Edward A. Birge, whose studies of Wisconsin lakes with Chancey Juday defined the field of freshwater ecology in America). The CZL was important for Brooks's classic studies of the American oyster, published in 1891 (Brooks 1996). Until 1879 knowledge of the breeding habits of oysters depended entirely on the study of European oysters. Brooks's research revealed that the American oyster, unlike its northern European relative, was not hermaphroditic, and this led to promising experiments on artificial oyster cultivation. Apart from this important practical application, Brooks saw the laboratory's work as contributing broadly to the study of the anatomy and development of marine organisms.

The CZL was not a fixed site but was convened in different locations around Chesapeake Bay, in North Carolina, and even further away in Bermuda and Jamaica. Brooks's description of the laboratory's early years conveys the opportunistic nature of the enterprise (Brooks 1884). In 1878 he took nine students to the seaside, hoping to occupy vacant buildings at Fort Wool, Virginia. The buildings were in poor repair, but the fort watchman vacated all but one room in his own house for the group to use. Spencer F. Baird, head of the U.S. Commission on Fish and Fisheries, supplied dredging apparatus and a couple of rowboats. In the second year the laboratory began its season at Crisfield, Maryland, where the group worked on three barges owned by the Maryland Fish Commission. When mosquitoes made the barges uninhabitable the laboratory moved back to Fort Wool, where the watchman again made his house available. For the next three years the university provided better support, allowing the laboratory to purchase a launch and sloop for collecting, and it moved 400 miles south of Baltimore to a rented house in Beaufort, North Carolina. That location was ideal for marine biology because of the richness of the fauna and the ease of collecting, and Brooks ran experiments on artificial oyster culture there (Keiner 1998). The laboratory moved then to a machine shop building at the Hampton Normal and Industrial Institute, Virginia, so that Brooks could study the condition of the oyster beds in Maryland. That location was poor for marine biology, as the collecting grounds were far away and the fauna lacked the richness of the Beaufort site.

At the time, the CZL was the only marine laboratory in the United States where a student could get an advanced education. The students who worked there later contributed to the development of the Marine Biological Laboratory at Woods Hole, Massachusetts, founded in 1888 on the model of the famed Naples Zoological Station (Benson 1987). In 1899 Brooks expressed dismay at the notion that the modern student might believe that biology was "nothing but" the study of the physical and chemical basis of life. Taking issue with Thomas Henry Huxley's essay "The Physical Basis of Life," which seemed to lend credence to this narrow view of biology, Brooks imagined how Aristotle would have answered Huxley had he been alive: "Is not the biological laboratory which leaves out the ocean and the mountains and meadows a monstrous absurdity? Was not the greatest scientific generalization of your times reached independently by two men who were eminent in their familiarity with living things in their homes?" (Brooks 1899, p. 41). Field research was a key component of the American vision of what biology should be, its significance underscored by the role that field observation played in Darwin's and Wallace's great contributions to biological theory.

Charles Otis Whitman, the first director of the Marine Biological Laboratory (MBL) at Woods Hole, argued in 1898 that biological stations should embrace all the important divisions of biology and not be artificially restricted to certain subjects or approaches. He advocated the introduction of "experimental biology," but he meant this term in a very special sense: not confined to work in the laboratory, but also including "experimental natural history, or biology, in its more general and comprehensive sense" (Whitman 1898, p. 42). This was "modern natural history," whose foundation was laid by Darwin and whose goal was to unify physiology and morphology within an evolutionary framework. Whitman hoped that the MBL would evolve toward this ideal, but was not able to translate his vision into reality (Pauly 1988, Maienschein 1988). Defeated by struggles over money and the control of the laboratory, he withdrew from active control of the MBL after 1902. Biology at the Laboratory was never the unified comprehensive science that Whitman hoped it would be (Maienschein 1986).

While neither Brooks nor Whitman spoke explicitly about "ecology," these ideas about biology as a modern science that did not spurn its natural history traditions also underlay the development of ecology in the late nineteenth century. Ecology, much like Brooks's and Whitman's concepts of general biology, represented resistance to narrow specialization. When Americans began to call themselves "ecologists" in the 1890s, they were signaling a rejection of the modern trend toward laboratory study as the exclusive definition of life science. At the same time they accepted the need for experimental research and quantitative methods in natural history. An ecologist, while engaged in physiological research, was not as narrow as the laboratory scientist: ecology was an experimental science that included fieldwork, a kind of "outdoor physiology" (Cittadino 1980). This broader vision of life science required new places for research: field stations, marine stations, and experimental farms that would not simply be vehicles for training teachers, but would also be research centers. The formation of ecology as a new discipline went hand in hand with the creation of field stations (Kohler 2002).

As with any enterprise that required money, scientists had to justify their activities in ways that appealed to those who funded the research. An example of a successful institution builder was Stephen Alfred Forbes, an Illinois naturalist and one of the grand old men of American ecology. Forbes is known for his classic essay of 1887, "The Lake as a Microcosm," which provided one of the earliest definitions of the ecological community (Real and Brown 1991). His research interests included the food relations of predators and prey, the function of different organisms within biological systems, the study of rivers as dynamic systems, and general biological surveys. In 1894 Forbes started referring to his work as "oecology," but even before adopting this word Forbes clearly conceived of a field of research devoted to studying the relationship between groups of organisms and their organic and inorganic surroundings.

The inspiration for this conception was likely a combined result of his interest in the new evolutionary biology, his own experiences as a naturalist, and possibly a religious sensibility that predisposed him toward a particular view of nature. The theoretical framework that guided Forbes came from Darwin and from the evolutionary speculations of Herbert Spencer, a philosopher whose writings were much in vogue in late-nineteenth-century America. But the way Forbes tackled these theoretical problems through detailed study of food relations in fishes and birds certainly reflected the local concerns of Illinois farmers and fishermen. One could argue that his success in defining a new field of research—eventually to be named "ecology"—was the result of the creative way in which general theory, imported from elsewhere, was brought to bear on specific problems encountered in his agricultural environment (Schneider 2000).

Forbes's status as a founding father of American ecology stems from his success in creating a long-lived research program that was distinctively ecological. Bocking's (1990) analysis of Forbes's career shows how he furthered the development of ecology through a pragmatic strategy that balanced the need for pure and applied research, ultimately helping to guarantee institutional support for "pure" ecology. The institutional basis of natural history at the time that Forbes began his career in the 1870s was meager, but by the 1880s he had a research program that combined interest in general biological problems with applied problems focused on economic entomology. He assembled a small group of researchers who made expeditions to nearby lakes and river systems, studied fish and their food supplies, and advanced the study of natural history as a scientific discipline with clear relevance to economic problems. His strategy involved merging general and applied ecology so that the two were inseparable; in his view economic entomology was a special division of ecology. In this way he obtained support not only for practical research, but also for research on general ecological problems.

By 1894, the year that Forbes adopted the term "oecology," he opened the University of Illinois Biological Research Station at Havana on the Illinois River, where biological studies had been underway for a couple of decades. This was to be a permanent research base for the university, but it was also meant to counter the trend toward the kind of mechanical laboratory study common in college education (Bocking 1990, p. 473). Forbes had in mind a research facility that would have direct economic applications, especially in fish culture, but would also open up a neglected area of biological study, namely ecology, which Forbes thought should be part of higher education. The research facility itself was a sixty-foot houseboat towed by a steamer and the biological station included thirteen substations on the river and surrounding lakes, all supporting research in ecology and systematics. The University of Illinois and the State Laboratory of Natural History, which in 1917 combined with the Office of State Entomologist to create the State Natural History Survey, became one of the growth centers for ecology, contributing to the definition of ecology as a new discipline and helping to ensure its disciplinary identity. As Forbes argued, the biological station would "restore to us what was best in the naturalist of the old school united to what is best in the laboratory student of the new" (Bocking 1990, p. 478). This union defined ecology as a subject and discipline.

Forbes's success in establishing institutions for the support of ecology encouraged those who worked with him to think about what the aim of ecology should be and to write works that would attract students to the subject. One such "ecology booster" was Charles Adams, Forbes's entomological assistant in the late 1890s, who finished his PhD at Chicago in 1908 and then joined the faculty at the University of Illinois. He became involved in a cooperative project between Forbes's State Laboratory of Natural History and an Ecological Survey Committee of the Illinois Academy of Science. One spin-off was a short book, published in 1913, in which Adams promoted ecology and explained its importance to biological work (Adams 1913). Adams argued that ecology was a dynamic science, a study of process. For this reason, he argued, the ecologist had to learn how to study in the field, not simply how to collect, and therefore it was important to select an area for repeated, prolonged visits. Field study, he pointed out, was not a matter of random observation but of informed study mediated by theory or by a general hypothesis. He argued that the act of reflection should not be thought of as something that happened only when scientists returned to their laboratories or their studies, but as something done in the field itself (Adams, 1913, p. 42). The creation of biological stations was vital, not just to aid the development of ecology but to balance the modern tendency to overemphasize laboratory work. Into the 1940s Adams continued to champion the need for well-supported field stations (Adams, 1946). His arguments illustrate how the creation of field stations, allowing for thorough immersion in the natural environment, was seen as essential to creative thinking in ecology: in effect such research environments were prerequisites for an independent discipline of ecology.

Another of Forbes's coworkers, Charles A. Kofoid, also saw the potential of establishing biological research stations to improve biological education as well as to address economic problems. Kofoid was the first superintendent of the Havana station in Illinois, but he left in 1900 for the University of California, Berkeley. He and his colleague William E. Ritter hoped to create a marine station near San Diego. Edward W. Scripps and Ellen Browning Scripps provided funds for a modest biological station that moved to its permanent home in La Jolla in 1907. Kofoid then embarked on a two-year tour of biological stations in Europe to make the case for government support of aquatic research in the United States (Kofoid 1910). His exhaustive survey, including photographs and floor-plans of laboratories, began with the Naples Zoological Station and went on to describe both marine and freshwater stations in France, Monaco, Great Britain, Germany, Austria-Hungary, Scandinavia, other European countries, and Russia. The European custom of integrating the research of such stations with higher education in biology seemed to him a laudable strategy for ensuring government support of science.

In America he drew attention to the pollution problems of waterways created by the growth of cities and industrialization and to the ill effects of overfishing that were already evident. He envisioned a research facility that would be not merely a biological station but "a station equipped for the solution of biological problems with the aid of all pertinent sciences," including chemistry, physics, meteorology, and hydrology (Kofoid 1910, p. 2). He imagined a large station with an international clientele and with many laboratories for individual researchers, who would have the opportunity for "long continued and close observation and experiment upon abundant and varied material" (Kofoid 1910, p. 6). Here is yet another way of envisioning a field station: as an international center for multidisciplinary study of particular places. Unfortunately, at La Jolla the scale of government largesse that he hoped for did not materialize and the fledgling marine station continued to depend on the patronage of the Scripps family. It did, however, affiliate with the University of California in 1912, when it was renamed the Scripps Institution for Biological Research, which was formally dedicated in 1916, marking the beginning of its evolution into a world-class oceanographic research institution (Raitt and Moulton 1967).

Some biologists fought uphill battles to establish field stations for their entire careers. Bocking (1990) contrasted Forbes's marked success with the very different outcome of Jacob Reighard's contemporaneous efforts to develop freshwater ecology in Michigan. As zoologist at the University of Michigan at Ann Arbor, Reighard hoped in the 1890s to develop aquatic biology, but his initial plans to create a portable laboratory fell through when the university did not support the enterprise. He then hoped for support from the Michigan Fish Commission, but from the start there was conflict between Reighard's personal interest in pure science and the practical concerns of the commission, which expected that studies would be relevant to commercial fisheries on the Great Lakes and was not willing to fund research in general biology or ecology. In 1898 Reighard proposed a permanent research station on the Great Lakes, received only a tenth of the funding he requested, and proceeded to do general biological research that had little connection either to regional needs in Michigan or to the practical needs of the Fish Commission. This station received support for only two seasons.

Forbes and Reighard operated within different institutional ecologies and each adopted a different adaptive strategy. Forbes succeeded in building the discipline of ecology on the foundation of field stations and state surveys, with the support of the University of Illinois. Reighard, despite efforts to establish field stations on the Great Lakes, could not get support from the

University of Michigan or from government institutions to operate research over the long term. Consequently he contributed little to the definition of the nascent discipline of ecology.

As an example of how a research station could contribute to the development of ecology as a new science, consider the Desert Botanical Laboratory of the Carnegie Institution of Washington, which became a major patron of scientific research after its incorporation in 1902 (Kingsland 2005, Kohler 1991). The Carnegie Institution wanted to develop its own facilities in areas of science thought to be important both for their theoretical problems and for the economic growth of the country. In biology the experimental study of evolution was considered to be an important new direction for research. Darwinian selection theory was being challenged by other theories, notably by the new mutation theory proposed by Dutch botanist Hugo de Vries. This theory suggested that evolution occurred rapidly and could be observed; moreover, if mutations could be induced experimentally, then there were exciting possibilities for producing new kinds of plants and animals. This kind of experimental work on evolution attracted the attention of the Carnegie officers.

The Carnegie Institution created two communities for research on evolution and related biological problems: the Desert Botanical Laboratory (founded in 1903) located outside of Tucson, and the Station for Experimental Evolution (founded in 1904) at Cold Spring Harbor, Long Island, New York. In both locations, biologists explored a wide variety of biological problems, combining laboratory experiment and field research. The Cold Spring Harbor station eventually led in the development of genetics and eugenics, whereas the Tucson laboratory, under the dynamic leadership of botanist Daniel Trembly MacDougal, emphasized research in plant physiology and ecology.

The physiological and ecological focus of the Desert Botanical Laboratory in part reflected its desert location, which was seen as particularly suitable for the study of general biological problems involving organism-environment relations. Ecology at this time was defined in part as the experimental study of adaptation. As Schimper, Cowles, and other proponents of ecology had realized, the analysis of adaptation involved physiological study, but also demanded field research. Because the desert was an extreme environment, it was thought that this would be an ideal place to test central ideas about how evolution occurred—for instance, comparing the role of competition between organisms to direct adaptation to the physical environment. The desert location provided an environment where different hypotheses about the causes of evolutionary change might be tested and where experiments on adaptation could be performed along with long-term observation of species distribution and abundance.

The importance of the Desert Botanical Laboratory was that it was devoted purely to research; resident scientists had the freedom to pursue any kind of research without administrative or teaching duties. The desert location made it easy for scientists to combine laboratory and field research and to move between different kinds of projects. Daniel MacDougal, whose background was in plant physiology, was just as interested in field studies and invited ecologists to work at the laboratory. Ecological studies began in the surrounding area and also in more distant locations. The accidental creation of the Salton Sea in California, caused by flooding of an irrigation channel of the Colorado River in 1905, provided a unique opportunity to study a region undergoing rapid changes. MacDougal along with colleagues from the University of Arizona spent several years studying the Salton "experiment," producing a multidisciplinary report on the region (MacDougal et al. 1914). Some of the research at the laboratory involved physiological ecology, and some of it was observational, looking at plant distribution and abundance and compiling climatic data. Forrest Shreve and Burton Livingston started long-term studies of desert vegetation and climate (Bowers 1988). The possibilities of long-term study, perhaps not fully appreciated at the time, were exploited by later generations of ecologists, who have continued to monitor the changes in the same landscapes that MacDougal, Shreve and Livingston studied (Hastings and Turner 1965, Turner et al. 2003). Of great value was the fact that MacDougal and others took many photographs of the areas they investigated, so that it was possible decades later to return to the same locations, identify some of the same plants from the original photographs, and chart exactly what changes had occurred in the interval.

The desert lab also contributed to the development of ecological theory, not so much by virtue of its desert location but by virtue of its role as a research sanctuary. Frederic Clements was invited to the laboratory, where he wrote his magnum opus, *Plant Succession: An Analysis of the Development of Vegetation* (1916), and thereby established himself as the leading theoretician of American ecology. Clements subsequently became a full-time Carnegie employee, working partly at Tucson, partly at an alpine laboratory on Pikes Peak in the Rocky Mountains, and later at Santa Barbara, California. His approach could be described as an effort to transcend place in order to generate theories that could apply to any landscape, and Shreve and Livingston detected many mistakes in his descriptions of desert vegetation (Bowers 1988, 59–60). Clements characterized vegetation in a broad way and was less interested in the specific details that defined particular places, such details being equivalent to "noise" that could be ignored. The desert laboratory's research embodied two antithetical approaches to ecology: Clements sought generality at the expense of accuracy, while Shreve, being more interested in the specific ecological problems of the desert, undertook detailed, descriptive studies which provided checks to Clements's errors.

All this work helped to establish ecology by giving it a firm foundation in research. Many of the laboratory scientists were charter members of the Ecological Society of America, founded in December 1915. The journal published by the Desert Botanical Laboratory, *Plant World*, was later handed to the Ecological Society and became *Ecology*, the first American journal devoted to ecological research. Here we can see the important connection between the creation of research laboratories and the development of ecology as a discipline. The history of American ecology is in part the history of the various struggles by scientists to attract funds for the creation of new kinds of research enterprises and biological stations. As the infrastructure for biological research gradually built up, so the discipline of ecology took on permanent form, with some ventures being more successful and long-lived than others.

The different research lines pursued in these places were subject to the judgments of the patrons of science. We have already noticed the need to balance practical needs with general biological research in places that received government support. At private research stations the interests of the patrons were also important. At the Tucson laboratory, for instance, scientists promoted the importance of basic research above applied work on the grounds that applied work was more the domain of government institutions such as agricultural experiment stations. To please the Carnegie officers, however, that research had to appear pathbreaking; hence, the experimental investigation of basic life processes was a crucial component. When the Carnegie Institution moved its experimental program away from Tucson and built a new laboratory at Stanford University, the observational ecology done at Tucson lost the support of the experimental scientists who had been there. Shreve, who took charge of the Tucson operations, found it difficult to convince the Carnegie officers that his observational work on vegetation and climate justified keeping the laboratory open during the Great Depression. In 1940, after a fire destroyed one building, the Carnegie Institution closed the laboratory and handed its buildings to the U. S. Forest Service, which later sold it to the University of Arizona.

Robert Kohler (2002) views the relationship between laboratory and field study as a competitive territorial struggle between two quite different cultures with different status. Ecology represents a hybrid of these two cultures, but Kohler argues that field science was for a long time viewed as secondary to the more rigorous experimental research pursued in the laboratory. Development of more sophisticated methods of field research took several decades. Kohler depicts those people who worked on the border between laboratory and field—the position of many ecologists—as having an ambiguous identity because they lacked a secure professional niche. He argues that it took about fifty years, until about mid-century, for this "border" region to become a more secure location. By that time the disciplines that mixed field and laboratory practices were more mature, and the status relations between the laboratory and the field were less one-sided.

While acknowledging the reality of the biases that Kohler describes, we are still left with the problem of explaining how such sciences as ecology managed to take hold in the early years of the century, when life on the "borderland" was supposedly at its most precarious. I would emphasize that in order to understand the formation of the discipline, we have to look to its material underpinnings. By the late nineteenth century ecology was recognized to be both an experimental science and a field science, but traditional laboratory-based approaches were viewed as inadequate to address broader problems concerning adaptation or concerning distribution and abundance of species. As a result, leading biologists stressed the importance of field study and encouraged patrons to create infrastructure to support research, albeit with mixed success. This was an uneven process requiring determined effort, but where it succeeded, as in the desert laboratory, it helped to launch the discipline of ecology in the United States.

Conclusion

In addition to exploring some ideas about the role of place in generating or challenging ecological theory, we have also considered the important role that field stations and field laboratories have played in the development of the discipline. In the European cases that have been touched on, colonial networks were important for creating infrastructure in exotic environments, thereby contributing to ecology as well as to other sciences. Within Europe, linking the research function of biological stations to higher education was a key step to ensuring government support, as Charles Kofoid found. Taking a closer look at the American context, success in developing a research school in ecology depended on showing that practical needs would be met, but also that basic research could serve practical interests. At the desert lab, studies aimed at understanding the mechanisms behind evolutionary changes—and perhaps one day controlling them—attracted the Carnegie Institution's support. Taken together, these biological stations and laboratories enabled scientists to explore a wide range of ecological problems, ranging from physiological ecology at the experimental end to descriptive studies that were more like natural history, but with a quantitative slant. That cumulative research took ecology over the threshold from passing fad to legitimate discipline.

This accumulation of a "critical mass" of research depended on the creation of research stations and the close study of environments that were particularly well suited to ecological research, whether it was the tropics of Java, the arctic environments of Alaska and Greenland, or the deserts of the Sahara and the United States. Ecology came into its own when scientists began to understand the value of place-based research: when they realized that collecting specimens from the field was inadequate and needed to be supplanted by prolonged study in the field. These scientists had a strong conviction that it was important to study problems in evolution, adaptation, and biogeography—not to reject natural history as old-fashioned and useless, but to transform it into something modern and vital. For European ecologists like Schimper, remaining within the confines of European society simply did not trigger the same questions about nature that one would ask in tropical, desert, or arctic regions.

Ecological problems were not only part of basic biology, but could have applications to agriculture, forestry, and fisheries, and were therefore tied to broader agendas concerning economic growth, development, and colonialism. In addition, field stations and laboratories could foster multidisciplinary studies of particular places and bring together scientists from around the world. The new research infrastructure also provided support for opportunistic studies of areas undergoing rapid transition, as on Krakatoa or at the Salton Sea. Thus the emergence of ecology as a discipline depended in part on intellectual developments (such as evolutionary theory and the theory of succession), in part on the conviction that ecology represented a distinctive approach that required its own disciplinary niche, and in no small measure on the success of early biologists in creating field stations to foster place-based research and educate the next generation. The fact that it was sometimes difficult to find funding for such enterprises underscores the need, still important today, to articulate the value of place-based and long-term research, and perhaps also to justify such research in terms of pragmatic societal needs beyond the basic desire to advance scientific understanding of the world.

References

Adams, C. C. 1913. *Guide to the Study of Animal Ecology*. Macmillan, New York.
———. 1946. Biological field stations. *Scientific Monthly* 62:469–70.
Adams, C. C., and G. D. Fuller. 1940. Henry Chandler Cowles, physiographic plant ecologist. *Annals of Association of American Geographers* 30:39–43.
Benson, K. R. 1987. H. Newell Martin, W. K. Brooks, and the reformation of American biology. *American Zoologist* 27:759–71.
———. 1988. From museum research to laboratory research: The transformation of natural history into academic biology. Pages 49–83 *in* R. Rainger, K. R. Benson, and J. Maien-

schein, eds., *The American Development of Biology*. University of Pennsylvania Press, Philadelphia.

Bocking, S. 1990. Stephen Forbes, Jacob Reighard, and the emergence of aquatic ecology in the Great Lakes region. *Journal of the History of Biology* 23:171–98.

Bowers, J. E. 1988. *A Sense of Place: The Life and Work of Forrest Shreve*. University of Arizona Press, Tucson.

Brooks, W. K. 1884. Chesapeake Zoological Laboratory: Report of the Director for its first six years, 1878–1883. *Johns Hopkins University Circulars, 1883–1884*. John Murphy and Company, Baltimore, pp. 91–94.

———. 1899. *The Foundations of Zoology*. Macmillan, New York.

———. 1996. *The Oyster*. Johns Hopkins University Press, Baltimore.

Browne, J. 1995. *Charles Darwin: Voyaging*. Princeton University Press, Princeton, NJ.

Cittadino, E. 1980. Ecology and the professionalization of botany in America, 1890–1905. *Studies in History of Biology* 4:171–98.

———. 1990. *Nature as the Laboratory: Darwinian Plant Ecology in the German Empire, 1880–1900*. Cambridge University Press, Cambridge.

1993. A 'marvelous cosmopolitan preserve': The dunes, Chicago, and the dynamic ecology of Henry Cowles. *Perspectives on Science* 3:520–63.

Clements, F. 1916. *Plant Succession: An Analysis of the Development of Vegetation*. Carnegie Institution of Washington, Washington, DC.

Cook, R. E. 1977. Raymond Lindeman and the trophic-dynamic concept in ecology. *Science* 198:22–26.

Cooper, W. S. 1913. The climax forest of Isle Royale, Lake Superior, and its development. I. *Botanical Gazette* 55:1–44.

———. 1923. The recent ecological history of Glacier Bay, Alaska: The interglacial forests of Glacier Bay. *Ecology* 4:93–128.

———. 1926. The fundamentals of vegetational change. *Ecology* 7:391–413.

Cowles, H. C. 1899. The ecological relations of the vegetation on the sand dunes of Lake Michigan. *Botanical Gazette* 27:95–117, 167–202, 281–308, and 361–91.

———. 1904. The work of the year 1903 in ecology. *Science* 19:879–85.

Darwin, C. 1964. *On the Origin of Species: A Facsimile of the First Edition*. Harvard University Press, Cambridge, MA.

Engel, J. R. 1983. *Sacred Sands: The Struggle for Community in the Indiana Dunes*. Wesleyan University Press, Middletown.

Evans, F. C. 1956. Ecosystem as the basic unit in ecology. *Science* 123:1127–28.

Gleason, H. A. 1926. The individualistic concept of the plant association. *Bulletin of the Torrey Botanical Club* 53:7–26.

Hastings, J. R., and R. M. Turner. 1965. *The Changing Mile: An Ecological Study of Vegetation Change with Time in the Lower Mile of an Arid and Semiarid Region*. University of Arizona Press, Tucson.

Keiner, C. 1998. W. K. Brooks and the oyster question: Science, politics, and resource management in Maryland, 1880–1930. *Journal of the History of Biology* 31:383–424.

Kingsland, S. E. 2005. *The Evolution of American Ecology, 1890–2000*. Johns Hopkins University Press, Baltimore.

Kofoid, C. A. 1910. *The Biological Stations of Europe*. Government Printing Office, Washington, DC.

Kohler, R. E. 1991. *Partners in Science: Foundations and Natural Scientists, 1900–1945*. University of Chicago Press, Chicago.

———. 2002. *Landscapes and Labscapes: Exploring the Lab-Field Border in Biology.* University of Chicago Press, Chicago.

Lindeman, R. L. 1942. The trophic-dynamic aspect of ecology. *Ecology* 23:399–418.

MacDougal, D. T., et al. 1914. *The Salton Sea: A Study of the Geography, the Geology, the Floristics, and the Ecology of a Desert Basin.* Carnegie Institution of Washington, Washington, DC.

Maienschein, J. 1986. Introduction. Pages 3–50 *in* J. Maienschein, ed., *Defining Biology: Lectures from the 1890s.* Harvard University Press, Cambridge, MA.

———. 1988. Whitman at Chicago: establishing a Chicago style of biology? Pages 151–82 *in* R. Rainger, K. R. Benson, and J. Maienschein, eds., *The American Development of Biology.* University of Pennsylvania Press, Philadelphia.

Margalef, R. 1968. *Perspectives in Ecological Theory.* University of Chicago Press, Chicago.

McIntosh, R. P. 1975. H. A. Gleason, 'Individualistic Ecologist,' 1882–1975: His contributions to ecological theory. *Bulletin of the Torrey Botanical Club* 102:253–73.

Nicolson, M. 1990. Henry Allan Gleason and the individualistic hypothesis. *Botanical Review* 56:91–161.

Olsson-Seffer, P. 1905. A biological station in Greenland. *Science* 21:180–91.

———. 1911. A plea for organized research in the tropics. *Science* 33:411–16.

Pauly, P. J. 1988. Summer resort and scientific discipline: Woods Hole and the structure of American biology, 1882–1925. Pages 121–50 *in* R. Rainger, K. R. Benson, and J. Maienschein, eds., *The American Development of Biology.* University of Pennsylvania Press, Philadelphia.

Raitt, H., and B. Moulton. 1967. *Scripps Institution of Oceanography: First Fifty Years.* Ward Ritchie Press, Los Angeles.

Real, L. A., and J. H. Brown, eds. 1991. *Foundations of Ecology: Classic Papers with Commentaries.* University of Chicago Press, Chicago and London.

Rumore, G. M. 2009. *A National Laboratory, a National Monument: Carving Out a Place for Science in Glacier Bay, Alaska, 1879–1959.* PhD dissertation, University of Minnesota.

Schimper, A. F. W. 1903. *Plant-Geography upon a Physiological Basis,* 2 volumes, revised and edited by Percy Groom and Isaac B. Balfour. Clarendon Press, Oxford.

Schneider, D. W. 2000. Local knowledge, environmental politics, and the founding of ecology in the United States: Stephen Alfred Forbes and "The lake as a microcosm" (1887). *Isis* 4:681–705.

Söderqvist, T. 1986. *The Ecologists: From Merry Naturalists to Saviours of the Nation: A Sociologically Informed Narrative Survey of the Ecologization of Sweden, 1895–1975.* Almqvist and Wiksell International, Stockholm.

Tansley, A. G. 1904. The problems of ecology. *New Phytologist* 3: 191–200.

———. 1926. Succession: the concept and its value. Pages 677–86 *in* B. M. Duggar, ed., *Proceedings of International Congress of Plant Sciences, Ithaca, New York.* George Banta, Menasha, Wisconsin.

———. 1935. The use and abuse of vegetational concepts and terms. *Ecology* 16:284–307.

Tobey, R. C. 1981. *Saving the Prairies: The Life Cycle of the Founding School of American Plant Ecology, 1895–1955.* University of California Press, Berkeley.

Turner, R. M., R. H. Webb, J. E. Bowers, and J. R. Hastings. 2003. *The Changing Mile Revisited: An Ecological Study of Vegetation Change with Time in the Lower Mile of an Arid and Semiarid Region.* University of Arizona Press, Tucson.

Vitousek, P. M., L. L. Loope, and H. Adsersen, H., eds. 1995. *Islands: Biological Diversity and Ecosystem Function.* Springer, Berlin.

Warming, E. 1899. On the vegetation of tropical America. *Botanical Gazette* 27:1–18.

Whitman, C. O. 1898. Some of the functions and features of a biological station. *Science* 7: 37–44.

Winsor, M. P. 1991. *Reading the Shape of Nature: Comparative Zoology at the Agassiz Museum.* University of Chicago Press, Chicago.

3 *Leopold's Legacy*

AN ECOLOGY OF PLACE

Donald M. Waller and Susan Flader

Figure 3.0. Aldo Leopold planting pines at the shack, 1936. Photograph by Carl Leopold. Reprinted courtesy of the Aldo Leopold Foundation.

Abstract

Trained as a forester, Aldo Leopold (1887–1948) was first by avocation and later by vocation an ecologist of remarkably broad vision and understanding. At each stage in his life—from his boyhood along the Mississippi River in Iowa and his education in the East, through his career with the U.S. Forest Service in the Southwest, to his return to the Midwest in Wisconsin—he sought out jobs, travel, and experiences to expand his understanding of landscapes and their biota. The trajectory of his intellectual growth reflects the lessons he learned in each region, amplified by trips he took to Germany and Mexico. The sense of place that informed and inspired his thinking emerges most insistently in *A Sand County Almanac*, centered on an abandoned farm he restored with his family along the Wisconsin River. This book and his community-based concepts of ecology and land health, esthetics, and ethics emerged from a lifetime of close and patient observation leavened by a deepening understanding of how humans affect ecosystems and how, in turn, people relate to landscapes. After exploring the sense of place Leopold experienced in various landscapes, we discuss his role in pioneering three approaches to ecological science: the comparison of different places, the use of history to analyze environmental change in a particular place, and the use of experiments. As Leopold pointed out, detailed knowledge of particular places gives us two rewards: the scientific understanding we need to manage land and an aesthetic appreciation for natural places that can inspire us to conserve and restore land. Leopold's land ethic is thus grounded in both ecological understanding and love of place. His lessons continue to resonate with ecologists and inspire conservationists.

Introduction

> I know a painting so evanescent that it is seldom viewed at all. . . . The artist has now laid his colors and sprayed them with dew. The Eleocharis sod, greener than ever, is now spangled with blue Mimulus, pink dragon-head, and the milk-white blooms of Sagittaria. Here and there a cardinal flower thrusts a red spear skyward. At the head of the bar, purple ironweeds and pale pink joe-pyes stand tall against the wall of willows. And if you have come quietly and humbly, as you should to any spot that can be beautiful only once, you may surprise a fox-red deer, standing knee-high in the garden of his delight (Leopold 1949, pp. 51–52).

Today, many Americans gain their graphic sense of nature from television documentaries or images beamed down from satellites. Aldo Leopold would have marveled at the power of these technologies to bring us so much information. His wonder, however, would be tempered by deep concern about how our experiences are becoming so tamed, circumscribed, and remote. Can any TV or digital image from space substitute for the intricate, intimate, and evanescent view of a Wisconsin River sandbar at dawn? He might be particularly concerned with how children are losing opportunities to experience nature directly. Richard Louv (2005) describes these serious declines in outdoor free-form play and the matching increases in what he terms "nature deficit disorder."

As a scientist, Leopold would delight in the rich data on land cover and ecological conditions that are now streamed to our desktops at the push of a button. But despite this largesse, few of us abandon fieldwork. Despite impediments, and despite sideways glances from some colleagues, we continue to send our students out into the field to ground-truth their data. Why? Most of us agree with Leopold that our personal experience of particular places has special value as an immediate, unfiltered, and primary source. It also reminds us that we must always integrate our field data with our broader knowledge of the historical, physical, and cultural factors that impinge on these places. Field experiences are simply more vivid than remotely sensed images, and more likely to lead us to new ideas.

Few knew, cared, or thought more about the places where they lived and worked than Aldo Leopold. In this chapter, we explore how he came to develop his sense of place and how particular places deepened his esthetic sensitivity, informed his science, shaped his approaches to conservation, and ultimately forged his ethics. Although we remember him now mostly for his mature and lyrical descriptions of nature and his pioneering contributions to conservation science, these emerged gradually from a lifetime's work and play in diverse wild and domesticated landscapes. Leopold the forester and Leopold the game manager trained and framed Leopold the scientist and Leopold the conservation sage. But Leopold remained throughout his career an intent observer. His fascination with place and the natural world germinated in his boyhood along the Mississippi River, pushed outward during his school years in New Jersey and at Yale, and unfolded during his tour of duty with the U.S. Forest Service in Arizona and New Mexico before flowering and bearing the fruit that we know upon his return to the native Midwest. Here, he reestablished roots, raised a family, and laid the foundations for a new profession and department while exploring the marshes, fields, and forests of the region and working with landowners and policymakers to protect and restore them.

Leopold's thinking was perhaps most penetrating as he pondered human relationships to the natural world at his farm along the Wisconsin River. Here, he and his family fashioned an old chicken coop into "the shack," planted trees, nursed abused land back to health, and sought a "weekend refuge from too much modernity" (Leopold 1949, p. viii). This place embodied and reflected Leopold's love for the land and became the ecological, esthetic and ethical compass for his conservation thinking. The shack, the farm, and the nearby countryside incubated the young field of game management (now wildlife biology), the embryonic fields of conservation biology and restoration ecology, and the germ of Leopold's land ethic. All three emerge in his landmark book, *A Sand County Almanac*—often in the deceptively simple form of his reflecting on the natural places around him. Leopold's sense of place underlay both his science and his land ethic. Places attracted and motivated him to explore the natural world. No wonder the titles he chose for his book and for his many essays evoke places: "The Choral Copse," "A Mighty Fortress," "Marshland Elegy," "Escudilla," "Song of the Gavilan."

Here, we recount episodes in Leopold's life to illuminate how particular places inspired and informed his community-based concepts of ecology and land health and their relation to esthetics and ethics. We then discuss his role in pioneering modern approaches to ecological science. We end by discussing his philosophical legacy to contemporary conservation and the ecology of place.[1]

Odyssey

Trained as a forester, Leopold developed into an adept applied scientist who actively sought and gained knowledge throughout his life. He particularly appreciated the opportunity to gain direct knowledge of nature via field observations in particular places. These observations began when he was a boy, eagerly exploring the bluffs and bottoms of the Mississippi River near Burlington, Iowa. The earliest materials we have in his own hand are ornithological notes and drawings of his observations. Leopold's grandfather had trained in Germany as a landscape architect and he imparted his esthetic sensitivity to his daughter and grandson. The German name for their home—*Lug-ins-Land*, or look into the land—reflected their strong attachment to both place and observation.

When Leopold went off to preparatory school at Lawrenceville, New Jersey, in 1904, he began a remarkable correspondence with his mother, comparing his own observations of bird arrivals, the blooming of plants, and other encounters on frequent "tramps" in the woods and streams of central New Jersey—and later in Connecticut during his years at Yale—with hers near their home on the Mississippi bluffs. In retrospect we can see these observations as the genesis of his lifelong fascination with phenology, the science of recurring phenomena in nature.

After college and forestry school at Yale, Leopold was eager to accept a field position with the U. S. Forest Service, which was then full of the excitement and esprit de corps of a young public agency. He began by leading a reconnaissance crew on the Apache National Forest in southeastern Arizona Territory. In part because he lacked familiarity with the place and the ways of seasoned locals under his supervision, he made some errors that triggered an official investigation. But he was given another chance and then advanced quickly, becoming supervisor of the Carson National Forest in New Mexico by age twenty-five and chief of operations for the twenty million acres in the southwestern district by age thirty-two. During these years he spent both professional and personal time exploring the varied landscapes of the Southwest. He also met and married a Hispanic woman, Estella Bergere of Santa Fe, who nurtured his aesthetic sensitivity and shared his attachment to the places they experienced together.

As an avid hunter and fisherman, Leopold learned the habits and habitats of game and fish. As a forest officer he organized citizen groups to help conserve them. He even served for a time as secretary of the Albuquerque

Chamber of Commerce, seeking to forge citizen consensus on improving the urban environment. As a grazing administrator and forest inspector he studied stocking levels, range conditions, watershed erosion, and how these changed over time. As a backcountry enthusiast he emphasized the special value of expansive public lands for recreation and wildlife, and proposed designating a large part of the Gila National Forest as the country's first wilderness area in 1922. Above all, he took every opportunity to gain an intimate understanding of how people, livestock, wildlife, vegetation, soils, waters, fire, and climate interacted within and across the landscapes of the Southwest. He also reflected on these in more than one hundred scientific and popular articles and Forest Service reports.

Hoping to advance both his own career and the role of science in his agency, Leopold transferred in 1924 to the Forest Products Laboratory in Madison, Wisconsin. Perhaps frustrated on both fronts, he left the Forest Service a few years later. To sustain himself, his growing family, and his continuing interest in wildlife, he persuaded the Sporting Arms and Ammunition Manufacturers' Institute to support his efforts to conduct an eight-state regional game survey. This afforded him opportunities to spend abundant time in the field, recording firsthand observations across a broad spectrum of Midwestern landscapes and discussing long-term change and conservation policy with hundreds of local farmers, sportsmen, and public officials. The results emerged in his report (1931) and in his landmark text *Game Management* (1933a).

In these travels and throughout his life, Leopold emphasized the value of personal observation and meticulous record-keeping. The tiny notebook he carried in his shirt pocket allowed him to record data in the field, and these he converted to careful field notes, often including sketches, as soon as he returned to camp, office, or home. Into his journals (figure 3.1) flowed a copious stream of details on weather, tracks, dens, vegetation, browse damage, and routes of travel. No two seasons or even days were exactly the same, but lessons were to be learned from each and particularly from the ensemble, which Leopold followed in all its moods and guises, analyzing changes and ruminating on their meaning.

Leopold's focus and eye for detail emerge in the observations we read in *A Sand County Almanac*. Over the years, he drew from his observations and journals to work and rework various themes, including the dramas and puzzles posed by wild creatures, the evanescent beauty of woods and streams, the rewards of personally reading the landscape, and the personal responsibilities of land stewardship. Importantly, he viewed a scientific understanding of natural systems as fully compatible with, and supported by, a personal emotional attachment to nature. He understood that both grow with direct and repeated contact with particular places. He wrote not in remote or ab-

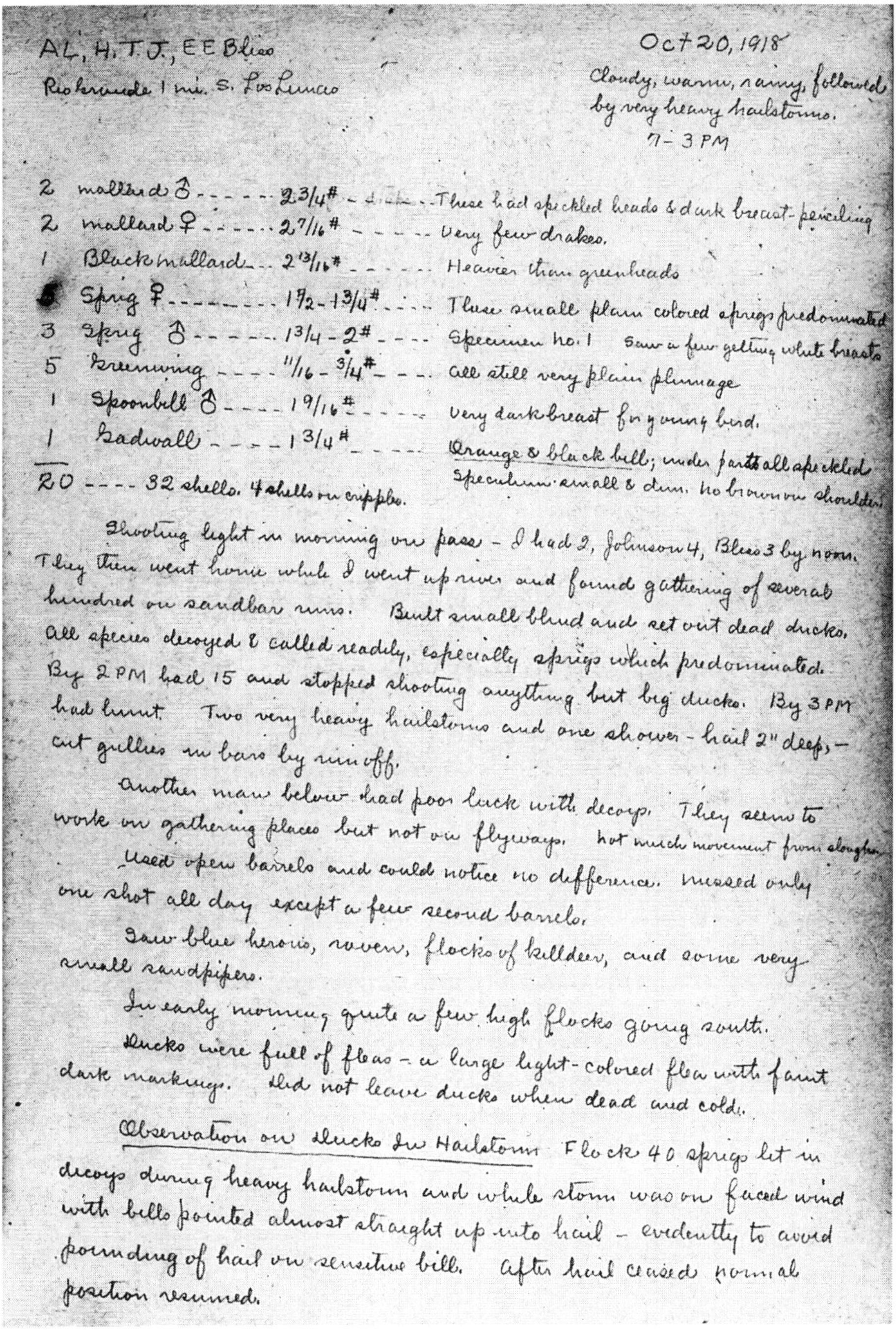

AL, H.T.J., E E Bliss   Oct 20, 1918

Rio Grande 1 mi. S. Los Lunas   Cloudy, warm, rainy, followed
                                               by very heavy hailstorms.
                                               7 - 3 PM

2   mallard ♂ ----- 2 3/4 # ---- ... These had speckled heads & dark breast-penciling
2   mallard ♀ ----- 2 7/16 # ---- ... Very few drakes.
1   Black mallard --- 2 13/16 # ---- ... Heavier than greenheads
    Sprig ♀ --------- 1 1/2 - 1 3/4 # ---- These small plain colored sprigs predominated
3   Sprig ♂ ------- 1 3/4 - 2 # ---- Specimen No. 1   saw a few getting white breasts
5   Greenwing ---- 11/16 - 3/4 # ---- all still very plain plumage
1   Spoonbill ♂ ---- 1 9/16 # ----- very dark breast for young bird.
1   Gadwall ---- 1 3/4 # ----- Orange & black bill; under parts all speckled
20 ---- 32 shells. 4 shells on cripples.   Specimen small & dim. No brown on shoulder

Shooting light in morning on pass - I had 2, Johnson 4, Bliss 3 by noon.
They then went home while I went up river and found gathering of several
hundred on sandbar runs.   Built small blind and set out dead ducks.
All species decoyed & called readily, especially sprigs which predominated.
By 2 PM had 15 and stopped shooting anything but big ducks. By 3 PM
had hunt.   Two very heavy hailstorms and one shower - hail 2" deep, -
cut gullies in bars by runoff.

Another man below had poor luck with decoys. They seem to
work on gathering places but not on flyways. Not much movement from slough.

Used open barrels and could notice no difference. Missed only
one shot all day except a few second barrels.

Saw blue herons, raven, flocks of killdeer, and some very
small sandpipers.

In early morning quite a few high flocks going south.

Ducks were full of fleas - a large light-colored flea with faint
dark markings. Did not leave ducks when dead and cold.

Observation on Ducks In Hailstorm   Flock 40 sprigs lit in
decoys during heavy hailstorm and while storm was on faced wind
with bills pointed almost straight up into hail - evidently to avoid
pounding of hail on sensitive bill. After hail ceased normal
position resumed.

Figure 3.1. Entry from Leopold's "New Mexico Journal" on October 20, 1918, showing his careful
field observation. Leopold Collection X25 2276, University of Wisconsin Archives. Reprinted
courtesy of the Aldo Leopold Foundation.

stract terms for other scientists but rather for a public that included fishermen ("The Alder Fork"), farmers ("Bur Oak" and "Sky Dance"), young men about to enter military service ("Flambeau"), and taxi drivers and bankers ("A Man's Leisure Time"). His lifelong odyssey of observation and reflection grounded in particular landscapes gave him the basis for synthesizing ecology, history, economics, and ethics so gracefully in his landmark essay "The Land Ethic."

Grounded in the Southwest

The value of extended observations of one place over time first struck Leopold in the Southwest, as epitomized in his riveting piece "Thinking Like a Mountain" (1949). This essay recounts a searing incident that likely occurred within months of his arrival on the Apache National Forest in 1909. The essay, not written until 1944, compresses into a few lyrical pages a wealth of experience and a revolution in how we think about ungulates and predators. The immediacy of Leopold's narrative, the simple tone of his writing, and the resonance of his conclusions make it clear that he sought an audience beyond his academic peers. He courageously tells the story of how he helped shoot a wolf (*Canis lupus*) with pups, watching "a fierce green fire dying in her eyes." In so doing, he admits his own complicity with conventional antipredator sentiments and shares observations and insights that actually occurred incrementally over succeeding decades (Flader 1974). These experiences led him to fundamentally alter his thinking about the role of predators. In particular, he uses this emotionally charged story to teach both an ecological and a moral lesson. The ecological lesson, reflecting years of observation, centers on the damaging effects of deer overabundance and overbrowsing and the role of wolves and other predators in limiting deer numbers and impacts. The moral lesson centers on the hubris of imagining that what we think is good for us here and now will sustain the systems we depend on over the long term. The power of the essay emerges from Leopold admitting his role in killing "varmints" and the progression of his own ideas and values, traits he sought to teach others. Interestingly, it was a former student who persuaded him to make this personal confession.

As his concerns about landscapes and watersheds grew, Leopold became convinced that local forest rangers needed to make regular detailed observations of flora, fauna, and other natural phenomena so they could compare these across years, units, and forests to assess the efficacy of their management. To facilitate this, he devised a new method to inspect forests complete with printed, notebook-sized tally sheets for recording the myriad relevant details (Leopold 1924a). This new system was viewed favorably by his superiors in Washington, and was considered a model for forest inspection nationwide. Leopold's emphasis on the importance of systematic long-term monitoring was later institutionalized by the U.S. Forest Service in the form of a standardized protocol for surveying forest conditions. This evolved into the Forest Inventory and Analysis (FIA) program (http://www.fia.fs.fed.us/). The National Park Service recently began a parallel "Vital Signs" monitoring program (Johnson et al. 2008).

Leopold was most open to making significant new observations and syntheses in areas where he did not have major professional responsibility. For example, his interest in southwestern watersheds developed informally in the early 1920s when he conducted official inspections of forests he had known intimately from his earlier years there. On these return visits he noted increased erosion, siltation, overgrazing, and associated shifts in vegetation. In his inspection reports and other writing, he puzzled over the causes and consequences of these changes. His observations and reflections came together in "Grass, Brush, Timber and Fire in Southern Arizona" (Leopold 1924b), which still stands as a landmark in ecological thinking. Here he synthesized a remarkable succession of observations and inferences to challenge Forest Service dogmas about how cattle grazing could prevent fires and about how trees prevent erosion. Reflecting on what he saw, Leopold realized that fire was not an unmitigated evil and could be beneficial, even necessary, in maintaining grasslands and preventing encroachment of brush. Fire-maintained grasses, in turn, often protected soils better than trees or brush.

Thinking hard about the practical problems posed by southwestern watersheds taught Leopold that disciplined observation, intellectual rigor, and an open mind formed the path to ecological insight. At the same time that he began to share his unorthodox ecological conclusions he realized that human biases and values presented even greater obstacles. Rather than accept erosion as the inevitable by-product of ranching and farming, Leopold worked to instill in both professional colleagues and private landowners a sense of responsibility for the land they managed. In fact, all his major efforts to articulate a land ethic—"Some Fundamentals of Conservation in the Southwest" (1923), "The Conservation Ethic" (1933b), "A Biotic View of Land" (1939a), "The Ecological Conscience" (1947a), and "The Land Ethic" (1949)—were grounded in his concerns with watershed protection (Flader 1987). However, in trying to teach these moral lessons, Leopold faced impediments beyond ecological misunderstanding. For him, gullied soils and clogged streams represented a failure to accept personal responsibility for maintaining and restoring the health of the land system.

Changing Places

Although Leopold clearly relished the deep understanding he gained by working in familiar places, he was also drawn to explore new places. He used the game survey of the late 1920s and early 1930s to gain a broader understanding of how wildlife species responded to differences in landscapes across the Midwest. Two other trips in the mid-1930s gave him opportunities to extend his interests in comparative ecology to foreign lands. Both of these trips allowed him to see his familiar landscapes in a new light, opening his eyes to how ungulate browsing could change landscapes, though in opposite ways.

Leopold traveled to Germany and Czechoslovakia in 1935 on a fellowship to study forests with colleagues. He looked forward to visiting the land of his ancestors and was eager to see the outcome of the Germans' centuries-long practice of forestry and wildlife management. But he was disturbed by what he found. Here, he observed directly what he later described as a style of forestry incompatible with wildlife, and a mode of wildlife management incompatible with forestry (Leopold 1936). Privileged hunters eager for abundant populations of red deer (*Cervus elaphus*) and roe deer (*Capreolus capreolus*) fostered densities of these ungulates so high that only seedlings of the least palatable tree species could survive. Foresters eager to pursue efficient forms of silviculture favored conifers like spruce (*Picea*) that were unable to provide nutritious browse for the deer and that led to long-lasting declines in soil fertility and understory diversity. Trees were planted in rows, few limbs or trunks were allowed to rot naturally, and fences were necessary to protect vulnerable seedlings. Instead of mimicking nature, Germanic forestry and game management were transforming forests to highly artificial, dysfunctional ecosystems.

Leopold had taught that the way to manage game was to manage habitat, but in Europe he realized that such efforts could run amok if they were not constrained and directed at emulating natural ecosystems. He was not alone. Backlash against overmanaged forests in Germany had led to the emergence of a "Dauerwald" (permanent woods) movement that emphasized restoring forests to a more natural mixture of species and age classes capable of supporting diverse populations of wildlife. Leopold saw the promise of Dauerwald, but the movement proved short-lived as the Nazis consolidated their control.

He returned from Europe a changed man. He realized that the Midwest was on the same path to overmanaging its forests and game populations to the detriment of wild species. In addition to publishing five papers on his German findings, he began to ruminate on the purposes of management in speeches and papers such as "Means and Ends in Wild Life Management,"

"Threatened Species," and "Conservation: In Whole or in Part?" (Flader and Callicott 1991). He realized that foresters and wildlife managers would have to adopt a broader vision of their activities if the full suite of "non-game" species and the wilder conditions they depend on were to have a chance to prevail. These were ecological lessons that emphasized, like his watershed work in the Southwest, how species interact in complex ecosystems in response to local and historical conditions. While not abandoning utilitarian values, Leopold began to stress the utility of balancing habitat needs among species (including humans) and preserving the intricate system as a whole.

Leopold experienced a complementary epiphany during a series of hunting trips into the Sierra Madre de Occidental in northern Mexico with his brother Carl, son Starker, and a friend. He wrote of the thick-billed parrot, which he regarded as the *numenon* or essence of the place, in "Guacamaja," and of the river's dancing riffles and fat rainbow trout in "Song of the Gavilan." What impressed him most, however, were the contrasts he saw between the denuded, overgrazed and eroded landscapes he was familiar with on the Gila and Apache just over the border and these wilder landscapes in northern Mexico where logging, cattle, fire suppression, predator control, and erosion had yet to take their toll. As in Germany just the year before, he confronted the contrast between wild lands and those that were highly managed but biotically impoverished. In the Sierra Madre, he found a naturally functioning system where deer still thrived but were kept in check by wolves and mountain lions, where wildfires burned but did no great or lasting damage, and where streams and rivers ran fast but clear between mossy banks.

Pondering these differences, Leopold came to realize that natural systems that maintain evolutionarily established abundances of predators and rhythms of disturbance to which native species are adapted also retain a fundamental natural resilience, or what he later termed "land health." With a shock, he realized that his eye had been conditioned to accept landscapes and ecological relationships as normal that he now recognized were degraded and out of balance. "It was here," he would reflect years later (Leopold 1947b), "that I first clearly realized . . . that all my life I had seen only sick land, whereas here was a biota still in perfect aboriginal health." The "sick land" included even the large expanse of the Gila National Forest that he had helped to designate as wilderness. The Sierra Madre, with its recurring fires and abundant predators, led him to reconsider what was normal and to recalibrate his ideas about what was ecologically appropriate. Goals for protecting and restoring lands in the Southwest would henceforth have to meet a higher standard.

Germany and Mexico gave Leopold new contexts for understanding ecological dynamics in places long familiar to him. By leaving the familiar places, he learned more about what they could become and how they might be judged. The similarity of the German and Mexican locations to these places allowed him to apply his ecological knowledge, while the differences allowed him to see the familiar places in a new light.

A meta-lesson that Leopold drew from these epiphanies was that ecologists cannot judge the system they are in—or trust the knowledge they think they have—unless they have an adequate basis for comparison. To properly interpret the ecological status and health of any landscape or ecosystem requires us to obtain a baseline of normalcy against which to judge it. This led Leopold on his return from Mexico to seek (unsuccessfully) to interest professional colleagues in an international research effort "to describe, and define, in actual ecological measurements, the lineaments and physiology of an unspoiled mountain landscape" (Leopold 1937). It also led him to realize that large expanses of wilderness, if not substantially degraded, could have great utility as yardsticks against which to measure more managed systems, a theme he developed in "Wilderness as a Land Laboratory" (Leopold 1941) —but with this caveat: the only healthy wilderness he knew south of the Canadian border as of 1941 was in the Sierra Madre. Ecologists increasingly refer to this "missing baseline" problem (Jackson 1997, Pandolfi et al. 2005). Many of us also worry that each succeeding generation of naturalists calibrates its eyes and expectations against reference systems that continue to decline in diversity, complexity, and distinctiveness.

The Special Roles of History

Just as Leopold learned to see landscapes with fresh eyes after changing places, familiarity with particular places also led him to recognize how landscapes change in response to historical and ongoing forces (figure 3.2). Past conditions, disturbance regimes, and management activities all have lasting effects in many ecosystems. Here again, Leopold was well ahead of his time in ecological thinking. He thought hard about conditions that prevailed in the past and how these might have influenced the landscapes he saw. His essay "Good Oak," in *A Sand County Almanac*, is remarkable in particular for the compelling metaphor it uses—sawing through the growth rings of an oak (probably *Quercus bicolor*)—to relate both the farm's history and broader trends in conservation history. Leopold expressed similar ideas in essays like "Marshland Elegy" and "On a Monument to the Pigeon," where he pondered the potential extirpation of the sandhill crane (*Grus canadensis*) from Wisconsin and the finality of the passenger pigeon's (*Ectopistes migratorius*) extinction. In these essays Leopold explores ecology lessons (such as the fact that pigeons will no longer disperse acorns across the land) and moral lessons: "The Cro-Magnon who slew the last mammoth thought

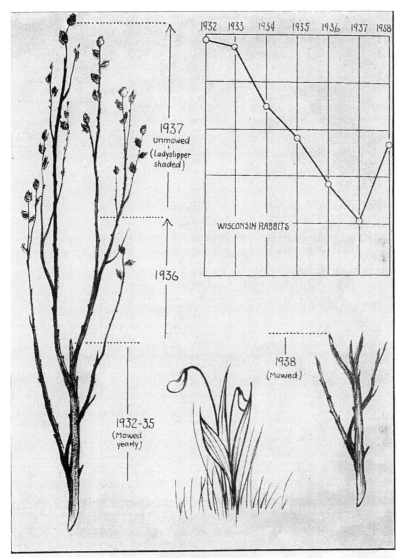

THE STORY OF A CYCLE

A mousy, unobtrusive, inconspicuous little bush, the bog-birch, plays an important role in the ups and downs of plant and animal life. Here is illustrated how it spells life or death to deer, grouse, rabbits, and ladyslippers in Wisconsin.
In 1932 to 1935 rabbits were abundant and ate down the bog-birches each winter, giving the ladyslippers the sun. During 1936 and 1937 the cycle decimated the rabbits and the bog-birches grew high and shaded out the ladyslippers. In 1938 the rabbits recovered, mowed down the birches and the ladyslippers regained their place in the sun.

Figure 3.2. "The story of a cycle." Leopold's systematic observation of rabbit population cycles allowed him to trace the impacts of their browsing on an entire ecosystem. Leopold's drawing was published in Leopold 1939b. Reprinted courtesy of the Aldo Leopold Foundation.

only of steaks." Knowing history and how we have affected other species so dramatically brings the specific moral lesson that we must pay attention to our impacts and work consciously to redirect and redesign our actions to be gentle to the land and the other species we share it with.

Understanding the historical effects of disturbance and land management on current plant and animal distributions and abundances has become an active area of research in ecology (Foster et al. 1997, 1998; Halpern and Spies 1995; Whitney and Foster 1988). In the southern Appalachians, forests clear-cut up to 87 years previously have recovered only half the species richness and one-third the understory of matched uncut forests (Duffy and Meier 1992). Smaller forest patches in this region retain a lower cover and diversity of herbaceous plants and are particularly prone to losing ant-dispersed herbs (Pearson et al. 1998). Lands that were tilled fields 100 years ago are now dominated by tulip poplars (*Liriodendron tulipifera*) and are far more likely to be invaded by exotic plants than adjacent untilled lands (T. Kuhman, unpublished data). The existence of such persistent historical legacies reinforces Leopold's concern about having adequate reference areas to serve as ecological benchmarks (Egan and Howell 2001).

Another aspect of using history to understand ecology involves our ability to track the changes that occur over time. In his journals Leopold always noted not only what and where but also when something occurred. He carefully recorded when plants leafed out, when they set flowers and fruit, and when they lost their leaves. He noted when migrating birds arrived in the spring and departed in the fall. He also noted when ice first covered a pond or river and when it broke up. These observations, which began in his boyhood ornithological journals and his correspondence with his mother, became an ingrained lifelong habit. He asked his students to keep such records, knowing that training and discipline in close observation would pay off in other aspects of their work. Collectively, he realized, such records also allow for a more systematic study of phenology—the study of the timing of various natural processes and occurrences. By examining changes in one place through time, we gain appreciation of both that which is regular and dependable about a place and how it varies from year to year. This regularity and variation parallel the similarities and differences we see among habitats within a region. When he was elected president of the Ecological Society of America, his highest scientific honor, Leopold submitted to *Ecological Monographs* a paper describing in detail his phenological observations (Leopold and Jones 1947). In an opening that is extraordinary for so august a journal, it begins: "Each year, after the midwinter blizzards, there comes a night of thaw when the tinkle of dripping water is heard in the land." Lovers of *A Sand County Almanac* will recognize this as the first sentence of that book's first essay, "January

Thaw," in which Leopold ruminates on his phenological observations at the shack.

Although Leopold appreciated the fact that his phenological studies had scientific value, he could not have anticipated how his data would be used to infer another kind of historical change at the close of the twentieth century. After moving back to Wisconsin to live near the family shack, his daughter Nina resumed the phenological observations that were so much a part of shack life. Each year, beginning in 1976, she collected parallel data to complement those collected from 1936 to 1948 by her father. These records span more than seven decades and still continue. They clearly document the creeping impacts of climate change on the birds and plants near the shack. We now know, for example, that spring arrives there a week or two earlier for many species than it did during Aldo's time (Bradley et al. 1999). Were he writing *A Sand County Almanac* today, Leopold would have to shift his July "Prairie Birthday" essay, as *Silphium* now blooms in June at the shack (Wright and Bradley 2008). We see how careful observations from one area, extended over many years, can provide unique insights into the global phenomenon of climate change. This could never have happened if Leopold had not believed in the value of making detailed place-based observations of seemingly trivial details.

A parallel story involves the eminent limnologist Gene Likens, whose systematic records on precipitation in the forested watershed of Hubbard Brook, New Hampshire, beginning in the 1960s led him to discover that rainwater was growing more sour, leading him to alert a continent to the growing menace of acid rain (Likens et al. 1979). Later measurements from the same forest would reveal how acid rain could arrest the growth of many trees, again demonstrating the value of accurate local baseline data. Likens chose to focus his own presidential address to the Ecological Society of America on the value of such long-term place-based data (Likens 1983).

The eminent plant ecologist John T. Curtis, Leopold's colleague at the University of Wisconsin during the 1940s (Meine 1988), also appreciated the value of comparative historical data. His chapter (Curtis 1956) in the renowned compendium *Man's Role in Changing the Face of the Earth* includes the now-famous figure of how forested habitats in southern Wisconsin's Cadiz Township became progressively smaller and more isolated as human activities—primarily farming—fragmented Midwestern landscapes. Curtis also undertook a broad and detailed survey of the remaining naturally vegetated landscapes in Wisconsin, resulting in his landmark book *The Vegetation of Wisconsin* (1959). These systematic quantitative data on community composition and species abundances were archived carefully and now provide an invaluable baseline for assessing ecological change in the latter twentieth century (Leach and Givnish 1996; Rooney et al. 2004; Rogers et al. 2008, 2009). It would have saddened Leopold to learn that species losses in northern Wisconsin forests reflect continuing problems with deer over-abundance (Wiegmann and Waller 2006).

Leopold also sought to teach his dynamic, holistic view of landscapes to students so that they too could read landscapes historically. At the time of his death he was working on a sequel to *Game Management*, a general textbook for his undergraduate survey course in wildlife ecology. The book would have focused on a series of place-based case studies situated in land-scapes as diverse as a prairie coulee, a northern Wisconsin forest, a horned-owl territory, a fencerow, a central Wisconsin marsh, and an Ozark farm. The approach was foreshadowed at least as early as *Game Management*, in which Leopold wrote (pp. 387–88):

> It is astonishing how few of those who have learned by rote rule or "nature study" the statics of the land's present inhabitants or condition, ever learn to read the dynamics of its past history and probable future. To see merely what a range is or has is to see nothing. To see why it is, how it became, and the direction and velocity of its changes—this is the great drama of the land, to which "educated" people too often turn an unseeing eye and a deaf ear. The stumps in a woodlot, the species age and form of fencerow trees, the plow-furrows in an abandoned field, . . . the fire-scars on a sawlog—these and a thousand other roadside objects spell out words of history, and of destiny, of game and of people.

We see that Leopold's interest in history went beyond predilection to en-compass an ecological comprehension of land. His interest in the dynamic—in history—reflected his interest in destiny, and in how we can redress the consequences of land degradation.

Ecology Discovers Experiments

In addition to acute observation and historical inference, Leopold's sense of place also led him to initiate experiments at a time when these were still unusual in ecology. His experimentation, however, was different from the rigorously controlled experiments then ascendant in laboratory biology. Instead, it built on the idea of "natural experiments" wherein two or more ecosystems are compared that differ conspicuously in some aspect such as land-use history or management (Diamond 1983). From there, Leopold found it a logical step to pursue the kind of "tweak-the-system-and-see-how-it-responds" field experiment that Peckarsky et al. (chapter 9) describe. For example, he was keen to understand what farmers could do to enhance habitat conditions for wildlife species, a goal he saw as integral to maintain-ing the biotic productivity and beauty of their farms. He therefore established cooperative agreements with groups of farmers and sent his students off to live in these communities. The most famous of these experiments was in the severely eroded Coon Valley watershed in southwestern Wisconsin. Hundreds of local farmers and dozens of trained technicians, including his own sons Starker and Luna, participated in this pioneering demonstration project to integrate pasturage, row crops, forestry, and wildlife, all while pursuing the New Deal goal of conserving soil. Leopold described these Coon Valley experiments in a 1935 article he subtitled "An Adventure in Cooperative Conservation."

Leopold also worked with John Curtis and others in the mid-1930s to establish the University of Wisconsin Arboretum as a "land laboratory" in Madison to study how to restore and manage natural ecological habitats. Rather than plant a wide variety of exotic trees and shrubs, these pioneers envisioned the UW Arboretum as a living museum, a laboratory, and a textbook in which scientists seeking to rebuild native plant communities could educate themselves as well as the general public in the art and science of what we now call restoration ecology. Here, scientists, managers, and teachers continue to grapple with the questions that Leopold and Curtis raised about natural systems and how best to restore their structure and function. In pursuing these goals, we now frequently apply multiple experimental treatments, monitor the outcomes, and revise our management based on these outcomes in a cycle that Leopold Chair of Restoration Ecology Joy Zedler refers to as "adaptive restoration" (Zedler 2005).

The shack itself was an arena for Leopold's ecological experiments in land restoration. He bought the run-down farm on the Wisconsin River in 1935 as a weekend retreat for his family, but it soon became a place for them to work together planting trees and prairie plants to restore wildlife habitats. Laboratory, backyard, research, and recreation all merged in efforts that also incubated the essays that became *A Sand County Almanac*. Here, Leopold learned valuable lessons in humility as his plantings failed repeatedly in the droughts of the 1930s. He came to appreciate the ofttimes inscrutable factors involved in life and death, growth and decay. His own direct struggles to rebuild a diverse, resilient, and esthetically satisfying biota led him to ponder the individual decisions he made every time he wielded an axe or shovel. Here he came to understand in a direct, personal way the imperative to restore what he called "land health" and the compelling need for a land ethic to guide daily management decisions.

The Upshot

Leopold's success as a scientist thus reflects his ability to exploit three distinct approaches to science. He was a keen observer of nature, sensitive to details and adept at making habitat comparisons; a student of both human and natural history, alert to how ecological systems change through the seasons and over time in response to disturbances; and a scientist who liked to tinker with management, laying the foundation for experimental restoration ecology. The existence and remarkable growth of both conservation biology and restoration ecology stand as testimony to his great influence. He also set an example of a scientist willing to roll up his sleeves to work on practical, community-based efforts in conservation, as at Coon Valley, and to engage the wider public in conservation issues. It is no accident that Leopold's own children became significant figures in several sciences, known for their keen field observations, place-based historical analysis, and commitment to experimentation and restoration (Carter 1980). Their time at the shack and their father's commitment to careful place-based ecological research clearly affected their relationships, development, and thought processes for life.

Today's World and Tomorrow's

One of the penalties of an ecological education is that one lives alone in a world of wounds. Much of the damage inflicted on land is quite invisible to laymen. An ecologist must either harden his shell and make believe that the consequences of science are none of his business, or he must be the doctor who sees the marks of death in a community that believes itself well and does not want to be told otherwise (Leopold 1953, p. 165).

The places we study are changing. Often this is painfully obvious. The forests and fields we may have played in as children or studied in college now house subdivisions instead of foxes and migrant songbirds. Other changes are more subtle. Even the natural areas that remain are unraveling as they become fragmented and more isolated. Rarer native species decline, and the species that thrive in disturbed and edge habitats steadily grow more common. These include habitat generalists like deer as well as opportunistic nest predators like skunks, crows, raccoons, and blue jays. The loss of top carnivores like wolves and cougars exacerbates this "meso-predator release." Cowbirds that also thrive in disturbed habitats parasitize migrant bird species by laying eggs in their nests. In tandem, these nest predators and parasites are decimating many native songbird populations, leading to general declines across eastern and midwestern North America (Robbins et al. 1989, Terborgh 1989).

Leopold taught that the way to manage game is to manage habitat. We now manage habitats so intensively that many game species, like white-tailed deer across the eastern United States, have rebounded with a vengeance. On his return from Germany and on other occasions, Leopold warned us specifically against this excess of success and generally against placing too much emphasis on any one output of natural systems. Despite this clear warning, the problem has grown worse across Wisconsin, with deer now well above the peak densities that so concerned him in the 1940s. Indeed, this is true across the country (Warren 1997, Augustine and DeCalesta 2003). Leopold would be sad and disappointed to see how deer are decimating understory cover, tree regeneration, and overall plant diversity (Alverson et al. 1988, Waller and Alverson 1997)—damage he so clearly warned about. As fewer baby oaks and hickories survive to adulthood, we see shifts in forest composition and reductions in the nut mast that help sustain wildlife populations and deer themselves. These shifts in species composition and interactions have become even more complex, with deer, squirrel, and oak mast dynamics contributing to outbreaks of human illness like Lyme disease (Jones et al. 1998, Allan et al. 2003; see also Ostfeld and Jones, chapter 10). What further proof could we ask for Leopold's dictum that conservation in whole must supersede conservation in part?

It is interesting to reflect on the fact that some of our best information on this degradation is coming from particular places like Rock Creek Park and Fairfax County, near Washington, D.C., where skilled "citizen scientists," often inspired by reading *A Sand County Almanac*, repeatedly resurvey the same site over many years. Leopold would also note, however, how local pictures are reinforced and given generality by systematic professional efforts. These include the detailed pictures of change emerging for the prairies and forests of Wisconsin from resurveys of the sites that colleague John Curtis and his students studied (e.g., Rogers et al. 2008, 2009). They also include long-term commitments to monitor habitats and wildlife populations, like the USDA–FIA program and the national Breeding Bird Survey. These results reinforce Leopold's belief in the importance of extensive wild areas. Songbird populations remain high in more heavily forested regions of the Midwest (Howell et al. 2000), and in expansive Western lands that remain in public ownership.

On the hopeful side of the ledger, Leopold would be delighted to see citizen scientists closely observing natural places, adding to science while gaining aesthetic pleasure and emotional ties in the process. He would be delighted also to learn that wolves and cougars are creeping back into the Rocky Mountains and the upper Midwest. More than 600 wolves now exist in Wisconsin. He would be dismayed, however, by pockets of local antagonism to wolf recovery both in Wisconsin and the Southwest (Dougherty 2007). His ideas about predators having significant "top-down" effects on their prey, limiting ungulate impacts, are gaining ground (Ripple and Beschta 2004a). Interestingly, this appears to occur both directly as ungulate populations are reduced and indirectly by inducing an "ecology of fear" that alters their behavior (Ripple and Beschta 2004b; see also Peckarsky et al., chapter 9). Carnivores can also help birds by consuming or chasing off the "meso-predators" mentioned above (Crooks and Soulé 1999). As our understanding of these "trophic cascades" improves, we gain support for Leopold's idea that all species should be sustained at ecologically functional densities.

Leopold would also be heartened by the tremendous interest in conservation biology and restoration ecology and the burgeoning community-based conservation movement, in which farmers and urbanites, ranchers and environmentalists join forces at the local level, often working in partnership with scientists and land managers to restore streams, watersheds, forests, ranges, and even urban prairies in a manner reminiscent of the Coon Valley Project. Some of these groups, such as the Quivira Coalition (White 2005) and the Malpai Borderlands Group (Sayre 2005; Curtin, chapter 12) in the Southwestern United States, take inspiration directly from Leopold's concepts of land health and a land ethic. One of the largest is the coalition of nearly two hundred groups known as Chicago Wilderness (http://www .chicagowilderness.org), working to preserve and restore ecosystem health and biodiversity in the Chicago metropolitan region. Hundreds of community-based organizations from across the nation came together for the White House Conference on Cooperative Conservation (2005), at which Leopold's commitment to place and his community-based land ethic were repeatedly invoked.[2] Paul Hawken (2007) has estimated that such groups worldwide number in the millions; he calls it "the largest movement in the world" (see also Feinsinger et al., chapter 18).

Although some of the more visionary schemes for reconnecting, restoring, and rewilding natural ecosystems across the continent, such as the Wildlands Project (http://www.twp.org) and the Rewilding Institute (http://www .rewilding.org), would have fascinated Leopold, he was above all a practical conservationist committed to restoring the health of working landscapes closer to home. He devoted his life to applying the lessons he learned about ecology to practical issues surrounding conservation. In *A Sand County Almanac*, his weekly columns for farmers, and other essays directed to the general public, he sought to reach people where they lived, enhance their esthetic appreciation and ecological understanding, rekindle their love of land, and inspire in them an ethical obligation to maintain and restore its health and beauty. As debates over how best to conserve, restore, and sustain natural ecosystems continue, we would be wise to share Leopold's parallel interests in esthetics and ethics. For Leopold ecology, esthetics, and ethics all emerged naturally from awareness of and love and respect for the land community. His sense of place, and the lessons he learned from the places where he lived and worked, infused all three. Eventually he would sum up these concerns in the simple moral statement: "A thing is right when it tends to preserve the integrity, stability, and beauty of the biotic community. It is wrong when it tends otherwise" (Leopold 1949, pp. 224–25). We can only make this judgment, however, in relation to particular places—and, ideally, in relation to the places we know best.

References

Allan, B. F., F. Keesing, and R. S. Ostfeld. 2003. Effect of forest fragmentation on Lyme disease risk. *Conservation Biology* 17:267–72.

Allombert, S., A. J. Gaston, and J.-L. Martin. 2005. A natural experiment on the impact of overabundant deer on songbird populations. *Biological Conservation* 126:1–13.

Alverson, W. S., D. M. Waller, and S. L. Solheim. 1988. Forests too deer: Edge effects in northern Wisconsin. *Conservation Biology* 2:348–58.

Aney, W. W. 1990. Deer and Dauerwald revisited: Aldo Leopold's 1935 visit to Germany in retrospect. Pages 14–19 *in* R. S. Mitchell, C. J. Sheviak, and D. J. Leopold, eds., *Ecosystem Management: Rare Species and Signficant Habitats.* Proceedings of the 15th Natural Areas Conference. New York State Museum, Albany.

Augustine, D. J., and D. DeCalesta. 2003. Defining deer overabundance and threats to forest communities: From individual plants to landscape structure. *Ecoscience* 10:472–86.

Bradley, N. L., A. C. Leopold, J. Ross, and W. Huffaker. 1999. Phenological changes reflect climate change in Wisconsin. *Proceedings of the National Academy of Sciences (USA)* 96: 9701–4.

Callicott, J. B. 1989. *In Defense of the Land Ethic: Essays in Environmental Philosophy.* State University of New York Press, Albany.

Carter, L. F. 1980. The Leopolds: A family of naturalists. *Science* 207:1051–55.

Crooks, K. R., and M. E. Soulé. 1999. Mesopredator release and avifaunal extinctions in a fragmented system. *Nature* 400:563–66.

Curtis, J. T. 1956. The modification of mid-latitude grasslands and forests by man. Pages 721–36 *in* W. L. Thomas, ed., *Man's Role in Changing the Face of the Earth.* University of Chicago Press, Chicago.

———.1959. *The Vegetation of Wisconsin.* University of Wisconsin Press, Madison.

DeCalestra, D. S. 1994. Effects of white-tailed deer on songbirds within managed forests in Pennsylvania. *Journal of Wildlife Management* 58:771–18.

Diamond, J. M. 1983. Laboratory, field, and natural experiments. *Nature* 304:586–87.

Dougherty, J. 2007. Last chance for the lobo. *High Country News* 39 (December): 10–17.

Duffy, D. C., and A. J. Meier. 1992. Do Appalachian herbaceous understories ever recover from clearcutting? *Conservation Biology* 6:196–201.

Egan, D., and E. A. Howell. 2001. *The Historical Ecology Handbook: A Restorationist's Guide to Reference Ecosystems.* Island Press, Washington, DC..

Flader, S. L. 1974. *Thinking Like a Mountain: Aldo Leopold and the Evolution of an Ecological Attitude toward Deer, Wolves, and Forests.* University of Wisconsin Press, Madison.

———. 1987. Aldo Leopold and the evolution of a land ethic. Pages 2–24 *in* T. Tanner, ed., *Aldo Leopold: The Man and His Legacy.* Soil Conservation Society of America, Ankeny, IA.

Flader, S. L., and J. B. Callicott. 1991. *The River of the Mother of God and other Essays by Aldo Leopold.* University of Wisconsin Press, Madison.

Foster, D. R., J. D. Aber, J. M. Melillo, R. D. Bowden, and F. A. Bazzaz. 1997. Forest response to disturbance and anthropogenic stress. *BioScience* 47:437–45.

Foster, D. R., D. H. Knight, and J. F. Franklin. 1998. Landscape patterns and legacies resulting from large, infrequent forest disturbances. *Ecosystems* 1:497–510.

Halpern, C. B., and T. A. Spies. 1995. Plant species diversity in natural and managed forests of the Pacific Northwest. *Ecological Applications* 5:913–34.

Hawken, P. 2007. *Blessed Unrest: How the Largest Movement in the World Came into Being and Why No One Saw it Coming.* Viking, New York.

Holling, C. S., and G. K. Meffe. 1996. Command and control and the pathology of natural resource management. *Conservation Biology* 10:328–37.

Howell, C. A., S. C. Latta, T. M. Donovan, P. A. Porneluzi, G. R. Parks, and J. Faaborg. 2000. Landscape effects mediate breeding bird abundance in midwestern forests. *Landscape Ecology* 15:547–62.

Jackson, J. B. C. 1997. Reefs since Columbus. *Coral Reefs* 16:S23–S32.

Johnson, S. E., E. L. Mudrak, and D. M. Waller. 2008. Comparing power among three sampling methods for monitoring forest vegetation. *Canadian Journal of Forest Research* 38:143–56.

Jones, C. G., R. S. Ostfeld, M. P. Richard, E. M. Schauber, and J. O. Wolff. 1998. Chain reactions linking acorns to Gypsy Moth outbreaks and Lyme disease risk. *Science* 279:1023–26.

Kuhlmann, W. 1997. Making the law more ecocentric: Responding to Leopold and conservation biology. *Duke Environmental Law and Policy Forum* 133:163–64.

Leach, M. K., and T. J. Givnish. 1996. Ecological determinants of species loss in remnant prairies. *Science* 273:1555–58.

Leopold, A. 1923. Some fundamentals of conservation in the Southwest. Pages 86–97 in *The River of the Mother of God and Other Essays*, S. L. Flader, and J. B. Callicott, eds., 1991. University of Wisconsin Press, Madison.

———. 1924a. Forest inspection as developed in the Southwest. Address to New York Forest Club, November 25. Leopold Papers.

———. 1924b. Grass, brush, timber and fire in southern Arizona. *Journal of Forestry* 22:1–10.

———. 1931. *Report on a Game Survey of the North Central States.* Sporting Arms and Ammunition Manufacturers' Institute, Madison, Wisconsin.

———. 1933a. *Game Management.* Charles Scribner's Sons, New York.

———. 1933b. The conservation ethic. *Journal of Forestry* 31:634–43.

———. 1935. Coon Valley: An adventure in cooperative conservation. *American Forests* 41:205–208.

———. 1936. Deer and *Dauerwald* in Germany. *Journal of Forestry* 34:366–75, 460–66.

———. 1937. Conservationist in Mexico. *American Forests* 43:118–20, 146.

———. 1939a. A biotic view of land. *Journal of Forestry* 37:727–30.

———. 1939b. The farmer as a conservationist. *American Forests* 45:294–99, 316, 323.

———. 1941. Wilderness as a land laboratory. *The Living Wilderness* 6 (July): 3.

———. 1947a. The ecological conscience. *Bulletin of the Garden Club of America* (September): 45–53.

———. 1947b. Foreword [to Great Possessions, an earlier version of *A Sand County Almanac*] *in* J. Baird Callicott, ed., 1987. *Companion to A Sand County Almanac: Interpretive and Critical Essays.* University of Wisconsin Press, Madison.

———. 1949. *A Sand County Almanac and Sketches Here and There.* Oxford University Press, New York.

Leopold, A., and S. E. Jones. 1947. A phenological record for Sauk and Dane Counties, Wisconsin, 1935–1945. *Ecological Monographs* 17:83–123.

Leopold, L. B., ed. 1953. *Round River: From the Journals of Aldo Leopold.* Oxford University Press, New York.

Likens, G. E., R. F. Wright, J. N. Galloway, and T. J. Butler. 1979. Acid rain. *Scientific American* 241:43–51.

———. 1983. A priority for ecological research. *Bulletin of the Ecological Society of America* 64:234–43.

Louv, R. 2005. *Last Child in the Woods: Saving our Children from Nature Deficit Disorder.* Workman Publishing, New York.

McShea, W. J., and J. H. Rappole. 2000. Managing the abundance and diversity of breeding bird populations through manipulation of deer populations. *Conservation Biology* 14: 1161–70.

Meine, C. 1988. *Aldo Leopold: His Life and Work.* University of Wisconsin Press, Madison.

Meine, C., and R. L. Knight, eds. 1999. *The Essential Aldo Leopold: Quotations and Commentaries.* University of Wisconsin Press, Madison.

Minteer, B. A., and R. E. Manning, eds. 2003. *Reconstructing Conservation: Finding Common Ground.* Island Press, Washington, DC.

Mitman, G. 2005. In search of health: Landscape and disease in American environmental history. *Environmental History* 10: 184–210.

Newton, J. N. 2006. *Aldo Leopold's Odyssey: Rediscovering the Author of a Sand County Almanac.* Island Press, Washington, DC.

Pandolfi, J. M., J. B. C. Jackson, N. Baron, R. H. Bradbury, H. M. Guzman, T. P. Hughes, C. V. Kappel, F. Micheli, J. C. Ogden, H. P. Possingham, and E. Sala. 2005. Are U.S. coral reefs on the slippery slope to slime? *Science* 307:1725–26.

Pearson, S. M., A. B. Smith, and M. G. Turner. 1998. Forest patch size, land use, and mesic forest herbs in the French Broad River Basin, North Carolina. *Castanea* 63:382–95.

Ripple, W. J., and R. L. Beschta. 2004a. Linking wolves and plants: Aldo Leopold on trophic cascades. *BioScience* 55:613–21.

———. 2004b. Wolves and the ecology of fear: Can predation risk structure ecosystems? *BioScience* 54:755–66.

Robbins, C. S., J. R. Sauer, R. S. Greenberg, and S. Droege. 1989. Population declines in North American birds that migrate to the neotropics. *Proceedings of the National Academy of Sciences (USA)* 86:7658–62.

Rogers, D. A., T. P. Rooney, and D. M. Waller. 2008. Fifty years of change in southern Wisconsin forests: Shifts in canopy and understory richness, composition and heterogeneity. *Ecology* 89:2482–92.

Rogers, D. A., T. P. Rooney, T. Hawbaker, V. Radcloff, and D. W. Waller. 2009. Paying the extinction debt in southern Wisconsin forest understories. *Conservation Biology* 23:1497–1506.

Rooney, T. P., S. M. Wiegmann, D. A. Rogers, and D. M. Waller. 2004. Biotic impoverishment and homogenization in unfragmented forest understory communities. *Conservation Biology* 18:787–98.

Sayre, N. F. 2005. *Working Wilderness: the Malpais Borderlands Group and Future of the Western Range.* Rio Nuevo Publishers, Tucson,

Terborgh, J. 1989. *Where Have All the Birds Gone?* Princeton University Press, Princeton, NJ.

Waller, D. M., and W. S. Alverson. 1997. The white-tailed deer: A keystone herbivore. *Wildlife Society Bulletin* 25:217–26.

Warren, R. J. 1997. Special Issue: Deer Overabundance. *Wildlife Society Bulletin* 25:209–577.

White, C. 2005. The working wilderness: A call for a land health movement. Pages 159–180 *in* W. Berry, ed., *The Way of Ignorance*. Shoemaker and Hoard, Emeryville, CA.

White House Conference on Cooperative Conservation. 2005. *Faces and Places of Cooperative Conservation: Profiles in Citizen Stewardship*. Executive Office of the President, Council on Environmental Quality. Washington, DC.

Whitney, G. G., and D. R. Foster. 1988. Overstorey composition and age as determinants of the understorey flora of woods of Central New England. *Journal of Ecology* 76:867–76.

Wiegmann, S. M., and D. M. Waller. 2006. Biotic homogenization in forest understories: Identity and traits of historical "winners" and "losers." *Biological Conservation* 129:109–23.

Wilson, E. O. 1984. *Biophilia: The Human Bond with Other Species*. Harvard University Press, Cambridge, MA.

Wright, S. D., and N. L. Bradley. 2008. Thinking like a flower: Phenology and climate change at the Leopold shack. Pages 41–53 *in* D. M. Waller and T. P. Rooney, eds., *The Vanishing Present: Wisconsin's Changing Lands, Waters, and Wildlife*. University of Chicago Press, Chicago.

Zedler, J. B. 2005. Restoring wetland plant diversity: A comparison of existing and adaptive approaches. *Wetlands Ecology and Management* 13:5–14.

Notes

1. We direct those interested in a fuller account of the events and ideas in Leopold's life to the biographical works by Flader (1974), Meine (1988), and Newton (2006). For broader tastes of Leopold's writing and its impact, see Flader and Callicott (1991) and Meine and Knight (1999). To learn more about Leopold's wider contributions, see Callicott's (1989) book on environmental ethics, Holling (1996) and Kuhlmann (1997) on environmental policy, Aney (1990) on the threats posed by overabundant deer (*Odocoileus virginianus*), Ripple (2004a) on the key role of predators in natural ecosystems, Minteer and Manning (2003) on community-based conservation, and Mitman (2005) and White (2005) on land health.

2. The very name for the conference may have been inspired by Leopold's (1935) description of his Coon Valley experiment in his article subtitled "An Adventure in Cooperative Conservation."

Part 2

THE IDIOSYNCRASY OF PLACE: CHALLENGES AND OPPORTUNITIES

Ian Billick and Mary V. Price

Nature is idiosyncratic. One's understanding of it as a scientist often will depend on exactly where and when one looks. This is no surprise. Ecological systems are complex, with many interconnected parts, and effects of a perturbation to one part of the system will propagate through it in different ways depending on the state of the system when the perturbation occurs. Hence, moving a field experiment even several meters, or conducting it a week or two later, can lead to substantially different outcomes. Such idiosyncrasy makes it difficult to understand the properties of a single place, much less extend that understanding to other places.

One response to the challenge of the idiosyncrasy of place, which Darwin (1859) so aptly described as an "entangled bank,"[1] is to run away very fast. There are undoubtedly a number of laboratory scientists whose introduction to the messiness and lack of certainty associated with fieldwork convinced them that their love of biology was best pursued with vials and PCR machines. Even ecological microcosms, which avoid much of the complexity of the natural environment by examining simplified components of a system in a controlled environment, are but a small step toward the ultimate goal of understanding the complex "system constituting a whole"[2] (Gause 1934). Because ecological systems are composed of multiple, interacting parts that often cannot be realistically replicated in a microcosm environment, experimental results run the risk of being an artifact of the microcosm itself. Additionally, many ecological processes occur on temporal and spatial scales that cannot captured within a microcosm. Microcosms can provide valuable insights into components of larger ecological questions (see, for example, Peckarsky et al., chapter 9), but cannot substitute in the end for field studies (Carpenter 1996).

Others escape the idiosyncrasies of place by looking across ecosystems, squinting so that the details disappear, and focusing primarily on the general patterns that emerge from a distant view. Some (e.g., Peters 1991, Martinez 1994) even claim that the primary or only path to ecological understanding is to uncover broad "macroecological" (Brown 1995) patterns that predict properties of ecological systems, such as invariant properties of food webs or geographic patterns in diversity. Such a stance argues that science should primarily focus on patterns that are both simple and true across sites and times.

The knowledge that comes from working in the laboratory, studying microcosms, or detecting macroecological patterns can be extremely valuable, and we could list many successes derived from these approaches. But such knowledge tells us nothing about a host of interesting questions. It does not tell us how ecological processes captured in microcosms will manifest themselves when embedded in a real place with all of its variability and complexity, nor why some systems deviate from macroecological patterns. Such approaches give us little practical ability to manage endangered species, restore damaged ecosystems, or predict the consequences of climate change for populations or food webs. That ability can only come from a detailed understanding of intact, idiosyncratic, and messy ecological systems. Some questions simply cannot be answered unless we open our eyes wide, wade into the tangled bank, and confront the complexity of the world head-on.

Turning Challenge into Opportunity

The challenge of idiosyncrasy can be turned on its head. Rather than a challenge, variation in time and space can be an opportunity. Pulliam and Waser (chapter 4) point out that research can explicitly take advantage of such variation. They start by analyzing two examples in which the results of field research are contingent on place. In the case of the wildflower *Ipomopsis aggregata*, ungulate browsing increased fecundity in some studies but decreased it in others. Similarly, density of the plant *Polygonatum biflorum* increased with soil moisture in one location but decreased in another. In both cases Pulliam and Waser indicate that broadening the focus of research to uncover the reasons for the differing outcomes has the potential to lead to a more general understanding. Particularly when it is informed by natural history intuition, research at multiple sites offers the opportunity to reveal mechanisms or patterns that are not apparent from a single site.

As Pulliam and Waser demonstrate, spatial variation provides an opportunity because such variation generates "natural experiments" (sensu Diamond 1983, 1986) in which independent and dependent variables vary naturally. Natural experiments provide opportunities to tease out cause and effect relationships on scales over which true experimentation is impractical. Rundel (chapter 5) illustrates this approach. In reviewing the history of comparative ecosystem studies, he notes that major Mediterranean-climate regions provide a replicated natural experiment that can be used to understand how temperate plant communities respond to winter rainfall and summer drought. He also takes this a step further, however, in emphasizing the substantial differences in species diversity among Mediterranean-climate ecosystems. These differences provide an opportunity to test ideas about how diversity responds to reliability and timing of rainfall and correlated variables like wildfire—that is, to aspects of the variance in conditions, rather than only to the average condition. Natural spatial variation can thus be used to generate insights into the mechanisms that control diversity under circumstances in which standard experimentation is impossible.

Variation through time also provides opportunities for natural experiments. Grant and Grant (chapter 6), by being in one place long enough, had the good fortune to observe a rare event: the founding of a new population of finches on one of the Galápagos Islands. They used this one-time event to better understand the role of founder events in speciation. Their long tenure also allowed them to observe repeated drought events. This repeated exposure helped them work out the mechanism by which drought affected the finches, and to appreciate that the evolutionary response depended critically on the state of the vegetation at the onset of the drought.

Armitage (chapter 7) also took advantage of variation through time and across sites to explore the link between social behavior and population dynamics in yellow-bellied marmots. Initially he observed multiple sites across ten years and saw that the population numbers at different sites did not vary concordantly, as one would expect if they were simply responding to large-scale fluctuations in weather. He therefore turned to another factor that could explain population dynamics: the behavior of these social animals. He spent more than four decades amassing data on the lifetime reproductive success of known individuals of known pedigree, and used natural variability in behavioral strategies to determine consequences of behavior for individual fitness. A simple assumption—that individuals act so as to maximize their individual fitness—went a long way toward explaining observed population dynamics.

The chapters by the Grants and by Armitage illustrate how watching an ecological system over extended periods leads to the accumulation of disparate pieces of information, all of which are potentially useful for understanding the system. One of the striking characteristics of the work by Grant and Grant is the extent to which they pulled together a range of information, including data on precipitation and seed size, as well as observations as anecdotal as the occurrence of a cactus spine in the throat of a single bird, in order to understand speciation of the finches. In similar fashion, Armitage combined quantitative estimates of the effects of individual behavioral strategies on fitness with observations of diet, dispersal, predators, and colony dynamics in order to develop a full understanding of the population ecology of marmots.

The final chapter in this part (Estes, chapter 8) further illustrates the value of synthesizing disparate pieces of information, an idea developed more fully by Peckarsky et al. (chapter 9) in part 3. The pieces Estes used will be familiar from the preceding chapters. He used natural experiments—variation across islands in sea otter and arctic fox abundance—to understand ways in which predators structure ecosystems. By observing the Aleutian archipelago for many years, he had the opportunity to use another natural experiment—a sudden decline in sea otter numbers—to provide independent confirmation of the role of sea otters in maintaining kelp forests. Estes did not limit himself to data from natural experiments. As he built working models of predation, he used information as diverse as whale energy requirements, the rate of observed otter deaths due to killer whale attacks, and isotope ratios in predator tissues. And while many of the details he used were idiosyncratic to place and focused on a specific ecosystem, his work has been broadly relevant for ecologists working on predation.

Indeed, there is an apparent paradox when we consider the relationship between the detail-oriented nature of the research described in this section and the general insights that result. How is it that the authors can work in specific locales, take advantage of place-specific natural history information, and yet deliver insights that are generally useful? In contrast to the phenomenological approaches (sensu Tilman 1987) used by individuals describing food webs (e.g., Cohen 1977, 1978), developing regressions (e.g., Peters 1991) or identifying macroecological patterns (Brown 1995), all of the authors in this section dug deeply enough to uncover mechanistic explanations for the empirical relationships they observed. And while the details of their systems are not exportable, the mechanisms they uncover, and the conceptual frameworks they use to frame their questions, are.

References

Brown, J. H. 1995. *Macroecology*. University of Chicago Press, Chicago.
Carpenter, S. R. 1996. Microcosm experiments have limited relevance for community and ecosystem ecology. *Ecology* 77:677–80.
Cohen, J. E. 1977. Ratio of predators to prey in community food webs. *Nature* 270:165–67.
———. 1978. *Food Webs and Niche Space*. Princeton University Press, Princeton, NJ.
Darwin, C. 1859. *On the Origin of Species by Means of Natural Selection*. Murray, London.
Diamond, J. M. 1983. Laboratory, field, and natural experiments. *Nature* 304:586–87.
———. 1986. Overview: Laboratory experiments, field experiments, and natural experiments. Pages 3–22 in J. Diamond and T. J. Case, eds., *Community Ecology*. Harper and Row, New York.

Gause, G. F. 1934. *The Struggle for Existence*. Williams and Wilkins, Baltimore.

Martinez, N. D. 1994. Scale-dependent constraints on food-web structure. *American Naturalist* 144:935–53.

Peters, R. H. 1991. *A Critique for Ecology*. Cambridge University Press, Cambridge.

Tilman, D. 1987. The importance of the mechanisms of interspecific competition. *American Naturalist* 129:769–74.

Notes

1. The "entangled bank" passage is in the concluding paragraph to the final chapter of *The Origin of Species*: "It is interesting to contemplate an entangled bank, clothed with many plants of many kinds, with birds singing on the bushes, with various insects flitting about, and with worms crawling through the damp earth, and to reflect that these elaborately constructed forms, so different from each other, and dependent upon each other in so complex a manner, have all been produced by laws acting around us."

2. G. F. Gause argued for microcosm studies using an analogy from physiology on page 3 of *The Struggle for Existence:* "First of all it had been necessary to show that separate elementary phenomena of vision, hearing, etc., can be fruitfully studied by physical and chemical methods, and thereupon only did the question arise of studying the organism as a system constituting a whole."

4 *Ecological Invariance and the Search for Generality in Ecology*

H. Ronald Pulliam and Nickolas M. Waser

Figure 4.0. The "Ipo Disciples" at work: field studies of scarlet gilia (*Ipomopsis aggregata*) in the Colorado Rocky Mountains. Photograph by Dennis Johns, 1981.

Abstract

Whereas the physical sciences seem to be characterized by universal constants and laws, which allow generalization from one system to another, it is questionable whether ecology enjoys any such invariants. The Hutchinsonian niche might be one ecological invariant, and it has been used as such in relating distribution and abundance of species to their niche requirements. On the other hand, ecologists are unsurprised to find variation in niche expression of a given species. We present two examples: geographic variation in tolerance to herbivory in the montane wildflower *Ipomopsis aggregata*, and geographic differences in the relationship between soil moisture and abundance of the understory herb *Polygonatum biflorum*. The first example illustrates the unfortunate possibility that perceived differences in niche expression derive from differences in method—"unfortunate" because it is difficult to detect unless researchers truly cooperate, and also because it can foster fruitless debate. Both examples illustrate two additional possibilities: that variable niche expression derives from location-specific differences in the "realized" niche of a species, and that it derives from differences in the "fundamental" niche due to genetic differentiation within species. Deducing the contributions of these two factors is beyond our current ability. Instead we discuss how to explore such possibilities by challenging multiple working hypotheses with increasingly sophisticated knowledge of natural history derived from long-term place-based observation and experimentation. This approach is an alternative to more common "null-hypothesis testing," and we assert that it can lead to a depth of mechanistic understanding at which one does in fact reach a level of ecological invariance.

Introduction

H e insisted that there is no Truth but only many truths; that Truth is not a colored bird to be chased . . . but a skeptical attitude toward life.

Sinclair Lewis (1925)

Consider the concept of niche.
If I knew what it meant I'd be rich.
Its dimensions are *n*
And a knowledge of Zen
Is essential to fathom the bitch!

Grant Cottam and David Parkhurst (unpublished, ca. 1969)

Some sciences—especially physics and chemistry—are known for their invariant physical constants and general laws. Just look in the back of any standard textbook: Planck's constant is $h = 6.626069 \times 10^{-34}$ Js; Newton's constant of gravity is $G = 6.6743 \times 10^{-11}$ m^3 kg^{-1} s^{-2}; the speed of light in a vacuum is $c = 299{,}792{,}458$ ms^{-1}. Furthermore, these constants can be used to quantify relationships among independent variables. For example, Newton's Law of Universal Gravitation states that the force with which one object attracts another is $F = G\frac{m_1 m_2}{r^2}$ where m_1 and m_2 are the masses of the two objects and r is the distance between them. Hence the physical sciences appear to enjoy laws, constants, and relationships—descriptions of the behavior of matter and energy—that apply universally, at least at macroscopic and nonrelativistic scales.

Does ecology have any similar universal laws or invariant constants, which allow us to generalize from one system to another, so as to predict a priori the behavior of systems that have not been studied? The exponential growth equation, $dN/dt = rN$, relating population growth rate (dN/dt) to population size (N), is sometimes called the Malthusian law (e.g., Gause 1934) and the parameter r is called the intrinsic rate of natural increase. But measuring *r* at one place and time may reveal little to nothing about its value in another place or time.

Ecological invariants known as "scaling and power laws," that describe power relationships across species or across ecological systems, also have been well documented and explored (e.g., Charnov 1993, Brown et al. 2000, Marquet et al. 2005). For example, the number of species (S) in a given land area (A) can be expressed as the species-area relationship, $S = cA^z$ (MacArthur and Wilson 1963). This is a pretty good law (description of a system's behavior), as it has been shown to apply for both true oceanic islands and for habitat "islands" in a "sea" of other habitats. But the exponent (z) varies, usually between 0.20 and 0.35, so the law hardly qualifies as a universal constant. Furthermore, we are concerned here with invariance of another sort—in particular, with whether or not the relationships among species, or between species and environment, documented at one place can be generalized and applied to other places.

Let us carry our musings further. The fact that *r* varies across species is somewhat akin to different chemical elements having different atomic mass. It could also be argued that *r* is really more like weight than mass, inasmuch as it varies as a function of biological species and environment rather as weight varies as a function of chemical element and environment. A more apt physical analogy to the species-area relationship or the exponential growth equation might therefore be a model for the rate of a certain chemical reaction at standard temperature and pressure: the rate of reaction applies only for specific environmental conditions. So far, so good; it seems that ecological and physical laws have some similarities. However, much of the predictive power of modern physical chemistry comes from the ability to extrapolate from standard conditions to other conditions. Alas, in ecology, a careful measurement of the rate of increase of a species in one environment might provide little information about how the same species will perform in other environments.

Is the Fundamental Niche an Ecological Invariant?

The "niche" of a species has served as a central concept in ecology, and is another candidate for an ecological invariant. Early ecologists (Grinnell 1917, Elton 1927) used the word to refer to a "recess" in the environment that has the potential to support a species. In this view the niche is an attribute of the environment; niches can be empty or full, and more than one species might vie for the same niche. G. E. Hutchinson (1957) turned the concept on its head, defining the niche as the environmental requirements of a species; in this view species, not environments, have niches. Hutchinson defined the "fundamental niche" of a species as an n-dimensional hypervolume, "every point in which corresponds to a state of the environment that would permit a species to exist indefinitely." Alternatively, the fundamental niche is the set of all environmental conditions for which the per capita population growth rate ($1/N \, dN/dt$) is greater than or equal to zero.

Is this mapping of fitness[1] on environment invariant? Standard texts often describe niche requirements as if they are fixed characteristics of species. In a typical example, *Bergey's Manual of Determinative Bacteriology* (Bergey et al. 1994) states that the bacterium *Norcardia farcinia* grows at temperatures between 20° and 40° C and at pH of from 6 to 10, and does best at 35° C and pH 7.5, but that *Treponema phagedenis* grows at temperatures of between 30° and 42° C and at pH of from 6.5 to 8, and does best at 37° C and pH 7. Should we treat such descriptions as invariants that determine the distribution and abundance of species in nature? This is commonly done, in fact, and has been done since the earliest days of ecology. For example, the distribution of trees along elevation and latitudinal gradients is often interpreted as a response to temperature and moisture requirements (e.g., Merriam 1890). In the southwestern United States, for example, Douglas fir (*Pseudotsuga menziesii*) is found at higher elevations where precipitation is high and temperatures are low, but descends to lower elevations in cool, moist canyons. The fixed relationship between climate and species distribution is a guiding principle of paleoecology, and is used extensively to reconstruct past climates.

In short, patterns in the distribution of terrestrial organisms (and analogous patterns in aquatic ecosystems) make sense in light of modern niche theory. The concept of fixed niche requirements provides a powerful tool for interpreting current distributions, reconstructing environmental conditions of past distributions, and predicting response to future conditions, including a changed global climate.

The Variable Niche

Although ecologists often treat the niche as an invariant, we also recognize that the observed niche may vary through space and time. Hutchinson himself distinguished between "fundamental" and "realized" niches. The fundamental niche characterizes the fundamental requirements of a species and, in this respect, might be thought of as invariant. The realized niche is molded by biotic interactions. Hutchinson argued that a species may be absent from a portion of its fundamental niche because it is displaced by competition from a superior competitor (figure 4.1). Imagine, for example, a laboratory study of the ability of a particular plant species to grow and reproduce. In a separate field study the species is present at low values of soil moisture but absent from wetter locations, even though it does well under

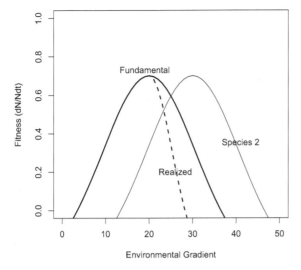

Figure 4.1. The realized niche may be a portion of the fundamental niche because a superior competitor (species 2) displaces the species from the rest of its fundamental niche.[2]

such conditions in the laboratory. Following Hutchinson, we postulate the presence of an unknown competitor that displaces the species from wetter locations. This seems akin to an astronomer postulating the presence of an unseen planet by observing some anomaly in the movement of a star.

The possibility of deducing the presence of a hitherto unknown competitor from the distribution of another species might lift the spirits of ecologists looking for generality and predictive power. Alas, there are other possible explanations for variable niche expression. For example, our plant might be able to resist fungal attack at low soil moisture, whereas the pathogen, not present in the laboratory studies, gains an upper hand in moister locations in nature. Indeed, ecological interactions are not limited to those that are antagonistic; mutualisms also might vary in space or time, leading to different observations of realized niches.[3]

Still other explanations are possible that may dash our hopes for the Hutchinsonian niche as an ecological invariant. Although it is not uncommon for ecologists to think typologically of a species, Charles Darwin introduced an alternative in chapter 2 of the *Origin of Species*: "No one supposes that all the individuals of the same species are cast in the same actual mold. These individual differences . . . afford materials for natural selection to act on and accumulate." If laboratory and field studies are done in different locations, they may compare genetically different entities within one species, the results of different histories of natural selection that have led to a different (in this example smaller) fundamental niche at one location than the other.

A Roadmap for What Follows

In what follows we present two examples of niche variability. First we consider the response of plants to a biotic dimension of the niche, the intensity of attack by herbivores. Reproduction of the wildflower *Ipomopsis aggregata* appears to respond differently to grazing by herbivores depending on geographic location. In most locations grazed plants suffer reduced flower and seed production, but in some locations grazing seems to *increase* fecundity. Tempting explanations include environmental or genetic differences that elicit different plant responses—that is, changes in realized or fundamental niches, respectively—but we cannot exclude the possibility that variable responses are artifacts of using different study methods. In the second example we explore why population density of the forest herb *Polygonatum biflorum* increases with increasing soil moisture at one site but declines in another. Since high population density is often taken as a reflection of suitable environmental conditions, we first consider this to be example of variable niche expression, but later raise the possibility that high density per se does not necessarily reflect suitable conditions. We end by recommending how ecologists can treat niche variability as an opportunity rather than a problem. We discuss some features of niche theory that reconcile the possibility of an invariant fundamental niche with a variable realized niche. More generally, we recommend ways in which place-based research, informed by knowledge of natural history, can contribute to an iterative process of model building that provides an alternative to classical "hypothesis testing" for gaining certainty about nature.

Variable Compensation by *Ipomopsis Aggregata* to Browsing of Inflorescences

Scarlet gilia, *Ipomopsis aggregata*, is an attractive wildflower common to mountains of western North America (Grant and Wilken 1986). Although often described as biennials, these plants are usually longer-lived semelparous perennials—that is, they flower once and die after spending multiple years (Waser et al. 2000) as vegetative rosettes of leaves. Flowers are both male, reproducing through export of pollen to other plants, and female, reproducing through import of pollen and setting of seeds. They are pollinated primarily by hummingbirds, but insects also play a role (Price et al. 2005). Throughout much of its geographic range, *I. aggregata* plants are browsed by ungulates (elk, *Cervus elaphus*, mule deer, *Odocoileus hemionus*, and/or white-tailed deer, *O. virginianus*) as they elongate their single flowering stalks. The usual response among surviving plants is to produce multiple secondary flowering stalks on lateral stems, eventually yielding a replacement crop of flowers.

The simplest expectation is that the unlucky browsed plant might produce a fraction of the seeds it would have otherwise—better than nothing, but not compensating for the loss of female reproductive success represented by the original inflorescence. In 1987, however, Paige and Whitham reported that naturally browsed plants of *I. aggregata* subspecies *formosissima* in northern Arizona, as well as plants clipped experimentally to mimic browsing, actually overcompensated. Browsed and clipped plants flowered virtually synchronously with undamaged plants and achieved the same total inflorescence height but, because of their multiple inflorescences, produced on average 2.8 times as many flowers and 3.0 times as many fruits and seeds as undamaged plants (in turn, overcompensation by experimentally-clipped plants averaged 1.9 times as many flowers and 2.4 times as many fruits and seeds). In subsequent work, Paige and colleagues reported overcompensation in other subspecies (in particular, *I. aggregata* subspecies *aggregata* in Colorado; Paige 1999), and in male reproductive success (success in siring seeds on other plants) in *I. aggregata* (Gronemeyer et al. 1997) and its relative *I. arizonica* (Paige et al. 2001).[4]

Overcompensation to browsing was not without precedent (e.g., McNaughton 1983). Still, overcompensation in *Ipomopsis* met with excitement because of its apparent magnitude. Several investigators attempted to confirm the result, without success.[5] Bergelson and Crawley (1992a, 1992b) summarized results of an extensive survey across a large portion of the geographic range of *I. aggregata* encompassing several subspecies, including those studied by Paige and colleagues. In comparing both naturally browsed and experimentally clipped plants to controls, they found the opposite of the previous result: dramatic undercompensation to browsing, with controls achieving up to 7.5 times the average female reproductive success of damaged plants.

Response to browsing by ungulates can be thought of as an expression of part of the Hutchinsonian niche, especially as extended by Maguire (1973), because it represents the mapping of a fitness component onto a biotic niche axis (figure 4.2). Contemplate a niche axis of increasing intensity of browsing. At some point it is easy to imagine that $1/N \, dN/dt$ becomes negative—browsing of reproductive structures is so intense that a plant species cannot replace itself and, in the absence of immigration, will go locally extinct. This represents a condition outside of the fundamental niche. But in scarlet gilia one confronts the possibility that ungulate herbivory may increase fecundity, so that ungulates act not as enemies but as mutualists.[6] This is a reversal of niche relationships in that one expects the species to be absent in areas of high ungulate density at one location, but to be concentrated in such areas

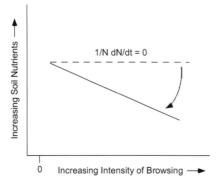

Figure 4.2. Hypothetical niche axes for *I. aggregata* in northern Arizona. If browsing by ungulates were absent, populations would achieve positive growth only if soil nutrients fell above
the level indicated by the dashed line, which is therefore the zero-net-growth isocline (ZNGI)
for this hypothetical situation. In fact, ungulate browsing does occur, and as browsing intensity
increases, so does fecundity. Thus the actual ZNGI has a negative slope, as indicated by the
solid line.

at another location. What could explain such variable niche expression, and
what does it say about the niche as an ecological invariant? We return to this
after the next section.

Variable Response of *Polygonatum Biflorum* to Soil Moisture

Polygonatum biflorum is a common understory herb in moist mid-elevation
forests of the southeastern United States. Pulliam and coworkers have analyzed survival, growth, and reproduction of *P. biflorum* of this and several
other species (Giladi 2006, Diez 2006, Diez and Pulliam 2007) in a region
extending from the Whitehall Forest (<150 m elevation) in the Piedmont
near Athens, Georgia, to the Coweeta Hydrologic Laboratory (750–1,500 m
elevation) in the Blue Ridge Mountains of North Carolina. This represents a
steep gradient in elevation, and thus in mean annual precipitation (≈120 cm
at Whitehall, >180 cm at Coweeta) and temperature (16.5° C at Whitehall,
<12° C at Coweeta). Plant populations at the southern end of the gradient
typically experience a three- to six-month period during the summer when
evaporation exceeds precipitation, whereas at higher elevations the period
of moisture stress is shorter and less severe.

If the niche is an ecological invariant, we might expect the abundance
of *P. biflorum* to be related to soil moisture in a consistent fashion. To explore this, Pulliam and coworkers marked individuals on six study grids at
Coweeta and six at Whitehall. On each grid, soil moisture was measured
throughout the growing season and the average was calculated. A prelimi-

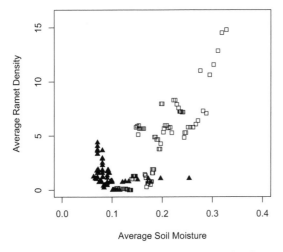

Figure 4.3. Relationships between *P. biflorum* ramet abundance and soil moisture at Coweeta (squares) and Whitehall (triangles). Average ramet abundance generally increased with increasing soil moisture at Coweeta, but not at Whitehall. The average ramet abundance shown is the nine-point running average with data arranged from the driest to moistest cell for each place.

nary look at the data (figure 4.3) suggests that *Polygonatum* abundance increases with moisture at Coweeta, and the opposite at Whitehall. We want to look behind this simple interpretation; in particular, we want to determine whether the relationship truly varies between the two locations approximately 100 km apart, and among grids separated by 0.5 to 2 km within each location. Our basic approach is to use a Bayesian "model selection" criterion to ask how the relationship varies between locations and among study grids within locations (box 4.1).

We anticipated an optimum moisture level, and expected that abundance might initially increase with increasing moisture and then eventually decrease. Parameter estimates from model 3a (box 4.1), the model that best fit the data, indeed suggest humped curves (figure 4.4) with intermediate moisture optima. But the overall trends were a decrease in abundance with increasing soil moisture at Whitehall and the opposite at Coweeta. We surmise that this is because none of the observed soil moisture values at Whitehall was less than 10%, so most observed abundances come from the descending right side of the Whitehall moisture curve. At Coweeta, only a few of the observed soil moisture values were above 35%, so most observed abundances come from the ascending left side of the Coweeta curve. The lack of observations in the lower part of the moisture curve at Whitehall may also partially explain the uncertainty about the parameter values for this location. Again, we return to these possibilities after the next section.

BOX 4.1. Analyzing responses of *Polygonatum biflorum* to soil moisture at two sites.

To compare the reactions to soil moisture of *Polygonatum biflorum* plants at Whitehall Forest, Georgia, and Coweeta, North Carolina, we used nested, hierarchical Bayesian models (Carlin and Louis 2000, Press 2003). Because we anticipated that abundance might initially increase with increasing moisture and then decrease, we included both first-order (moisture) and second-order (moisture-squared) terms in all models, and subsequently removed terms to see whether the model was improved without them. Bayesian regression analysis was performed with the WINBUGS statistical package (MRC Biostatistics Unit, Cambridge University, UK), with 30,000 iterations and a burn-in of 10,000 iterations.

The first set of models is nonhierarchical and assumes that regression coefficients relating density to moisture are the same at all sites:

$$D_{ijk} = \exp(a + b_1 M + b_2 M^2) + \varepsilon_{ijk}, \qquad \text{(model 1A)}$$

where D_{ijk} is the number of ramets on cell i of grid j at site k, a is the intercept, b_1 and b_2 are first- and second-order regression coefficients, and ε_{ijk} is the Poisson distributed error term. The second set of models is similar, but it allows the regression coefficients to vary between sites:

$$D_{ijk} = \exp(a_k + b_{1k} M + b_{2k} M^2) + \varepsilon_{ijk}, \qquad \text{(model 2A)}$$

where a_k, b_{1k}, and b_{2k} are the regression coefficients for place k. Finally, the third set is similar but allows each grid in each site to have a different intercept:

$$D_{ijk} = \exp(a_{jk} + b_{1jk} M + b_{2jk} M^2) + \varepsilon_{ijk}, \qquad \text{(model 3A)}$$

where a_{jk}, b_{1jk}, and b_{2jk} are the regression coefficients for grid j of place k. We also included a variant B of each model in which the moisture term is retained and the moisture-squared term is removed; and a variant C in which the moisture term is removed and the moisture-squared term is retained.

Bayesian regression models require specifying "prior distributions": probability distributions that represent the modeler's beliefs, based on previous studies or natural history intuition, about parameter values before new data are taken into account. Both the prior distributions and the data influence "posterior distributions" which specify the probability distribution for each parameter after the data have been considered. In the present case, we had no previous studies to inform the priors, so the resulting posterior distributions of the parameters were primarily determined by the data rather than by prior expectation.

Table 4.1 shows the results. In all cases the models were improved by retaining the regression coefficients for both moisture and moisture-squared, and allowing all coefficients to vary with place and grid. For the best model, 3a, all

BOX 4.1. *(continued)*

first-order (b_1) moisture coefficients (slopes) were significantly greater than zero at Coweeta, and all second-order (b_2) moisture-squared coefficients were significantly less than zero, indicating bell-shaped moisture curves. Whitehall showed the same trends, but most of the 95%-credible intervals for parameter values overlapped zero, indicating more uncertainty about the exact shape of the moisture curves than at Coweeta.

TABLE 4.1. Comparisons of nine models relating *P. biflorum* abundance to soil moisture. The models are compared using the Deviance Information Criterion (DIC), which takes into account the deviation of observations from model predictions and the effective number of parameters (pD) in each model (Spieglhalter et al. 2002). DIC is similar to the non-Bayesian Akaike's Information Criterion (AIC) inasmuch as a lower DIC implies a more parsimonious model.

Model	pD	DIC
1a. $D_{ijk} = \exp(a + b_1 M + b_2 M^2) + e_{ijk}$	3.1	1935.8
1b. $D_{ijk} = \exp(a + bM) + e_{ijk}$	2.0	2035.2
1c. $D_{ijk} = \exp(a + bM^2) + \varepsilon_{ijk}$	2.0	2130.2
2a. $D_{ijk} = \exp(a_k + b_{1k} M + b_{2k} M^2) + \varepsilon_{ijk}$	5.3	1852.8
2b. $D_{ijk} = \exp(a_k + b_k M) + \varepsilon_{ijk}$	3.9	1884.6
2c. $D_{ijk} = \exp(a_k + b_k M^2) + \varepsilon_{ijk}$	3.9	1911.8
3a. $D_{ijk} = \exp(a_{jk} + b_{1jk} M + b_{2jk} M^2) + \varepsilon_{ijk}$	23.4	902.1
3b. $D_{ijk} = \exp(a_{jk} + b_{jk} M) + \varepsilon_{ijk}$	17.4	950.8
3c. $D_{ijk} = \exp(a_{jk} + b_{jk} M^2) + \varepsilon_{ijk}$	19.3	954.3

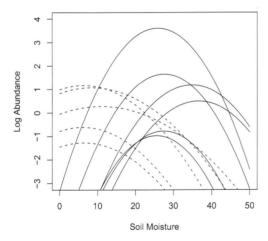

Figure 4.4. Moisture curves for *P. biflorum* abundance, drawn using the mean parameter estimates for model 3a (box 4.1). The six curves to the left (dashed lines) are for study grids at Whitehall, and the six curves to the right (solid lines) are for grids at Coweeta. The *y* axis is the natural log of abundance, allowing all values to fit on the same graph.

Possible General Reasons for Variable Niche Expression

We can imagine several hypotheses to explain an observation of variable niche expression (box 4.2). Our immediate goal is not to choose among these hypotheses, but to show for both examples given above that there are several explanations consistent with niche theory.

Different methods

Compensation to browsing in I. *aggregata* exemplifies the possibility that apparent differences in niche expression simply derive from different approaches used in different times or locations. Indeed, much discussion of this system in the literature has focused on differences in methodology. In responding to Bergelson and Crawley (1992a), Paige (1994) pointed to details of method, such as the precise method of mimicking browsing with experimental clipping, as critical to the results obtained. Bergelson et al. (1996) and Juenger and Bergelson (1997) copied as best they could the methods used by Paige, and still obtained undercompensation. Still, it is impossible to rule out subtle differences in method. For example, it may be that early reports of overcompensation failed to distinguish browsed individuals from those damaged during the years of vegetative growth that precede the flowering year. Vegetative damage can produce a multi-stalked flowering architecture resembling that of browsed plants, but it affects fitness in different ways—by lowering survival and delaying flowering (Brody et al. 2007). A team that confirmed the remains of a central inflorescence before assigning a plant as "browsed" might arrive at a very different accounting of fitness than a team that unknowingly included plants damaged in vegetative stage.

We can be much more confident that geographic differences in the *Polygonatum* system are not due to differences in method. All measurements in this case were done with the same method by the same investigators. The alert reader may see a potential flaw, since local abundance of *Polygonatum* may depend on additional soil properties that might vary between sites (just as it may depend on site-to-site variation along unmeasured biotic niche axes: e.g., abundance of a competitor). However, Pulliam and coworkers analyzed soil properties and found broad overlap between Coweeta and Whitehall, rather than differences that might explain the different *Polygonatum* abundance patterns.

Variation in the realized niche

Although the fundamental niche might be invariant, its realized expression may vary. In the *Ipomopsis* example, it is easy to erect location-specific

BOX 4.2. Three reasons for variable niche expression

1. Different methods
Apparent variation in niche expression may simply reflect different methods used by different investigators. Most ecological studies are done in a short time in one location by a single investigator. Each investigator chooses methods that he or she thinks appropriate for the question at hand. Often niche measurement, *per se*, is not a primary aim, and standardization of methodology may be given lower priority than cost and convenience. When a similar study is done by another investigator, different methods may lead to different results. The final outcome may be a contentious and futile debate in the literature, with each investigator defending his or her methods and results.[7]

2. Variation in the realized niche
Under this hypothesis, observed differences in niche expression are real, not an artifact of method. Even if the fundamental niche is invariant, its expression is likely to vary with ecological context: we may be looking at different realized niches, i.e., different parts of the fundamental niche. To see if this is so, we need to recognize other factors that vary across time and space, and to explore the possibility that they are responsible for differences in niche expression.

3. Variation in the fundamental niche
Finally, we can imagine that variable niche expression reflects genetic variation in the fundamental niche. Such variation seems especially likely over larger spatial or temporal scales. A species is not monolithic; rather it can be thought of as a cluster in phenotype space, corresponding to a cluster in genotype space.[8] If natural selection or other agents of genetic change produce geographic races or subspecies with different fundamental niches, then we expect variable niche expression.

hypotheses for the reversal of the relationship between browsing and fitness. Sharaf and Price (2004) showed that most of the undercompensation at the Rocky Mountain Biological Laboratory (RMBL) in Colorado was due to direct loss of biomass to browsers, rather than to indirect changes in pollination success resulting from delayed flowering. Furthermore, Campbell and Halama (1993) reported that flower production at the RMBL was limited by availability of water and nutrients. Combining these results suggests that compensation might be limited in dry, nutrient-poor soils. In northern Arizona, where most reports of overcompensation arose, the growing season is longer and warmer than at the RMBL. Assuming that the Arizona soils are relatively rich in nutrients (Whitham et al. 1991), and that summer rains provide adequate moisture to mobilize these nutrients, one can imagine that browsers open the possibility for plants to use new nutrients (rather than those stored in roots) to produce more flowers than could be produced on a single original inflorescence. Furthermore, in Arizona the southward migration of hummingbirds may span a longer period than in Colorado (e.g., Calder 1987). Under such combined conditions, browsing might push plants into a region of the fundamental niche of lower fitness in Colorado but of higher fitness in Arizona (figure 4.2). Evaluating this scenario— or others that can be erected to explain location-specific tolerance to browsing without invoking change in the fundamental niche (see Aarssen 1995 for examples)—requires an understanding of mechanism beyond that presented in most of the papers we have cited. The study by Sharaf and Price (2004) exemplifies the value of peeling back the onion of natural history sufficiently to understand mechanism, a point to which we will return.

Pulliam and coworkers are exploring several alternative hypotheses for possible differences in the realized niche of *Polygonatum*. One of these (figure 4.5a) assumes that population density of the species is not an adequate measure of fitness. On average, Whitehall is drier than Coweeta. Suppose that asexual reproduction (cloning), which is common in this species, is more likely in drier soils, resulting in a secondary peak in ramet abundance (but not fitness) under dry conditions. Higher cloning frequency at Whitehall, for which there is some evidence, might then explain the bimodal pattern of densities in figures 4.3 and 4.4. A second hypothesis (figure 4.5b) assumes that a competitor, herbivore, or pathogen at Whitehall prevents plants from fully using moister sites, while a different enemy at Coweeta prevents them from fully using drier sites. Potential competitors seem numerous at Coweeta, where the understory herb layer is dense, but less likely at Whitehall, where the understory is sparse. Fungal pathogens might reduce abundance in moister sites at Whitehall, although we have no evidence for this at present.

Variation in the fundamental niche

The scenario of genetic variation in the species, and in its fundamental niche, is easy to apply to *Ipomopsis*. Based on the initial report of overcompensation in *I. aggregata*, Vail (1992) developed bet-hedging models describing conditions under which different degrees of compensation might be observed in a semelparous plant species. The basic logic is that a sufficiently high probability of herbivore attack can favor an individual that invests less than half of its available resources into an initial reproductive effort. Following

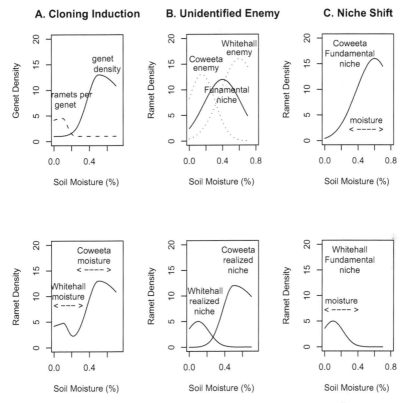

Figure 4.5. Three competing hypotheses to explain differences in the relationship between *Polygonatum* abundance and soil moisture. See text for explanations.

attack, such an individual will be able to surpass the reproductive success of an individual that has not been attacked, under a number of simplifying assumptions. These early models have been refined (see Stowe et al. 2000 and citations therein), but the basic prediction remains. In populations with a history of severe browsing one might observe overcompensation because of an evolved bet-hedging system, whereas in regions where browsing is less frequent one might observe undercompensation. Aarssen (1995) provides additional reasons why removal of the apical meristem and activation of lateral meristems would lead to overcompensation.

The hypothesis of different fundamental niches might likewise explain the observed responses of *Polygonatum* to soil moisture. Imagine an evolutionary change in the fundamental niche (figure 4.5c) such that the optimum condition for growth occurs in drier soils at some sites and moister soils at others. This is a reasonable hypothesis since soil moisture at Whitehall is lower on average (5–10%) than at Coweeta (15–25%), and natural selection might favor characteristics that allow plants to grow best under the most common moisture conditions. The *Polygonatum* observations are suggestive in another respect: the lower maximum abundance at Whitehall is what one might expect if it were more difficult to adapt morphologically or physiologically to low than to high soil moisture.

How Ecologists Might Best Proceed

We have presented evidence for two species that the magnitude of a niche response and even the sign of that response vary between locations. Furthermore, we have outlined three general hypotheses for such variable niche expression (box 4.2). Two of these hypotheses are consistent with the proposition that the *fundamental* niche is an ecological invariant, remaining sufficiently constant that any location-specific differences result from differences among researchers in methods, or (our second and far more interesting hypothesis) from differences in the niche a species actually realizes in each location. The third hypothesis posits that the fundamental niche evolves differently from site to site. In the remainder of this chapter we do not attempt to choose among these hypotheses for the specific species discussed above, but instead we consider what ecologists can do, in general, to differentiate among the hypotheses.

Standardization of method

How might we determine whether apparent niche variation is an artifact of different methods? The obvious answer is to standardize methods and see whether differences disappear. The Long Term Ecological Research (LTER) initiative of the U.S. National Science Foundation (NSF) provides an example of standardization. Recognizing that many ecological questions are too big to be addressed by individual workers, LTER sought to provide sites for intensive collaborations. As the program grew, so did the desire to compare across these sites, and it was recognized that lack of standardized methods made this difficult (National Science Foundation 2002). The NSF responded by encouraging standardization, especially in data and metadata protocols, and by funding cross-site initiatives that required investigators to use the same methods. At present it is launching an even bigger and more ambitious program, the National Ecological Observatory Network (NEON), to facilitate cross-site comparisons and generalizations.

There are dangers, however, in methods being too rigidly dictated (see, e.g., chapter 5 in Golley 1993). Often progress toward an answer to a puzzling scientific question is fastest if the methods used are allowed to evolve as knowledge is gained. More generally, excessive regimentation can stifle the creativity, akin to artistic creativity, that represents the most mysterious and productive aspect of the scientific endeavor.[9] Although "wild ideas" are usually just that, it is equally true that forward leaps in understanding usually are culled from such unorthodoxy, so we would be ill-advised to impose any orthodoxy too strongly.

The ecology of place, natural history intuition, and multiple working hypotheses

Suppose that we show that spatial variation in niche expression is not a result of different methods. Two possibilities remain: that the realized niche varies, or that the fundamental niche does (box 4.2). To go further we need to erect and explore hypotheses for as yet unrecognized influences that might now be molding the realized niche, or that in the past might have caused the evolution of different fundamental niches. To detect such influences we may employ classical experiments, such as reciprocal transplantation among sites,[10] and new technologies such as those that unravel the genetic architecture of quantitative traits involved in niche expression (e.g., Juenger and Bergelson 2000). But none of these approaches replaces "natural history intuition" in getting to a deeper understanding of causes—those differences in ecological context that must ultimately explain differences in realized or fundamental niches.

Natural history intuition is difficult to describe or prescribe (see also Waller and Flader, chapter 3). We propose that it involves an iterative process of observation of nature, attempts at explaining the observations, and modification of one's list of possible explanations based on new observations (see also Price and Billick, introductory remarks to part 3). This process of model building requires time and patience since many natural events are uncommon and, as such, undetectable at first glance. We therefore emphasize the value of the ecology of place, as this term is used throughout this volume: the value of long-term study in a single place, using diverse methods, which allows the development of natural history intuition. Of course, once such detailed understanding is achieved at one place, there is great additional value in testing the generality of that understanding through similar studies at additional places, using common methods (see also Waller and Flader, chapter 3; Rundel, chapter 5; Estes, chapter 8).

We have emphasized above the need to keep an open mind to multiple possible explanations. This is another way of saying that the best way to understand many ecological phenomena (including, but not limited to, the niche) is to develop multiple working hypotheses, rather than the traditional Fisherian approach which compares a null hypothesis to a single alternative hypothesis. The approach of multiple working hypotheses involves building multiple alternative causal models for the underlying ecological phenomena one has observed. Advocacy of multiple working hypotheses dates at least to the geologist T. C. Chamberlin (1890; see also Railsback 2004), who proposed that hypotheses be erected a priori. We instead imagine them evolving to some extent sequentially, as the ecological detective (sensu Hilborn and Mangel 1997; see also the quotation from Sinclair Lewis at the head of this chapter) accumulates natural history knowledge and recognizes previously ignored factors to be important. In either case, this approach reduces the risk of the investigator becoming attached to a favored hypothesis, and instead increases the likelihood that he or she remains a "skeptical inquirer" prepared to admit that earlier hypotheses were mistaken.

Adopting multiple hypotheses takes us away from the philosophical position that causation can be viewed as an "either/or" dichotomy, and that progress is made by rejecting one possibility in favor of an alternative null hypothesis. Instead, we view causation as potentially more complex (Hilborn and Stearns 1982), although we may still sometimes be able to group hypotheses into classes and make observations that allow us to eliminate entire classes (Platt 1964; for examples in ecology see Waser and Price 1985, Harmon and Franklin 1989). Also, carrying forward a given study across time in one place moves us away in another sense from an either/or perspective on ecological phenomena, by allowing us to say *how often* one finds each of several outcomes. Refining a causal model of ecological phenomena using the approach just described is implicitly a Bayesian process (Hilborn and Mangel 1997, Ellison 2004): through successive observations and experiments we become increasingly confident of a given causal model, using the confidence at each step to erect prior probabilities that affect our interpretation of the next round of observations.

The approach taken so far to understanding how *Ipomopsis* responds to inflorescence browsing, for example, has largely been to test the hypothesis of overcompensation against some null such as no compensation or exact compensation, with a decision being based on some arbitrary (i.e., 5%) chance of accepting a false alternative hypothesis. Faster progress might be made by developing causal hypotheses about how drought conditions, soil nutrient availability, phenology of browsing, phenology of pollinators, and so on interact to determine the expected life histories and final fecundities of browsed and unbrowsed plants. Ultimately one may be able to build a causal model that explains the observed results with enough fidelity, and in enough mechanistic detail, that one has pinpointed likely explanations for variable niche expression. Similarly, in exploring how *Polygonatum* responds to soil moisture we need to answer basic who, what, when, and where questions. Is *Polygonatum* at Whitehall genetically the same as the biological entity at Coweeta? Is the abundance of *P. biflorum* greater where soil moisture is higher? Does *P. biflorum* produce more clones in drier habitats? The analysis presented earlier gives an example of using multiple competing hypotheses and Bayesian analysis to characterize the functional form of the relationship between abundance and soil moisture at different locations. Once such basic patterns are established, the Bayesian approach can be used again to differentiate among multiple competing hypotheses.

Conclusions

We end where we started—with a comparison to the physical sciences—and a simple example from chemistry that not only the rate but also the products of a reaction can vary with the environment. There is no general law that requires the reaction of chemicals X and Y to always yield product Z ($X + Y \rightarrow Z$), but rather a set of conditions (an environment) for which this happens at a predictable rate, and other conditions under which $X + Y \leftarrow Z$ or $X + Y \rightarrow Q$[11]. There is a set of rules to be discovered by observation and experimentation that specify the direction and rate of the reaction. Similarly, in ecology there is no fixed response of a species to an environmental variable, but rather a putative set of rules that predict how the response will change as other environmental variables also change. Accordingly, we should take variation in niche expression, or in any other ecological phenomenon, not as an annoyance but as an opportunity to discover deeper levels of mechanism that explain the observation at a more general level (see also Levin 1981, Tilman 1987).

Said differently, we should not be surprised by location-specific patterns in niche expression—and in other ecological phenomena—but we should persist in searching for their predictability (and here our conclusion seems to differ from that of Feinsinger et al., chapter 18). The key word—and the bugbear—is predictability. Whereas some ecologists interpret nature's variability as pure idiosyncrasy, and despair that conclusions from one location can apply to another, we counter that ecological patterns must be complex realizations of an understandable set of rules governing responses of species across space and time. If one is able to reach this level of understanding by the methods of natural history intuition we have sketched or by some other route, so that one can explain and predict variation in an ecological phenomenon based on a deeper perception of shared mechanisms, then one will have achieved a most crucial form of "ecological invariance."

Acknowledgments

We thank Mary Price for encouragement, Alison Brody and Mary Price for discussing compensation in *Ipomopsis*, and Dave Parkhurst for permission to use his limerick. The *Polygonatum* work has benefited from discussion with students in the Pulliam lab, especially Jeff Diez, Robert Warren, and Itamar Giladi, and was supported by the United States National Science Foundation (DEB-0235371 and DEB-9632854, the Coweeta LTER). Finally, we are indebted to Ian Billick, Emily Jones, Charley Krebs, Michele Lanan, and Mary Price for comments on earlier drafts.

References

Aarssen, L. W. 1995. Hypotheses for the evolution of apical dominance in plants: Implications for the interpretation of overcompensation. *Oikos* 74:149–56.

Austin, M. P., A. O. Nicholls, and C. R. Margules. 1990. Measurement of the realized qualitative niche: Environmental niches of five *Eucalyptus* species. *Ecological Monographs* 60:161–77.

Austin, M. P., and T. M. Smith. 1989. A new model for the continuum concept. *Vegetatio* 83:35–47.

Bergelson, J., and M. J. Crawley. 1992a. Herbivory and *Ipomopsis aggregata*: The disadvantages of being eaten. *American Naturalist* 139:870–82.

———. 1992b. The effects of grazers on the performance of individuals and populations of scarlet gilia, *Ipomopsis aggregata*. *Oecologia* 90:435–44.

Bergelson, J., T. Juenger, and M. J. Crawley. 1996. Regrowth following herbivory in *Ipomopsis aggregata*: Compensation but not overcompensation. *American Naturalist* 148:744–55.

Bergey, D. H., J. G. Holt, N. R. Krieg, and P. H. A. Sneath. 1994. *Bergey's Manual of Determinative Bacteriology*. Williams and Wilkens, Baltimore.

Brody, A. K., M. V. Price, and N. M. Waser. 2007. Life-history consequences of vegetative damage in scarlet gilia, a monocarpic plant. *Oikos* 116:975–85.

Brown, J. H., G. B. West, and B. J. Enquist. 2000. Scaling in biology: Process, causes and consequences. Pages 1–24 *in* J. H. Brown and G. B. West, eds., *Scaling in Biology*. Oxford University Press, New York.

Brumfiel, G. 2006. Outrageous fortune. *Nature* 439:10–12.

Calder, W. A. 1987. Southbound through Colorado: Migration of rufous hummingbirds. *National Geographic Research* 3:40–51.

Campbell, D. R., and K. J. Halama. 1993. Resource and pollen limitations to lifetime seed production in a natural plant population. *Ecology* 74:1043–51.

Chamberlin, T. C., 1890. The method of multiple working hypotheses. *Science* (old series) 1592–96 (reprint in *Science* 148:754–59, 1965).

Carlin, B. P., and T. A. Louis. 2000. *Bayes and Empirical Bayes Methods for Data Analysis, 2nd Edition*. Chapman and Hall, Boca Raton, FL.

Charnov, E. L. 1993. *Life History Invariants*. Oxford University Press, Oxford.

Diez, J. M. 2006. Hierarchical patterns of symbiotic orchid germination linked to adult proximity and environmental gradients. *Journal of Ecology* 95:159–70.

Diez, J. M., and H. R. Pulliam. 2007. Hierarchical analysis of species distributions and abundance across environmental gradients. *Ecology* 88:3144–52.

Ellison, A. M. 2004. Bayesian inference in ecology. *Ecology Letters* 7:509–20.

Elton, C. 1927. *Animal Ecology*. Sedgwick and Jackson, London.

Gause, G. F. 1934. *The Struggle for Existence*. Williams and Wilkens, Baltimore.

Giladi, I. 2006. Choosing benefits or partners: A review of the evidence for the evolution of myrmecochory. *Oikos* 112:481–92.

Golley, F. B. 1993. *A History of the Ecosystem Concept in Ecology: More than the Sum of the Parts*. Yale University Press, New Haven, CT.

Grant, V., and D. H. Wilken. 1986. Taxonomy of the *Ipomopsis aggregata* group (Polemoniaceae). *Botanical Gazette* 147:359–71.

Grinnell, J. 1917. The niche-relationships of the California Thrasher. *Auk* 34:427–33.

Gronemeyer, P. A., B. J. Dilger, J. L. Bouzat, and K. N. Paige 1997. The effects of herbivory on paternal fitness in scarlet gilia: Better moms also make better pops. *American Naturalist* 150:592–602.

Harmon, M. E., and J. F. Franklin. 1989. Tree seedlings on logs in *Picea-Tsuga* forests of Oregon and Washington. *Ecology* 70:48–59.

Hilborn, R., and M. Mangel. 1997. *The Ecological Detective: Confronting Models with Data*. Princeton University Press, Princeton, NJ.

Hilborn, R., and S. C. Stearns. 1982. On inference in ecology and evolutionary biology: The problem of multiple causes. *Acta Biotheoretica* 31:145–64.

Hutchinson, G. E. 1957. Concluding remarks. Population studies: Animal ecology and demography. *Cold Spring Harbor Symposium on Quantitative Biology* 22:415–27.

Juenger, T., and J. Bergelson. 2000 The evolution of compensation to herbivory in scarlet gilia, *Ipomopsis aggregata*: Herbivore-imposed natural selection and the quantitative genetics of tolerance. *Evolution* 54:764–77.

Kemp, M. 2005. From science in art to the art of science. *Nature* 434:308–9.

Langlet, O. 1971. Two hundred years genecology. *Taxon* 20:653–721.

Levin, S. A. 1981. The role of theoretical ecology in the description and understanding of populations in heterogeneous environments. *American Zoologist* 21:865–75.

Lewis, S. 1925. *Arrowsmith*. Harcourt Brace and World, New York.

MacArthur, R. H., and E. O. Wilson 1963. An equilibrium theory of insular zoogeography. *Evolution* 17:373–87.

Maguire, B. Jr. 1973. Niche response structure and the analytical potentials of its relationship to habitat. *American Naturalist* 107:213–46.

Marquet, P. A., R. A. Quiñones, S. Abades, F. Labra, M.Tognelli, M. Arim, and M. Rivadeneira. 2005. Scaling and power laws in ecological systems. *Journal of Experimental Biology* 208:1749–69.

Maschinski, J., and T. G. Whitham. 1989. The continuum of plant responses to herbivory: The influence of plant association, nutrient availability, and timing. *American Naturalist* 134:1–19.

McNaughton, S. I. 1983. Compensatory plant responses to herbivory. *Oikos* 40:329–36.

Merriam, C. H. 1890. Results of a biological survey of the San Francisco Mountain region and desert of the Little Colorado, Arizona. *North American Fauna*, no. 3.

National Science Foundation. 2002. Long-term ecological research program: Twenty-year review. Online URL http://intranet.lternet.edu/archives/ documents/reports/20_yr_review/.

Paige, K. N. 1994. Herbivory and *Ipomopsis aggregata*—differences in response, differences in experimental protocol—a reply. *American Naturalist* 143:739–49.

———. 1999. Regrowth following ungulate herbivory in *Ipomopsis aggregata*: Geographic evidence for overcompensation. *Oecologia* 118:316–23.

Paige, K. N., and T. G. Whitham. 1987. Overcompensation in response to mammalian herbivory: The advantage of being eaten. *American Naturalist* 129:407–16.

Paige, K. N., B. Williams, and T. Hickox. 2001. Overcompensation through the paternal component of fitness in *Ipomopsis arizonica*. *Oecologia* 128:72–76.

Platt, J. R. 1964. Strong inference. *Science* 146:347–53.

Press, S. J. 2003. *Subjective and Objective Bayesean Statistics: Principles, Models, and Applications*. John Wiley and Sons, New York.

Price, M. V., N. M. Waser, R. E. Irwin, D. R. Campbell, and A. K. Brody. 2005. Temporal and spatial variation in pollination of a montane herb: A seven-year study. *Ecology* 86:2106–16.

Pulliam, H. R. 1988. Sources, sinks, and population regulation. *American Naturalist* 132:652–61.

Railsback, L. B. 2004. T. C. Chamberlin's "Method of Multiple Working Hypotheses": An encapsulation for modern students. Online URL www.gly.uga.edu/railsback/railsback_chamberlin.html.

Schluter, D., and P. R. Grant. 1984. Determinants of morphological patterns in communities of Darwin›s finches. *American Naturalist* 123:175–96.

Sharaf, K. E., and M. V. Price. 2004. Does pollination limit tolerance to browsing in *Ipomopsis aggregata*? *Oecologia* 138:396–404.

Spieglhalter, D. J., N. G. Best, B. P. Carlin, and A. van der Linde. 2002. Bayesean measures of model complexity and fit. *Journal of the Royal Statistical Society Series B* 64:583–639.

Stowe, K. A., R. J. Marquis, C. G. Hochwender, and E. L. Simms. 2000. The evolutionary ecology of tolerance to consumer damage. *Annual Review of Ecology and Systematics* 31:565–95.

Tilman, D. 1987. The importance of the mechanisms of interspecific competition. *American Naturalist* 129:769–74.

Turesson, G. 1922. The genotypical reponse of the plant species to the habitat. *Hereditas* 3:211–350.

Vail, S. G. 1992. Selection for overcompensatory plant responses to herbivory: A mechanism for the evolution of plant-herbivore mutualism. *American Naturalist* 139:1–8.

Waser, N. M. and M. V. Price. 1985. The effect of nectar guides on pollinator preference: Experimental studies with a montane herb. *Oecologia* 67:121–26.

Waser, N. M., M. V. Price, and R. G. Shaw. 2000. Outbreeding depression varies among cohorts of *Ipomopsis aggregata* planted in nature. *Evolution* 54:485–91.

Whitham, T. G., J. Maschinski, K. C. Larson, and K. N. Paige. 1991. Plant responses to herbivory: The continuum from negative to positive and underlying physiological mechanisms. Pages 227–56 *in* P. W. Price, T. M. Lewinsohn, G. W. Fernandes, and W. W. Benson, eds., *Plant-Animal Interactions: Evolutionary Ecology in Tropical and Temperate Regions*. John Wiley and Sons, New York.

Wolf, P. G., P. S. Soltis, and D. E. Soltis. 1993. Phylogenetic relationships of the *Ipomopsis aggregata* complex and related species (Polemoniaceae): Evidence from chloroplast DNA restriction site variation. *Systematic Botany* 18:652–62.

Wright, S. 1932. The roles of mutation, inbreeding, crossbreeding and selection in evolution. Pages 356–66 *in* D. F. Jones, ed. *Proceedings of the Sixth International Congress of Genetics*. Brooklyn Botanical Garden, Brooklyn, NY.

Notes

1. $1/N \, dN/dt$ is a measure of mean fitness. Maguire (1973) represented the Hutchinsonian fundamental niche as a mapping of fitness on environmental variables. The word "map" is appropriate because Maguire drew a three-dimensional topography, with isolines of positive fitness forming hills and mountains on a surface of two environmental dimensions: this is not the same as Wright's (1932) adaptive landscape, which has dimensions of gene frequencies instead of environmental factors, and valleys of negative fitness! Maguire even used his topographic approach to explore a variant of what are now called zero-net-growth isoclines (ZNGIs).

2. For simplicity we have drawn niches along a single axis as unimodal functions. However, fundamental niches might be multimodal because environmental or resource gradients at any location are discontinuous (e.g., discrete classes of seed size and hardness; Schluter and Grant 1984), and realized niches might be multimodal because of the action of other species at the same or different trophic levels. Austin and Smith (1989) and Austin et al. (1990) discuss the actual shapes of niche functions.

3. This brings up interesting thoughts. The mutualism might be critical for our hypothetical plant species to occur, or might simply increase fitness without being essential. In the former case which is the fundamental niche, and which realized? Is realized larger than fundamental? Might mutualists (e.g. mycorrhizae) add extensions to the fundamental niche, these extensions disappearing in their absence, representing a shrinkage to the realized niche (i.e., absence of a beneficial interaction resembles presence of a detrimental one)? These may seem like questions about noise made by trees falling in the forest, but perhaps they illustrate interesting elaborations (limitations?) of the Hutchinsonian niche (see also Pulliam 1988).

4. *Ipomopsis arizonica* is sometimes considered a subspecies of *I. aggregata*, although molecular evidence (Wolf et al. 1993) supports specific status. Undercompensation following browsing, or at best perfect compensation (seed production equal to that of unbrowsed plants), had been reported previously for female reproductive success of this species (Mashinski and Whitham 1989).

5. We personally know of one unpublished case. A. K. Brody, S. Louda, and N. M. Waser (unpublished data), working at the Rocky Mountain Biological Laboratory (RMBL) in western Colorado, found only undercompensation in repeated comparisons of artificially browsed plants to controls.

6. Various authors have objected to this use of "mutualist," because overcompensating plants may have evolved with herbivores to "make the best of a bad deal," and must have lower fecundity than their ancestors that lived without herbivores. For details see Stowe et al. 2000 (pp. 577–78).

7. Not all debates over results are sterile; at best they may resolve apparent differences in results and stimulate wider discussion about appropriate methodology. But at the worst they may generate more heat than light, and inflict substantial harm: this was the case with the "null-model debate" in community ecology, which, despite its merits, and by its acrimony, needlessly depopulated the field for several academic generations at a time when the emerging discipline of conservation biology was identifying burning applied questions at the community level.

8. Molecular phylogenetic methods are making it more possible to confirm the latter, sometimes yielding surprises that "species" are polyphyletic; indeed, this is so for some *Ipomopsis* "species" (Wolf et al. 1993).

9. See Kemp (2005) for one discussion of the links between art and science.

10. Langlet (1971) attempts to put to rest the perception that such studies began with Turesson (1922), tracing their history to silviculture in the 1700s.

11. Indeed, recent advances in string theory suggest the conclusion that values of the physical constants apply only in our own universe, and are effectively chance values among all values realized in an astounding number of parallel universes (Brumfiel 2006). In this regard physical constants and laws might be seen generally to be context-dependent, the complaint often leveled at ecological constants and laws.

5 *Convergence and Divergence in Mediterranean-Climate Ecosystems*

WHAT WE CAN LEARN BY COMPARING SIMILAR PLACES

Philip W. Rundel

Figure 5.0. Convergence in canopy structure of evergreen shrublands in Mediterranean-climate regions: (a) maquis community on Mount Carmel, northern Israel (photograph by P. Rundel, 2008); (b) chaparral community in the Santa Monica Mountains of southern California (National Park Service photograph, ca. 2005).

Abstract

The five Mediterranean-climate regions of the world—California, central Chile, the Mediterranean basin, the Cape Region of South Africa, and south and southwestern Australia—share a remarkable similarity in the structure of evergreen, leathery-leaved shrublands that dominate the landscape. This pattern has long evoked interest in comparative ecological studies among these five geographically isolated regions with their unique climatic regime. The focus of scientific research has progressed through time from initial generalizations about ecosystem convergence in the five Mediterranean-climate regions to a focus on multiple exceptions to patterns of convergence, and finally to a deeper understanding of processes and ecological relationships that both link and separate individual regions. Today there is expanding interest in the development of models for the evolution of plant life-history traits in Mediterranean-climate regions as an approach to better understanding the nature of both past and present selective forces in each of the five regions. This paradigm has developed as the broader community-level vision of Mediterranean-climate "places" has been replaced by an evolutionary approach to understanding plant functional groups and their role in structuring communities. Mediterranean-type ecosystems continue to provide many opportunities for comparative studies of the controlling factors in community structure and biodiversity, and of macroecological patterns in convergent community evolution.

Introduction

There has been a widespread opinion among ecologists and philosophers of science that because of the complexity of the biological world, ecology and evolution lack broad universal laws that underpin successful scientific prediction and explanation. This view was well articulated by Lawton (1999) who describes much of ecology as a "rummage-box of case histories." Other ecologists have differed sharply with Lawton, arguing that multiple case studies provide significant insights and that numerous quantitative "laws" exist which describe how populations grow and how natural selection operates (Murray 1999, Turchin 2001). There has, however, been less effort to dispute Lawton's claim with regard to community and ecosystem ecology, which he opined was so complex that the field had not produced much in the way of patterns or workable contingent rules.

A different view of the existence of universal laws in ecology was taken by O'Hara (2005), who argued that much of the controversy over whether ecology had laws stemmed from a failure to recognize that the concept of a "law" could be applied to two kinds of scientific generalizations. Philosophers of science typically consider causal laws, which are intended as descriptions of how some aspect of the universe operates. Newton's laws of motion and Malthusian principles of population growth are examples of such causal laws. There is a second form of "law," however: correlative laws, which describe patterns of predictability but have only weak explanatory power and thus do not provide a deep understanding of the causal processes that produce an observed pattern. By this usage, community and ecosystem ecology do have laws—as, for example, with Rapoport's rule for latitudinal ranges, or the species-area relationships of island biogeography. Even though correlative laws do have exceptions and are rarely universally true, they can form the basis for a strong science of ecology (O'Hara 2005). This is because they invite explanation, and thus a search for causal factors that are responsible for the observed patterns.

Unlike the causal laws, correlative laws are not invalidated by exceptions because they describe regularities rather than causal truths. Weber (1999) points out that with correlative laws it is often the exception that helps define the rule, in that exceptions allow one to better understand the limits of the law's applicability. Moreover, correlative laws often lead to expanded research to identify causal factors that produce predictable patterns in communities and ecosystems.

Correlative laws, and exceptions to them, play a productive role in ecology. In fact, the entire field of macroecology concerns itself with identifying and explaining ecological patterns on broad taxonomic, temporal, and geographical scales (Brown and Mauer 1989, Brown 1995). In this chapter I illustrate how correlative laws facilitate ecological understanding with a macroecological example—the correlation between a place's climate and the basic structure and functional attributes of its dominant vegetation. Because the example is a geographical one, it illustrates what can be learned from a comparison of multiple places. It also illustrates how natural-history knowledge of particular places informs understanding of the broader patterns of ecosystem distribution, as well as how exceptions to general patterns at one place can stimulate research and inform understanding at another.

My example tells the story of the long history of comparative ecological studies of the five Mediterranean-climate regions of the world. Ecologists more than a century ago noted the remarkable similarity in the structure of evergreen, sclerophyllous (leathery-leaved) shrublands that dominated the landscape in five Mediterranean-climate regions—California, central Chile, the Mediterranean basin, the Cape Region of South Africa, and south and southwestern Australia (figure 5.1). The concept of climatic influence on the evolution of vegetation structure and functional attributes began with simple natural-history observations. This early work has motivated a remarkable body of scientific research to test hypotheses of convergent evolution at both macroecological and regional levels.

My narrative charts the evolution of our thinking as scientific knowledge has progressed from initial generalizations about climate-vegetation relationships in Mediterranean-climate regions to a period of realization of the existence of multiple exceptions to patterns of convergence, and finally to a deeper understanding today of processes and ecological relationships that both link and separate individual regions. I also describe how differences in

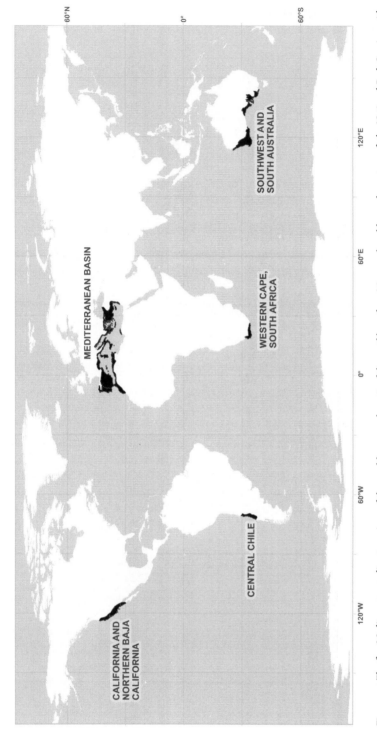

Figure 5.1. The five Mediterranean-climate regions of the world cover only 2.5% of the earth's surface. Map produced by, and courtesy of, the National Park Service, with continental boundaries taken from ESRI Data and Maps, 2007.

specific components of community structure and species attributes between these regions have not negated the value of the correlative laws of ecosystem convergence, but instead have provided the means to develop causal hypotheses to account for these dissimilarities and thus better understand the ecology of Mediterranean-climate ecosystems.

Convergent Evolution

Convergent evolution is generally defined as the emergence of biological traits that exhibit similar function and appearance but have evolved through independent evolutionary pathways. The similarities shared in the case of convergent evolution are not the result of evolution from a common ancestor, but rather are explained as shared adaptive solutions to similar environmental pressures. Textbook examples of convergent evolution include fundamental structural traits such as the multiple origins of wings (bats, birds), biochemical traits such as the evolution of functionally similar but distinct antifreeze proteins in divergent species of fish in the Antarctica and Arctic, and combined morphological and physiological adaptations to aridity seen in New World cacti and succulent African euphorbias.

Ecologists have extended the concept of evolutionary convergence to describe convergent ecosystem evolution between broad ecological biomes. The assumption in these studies has been that climatically analogous habitats produce a combination of environmental stress factors for which there exist only a limited number of successful evolutionary responses. Temperate conifer forests provide a good example in which forest communities on low-nutrient or skeletal montane soils are typically dominated by evergreen conifers with common adaptive traits of canopy architecture, nutrient-use efficiency, water relations, and cold tolerance. Temperate grasslands provide another example in which climate and edaphic factors strongly influence the dominance of grasses.

This concept of convergent ecosystem structure and function dates back to a tacit assumption of convergence in the global vegetation classifications developed at the end of the nineteenth century (Grisebach 1872, Drude 1890, Schimper 1903). Some of these early writings were remarkably prescient in developing modern concepts of ecological convergence, considering the limited data available at the time. Schimper (1903), for example, provided a broad albeit imperfect basis for understanding the adaptive significance of structural features in plants. In describing the vegetation of the five Mediterranean-climate regions of the world, he wrote, ". . . the vegetation bears essentially the same stamp, in spite of deep-seated differences in composition of the flora; it is dominated by sclerophyllous plants, and always, although to a subordinate extent, by tuberous and bulbous plants." In his comment on the fundamental floristic differences between these regions, Schimper was clearly focusing on what today would be called convergent ecosystems.

Other seeming examples of convergent ecosystems relate to climate, but lack a clear evolutionary context because separate geographic communities have species that share common ancestral roots. Broad-leaved trees form the dominant vegetation cover in east Asia, the eastern United States, and western Europe under similar climatic regimes. Winter-deciduous behavior, typically cued by photoperiod, adapts these trees to tolerate cold winter conditions when photosynthetic production would otherwise be limited. An arborescent growth form is critical in competing for solar irradiance during the favorable months for growth. These three forest areas, however, do not have independent evolutionary roots, but rather share a paleoecological component of origin in southern China in the early Tertiary.

Mediterranean-Climate Regions

Mediterranean-climate regions of the world have long evoked interest in comparative studies with their unique climatic regime and their shared trait of community dominance by evergreen, leathery-leaved shrubs (Cowling et al. 1996; Rundel 2000, 2004). The core Mediterranean-climate "places" are typically defined as temperate areas strongly dominated by winter rainfall and summer drought, but where precipitation exceeds potential evapotranspiration over the rainy season. While these regions amount to only about 2.2% of the world's land area (figure 5.1, table 5.1), they have tremendous conservation significance because they are home to approximately 20% of all of the world's vascular plant species (Cowling et al. 1996).

The modern communities of Mediterranean-climate regions are relatively recently evolved in the late Tertiary and Quaternary from largely independent ancestries. The contemporary floras of each of the Mediterranean-climate regions comprise admixtures of floras independently derived from relict subtropical forest and recent drought- and fire-adapted lineages

TABLE 5.1. Comparative area, topographic heterogeneity, climatic heterogeneity, and estimated natural fire frequency for the five Mediterranean-type climate regions of the world. Fire frequency ranges are typical values for shrublands rather than extreme ranges.

Region	Area ($\times 10^6$ km²)	Topographic heterogeneity	Climatic heterogeneity	Natural fire frequency (yrs.)
California	0.32	high	very high	40–60
Central Chile	0.14	very high	very high	fire-free
Mediterranean Basin	2.30	high	high	25–50
Cape Region	0.09	moderate	high	10–20
S.W. Australia	0.31	low	moderate	10–15

evolved in seasonal semiarid regions with a summer rainfall regime. Before the onset of global cooling and seasonal drying in the Pliocene, subtropical forests dominated all five regions. Drought-tolerant shrublands with evergreen leathery leaves and dry woodlands subsequently replaced these forests, most prominently during the Quaternary when the strongly seasonal Mediterranean-type climate developed, and fire became a significant factor in all these regions except Chile. Most of the regions support extensive areas of forest, woodland, and drought-deciduous shrubland in addition to evergreen shrublands, but comparative studies have not focused on these other systems that lack the same strong elements of apparent convergence.

Convergent Ecosystem Studies

Before the 1960s, a barrier to understanding the evolution of ecosystem structure and function had been that studies were difficult to put into context due to the absence of a proper standard of comparison or control. Principles of ecosystem adaptive function could not be generalized without a properly replicated ecosystem "place." In the Mediterranean-climate regions of the world, however, ecologists recognized a powerful natural experiment. Here were replicated opportunities to compare and contrast how evolution had adapted plant and animal species to unique climate conditions and stresses in five isolated areas of the world.

Modern comparative studies of Mediterranean-climate regions began in the late 1960s with ecological comparisons of California and Israel (Naveh 1967); Australia, France and California (Specht 1969a, 1969b); and California and Chile (Mooney and Dunn 1970a, 1970b). These studies pointed out similarities in climatic regime, vegetation structure, and ecological processes such as fire succession, nutrient dynamics, and hydrologic flow. With these publications and expanding interest in these regions, the U.S. National Science Foundation agreed to sponsor a 1971 symposium in Valdivia, Chile, on the origin and structure of Mediterranean-type ecosystems. This conference had a stated goal of providing a preliminary critical overview of the information available on the origin, evolution, and structure of Mediterranean-climate ecosystems. Its product was a synthetic volume offering review chapters for the Mediterranean-climate regions on themes of physical geography, vegetation, soil systems, animal ecology, and human impacts (di Castri and Mooney 1973). This volume was highly significant in stimulating a broad international program of study.

At the same time that the Valdivia symposium was being planned, the International Biological Program (IBP) was initiated with support from the U.S. National Science Foundation to begin major ecological studies with diverse ecosystem foci. Two programs within the U.S. IBP Program were centered on convergent evolution themes. One of these was a project comparing desert ecosystems in the Sonoran Desert of Arizona with that of the Monte Desert in Argentina. The other was a Mediterranean-climate ecosystems program comparing Chile and California (Golley 1993).

The guiding hypothesis of this IBP convergent-evolution theme was that similar physical environments produce a common set of physical stresses to which there exist only a limited number of successful adaptive strategies. Since each form of adaptive strategy provides energetic benefits but requires energetic costs, natural selection will choose a convergent set of structures and functions that maximize Darwinian fitness (Mooney 1977). At the end of the six years of IBP studies, the researchers concluded that the climatic similarities between Chile and California were responsible for a substantial degree of convergence in the structure of plant communities. This conclusion was reached on the basis of quantitative studies of plant growth forms, canopy architecture, and leaf phenology and physiology. At the level of consumers, however, they concluded that the direct adaptive links between climate and morphological types were weaker and difficult to demonstrate. Above the level of the community, convergences in ecosystem processes were more difficult to demonstrate in both plants and animals, as they were less obvious and obscured by chance and historical events (Mooney 1977).

Researchers in the Sonoran/Monte program reached similar conclusions (Orians and Solbrig 1977). They found what appeared to be striking convergences between the two regions in community structure, life-form distribution, and biomass. In addition, adaptive plant morphology and ecophysiology were very similar, but interpretation of these traits was confounded because many species pairs of woody plants across the two regions were congeners. Thus, parallel rather than convergent evolution was a possible explanation. Small vertebrates strongly affected by their physical environment, such as anurans, were reported as highly convergent in adaptive traits, with seemingly equivalent species pairs. With larger animals, as with most mammals, biotic factors of competition and predation, as well as historical factors, seemed to outweigh any convergence in adaptive traits.

An expansion of the convergent-ecosystem model for California and Chile was explored by Miller (1981) who used community theory to predict that there should be the evolution of optimal or maximal resource-use efficiency for light, water, and nutrients in a stable plant community. Extending this model, he predicted that evergreen shrublands in California and Chile should deploy similar patterns of resource utilization despite their differing phylogenetic histories. The conclusions of this study, however, showed limitations in this approach based on the methodologies and models used. Mature evergreen shrublands in the two regions failed to perform as expected by community theory, with uneven utilization of resources. Results indicated that there were resources of light, water, and nutrients that were not harvested efficiently by the existing mixes of community composition. Pole-facing communities in the two regions were found to be more alike in their efficiencies of resource utilization than were pole- and equator-facing slopes of chaparral in California, suggesting that some form of ecosystem convergence might be present. There was no evidence for convergence or ecological "equivalence" at the species level, with variable suites of adaptive traits present among species from both regions. Interestingly, Miller posited from his models that water-use efficiency was environmentally controlled while light and nutrient-use efficiencies were subject to biological adaptation. These conclusions suggested the possibility that water availability, with or without biological competition, was a strong factor of selection in both regions, while light and nutrients were less limiting.

Divergence as an Ecological Theme

By the late 1970s, the initial focus on the similarities of vegetation structure in the five Mediterranean-climate regions of the world had given way to interest in the subtle but sometimes obvious differences that also existed between certain of the Mediterranean-climate regions (table 5.1). Markedly lower soil fertility, greater amounts of summer rain, differential disturbance regimes, and more frequent fires were invoked to explain differences in plant traits and community structure between the Mediterranean-climate ecosystems of the Cape Region and southwestern Australia and those of the climatically analogous areas of California, Chile, and the Mediterranean Basin (Cody and Mooney, 1978, Barbour and Minnich 1990, Keeley 1992, Arroyo et al. 1994, Rundel et al. 1998, Rundel 1999). Prominent in these differences were contrasts between the extremely high levels of plant species diversity in the Cape Region and southwestern Australia and the lower levels of diversity at a community level in the other three regions (Cowling et al. 1994, 1996, Rundel 2000). For example, the Cape Region, about one-fourth the area of the California Mediterranean-climate region, has more than twice as many plant species. Interest in identifying the common denominator separating the Cape Region and southwestern Australia from the other regions focused on stable geologic position at the center of global plates, associated long history of soil weathering, and the presence of acidic low-nutrient soils in these two austral regions.

It was obvious to researchers that the low-nutrient and acidic soils typical of the Cape Region and southwestern Australia were correlated with high levels of plant species diversity. Was there a causal relationship of some kind between these two conditions? The theme of nutrient availability and biodiversity in Mediterranean-climate regions was addressed in a symposium held in Stellenbosch, South Africa, in 1980 to assess the role of nutrients in patterns of species diversity and ecosystem convergence, as well as the mechanisms of diversity maintenance in the low-nutrient Mediterranean-climate regions (Kruger et al. 1983). Although a published volume provided a greatly expanded view of inter- and intra-regional patterns of nutrient ecology, it failed to identify a clear model explaining the regional patterns of species diversity based on patterns of regional nutrient availability. Long histories of geologic stability and weathering in the Cape Region and southwestern Australia have each produced highly leached soils low in nutrients, but with very different geomorphic histories. The classic fynbos soils of the Cape Region are weathered quartzite, while soils in southwestern Australia include laterites, weathered sands over laterite, and younger sand plains. Moreover, the high species richness of fynbos shrublands on acidic nutrient-poor soils is matched by similarly high species richness of communities on richer soils derived from shales.

If nutrients were not a causal factor in explaining biodiversity, what other factors could be considered? It had long been recognized that despite the broad similarities of rainfall regime in Mediterranean-climate places, subtle differences in rainfall regimes were also present (di Castri 1981, Barbour and Minnich 1990). Most research on convergence in Mediterranean-climate regions has used mean annual rainfall as the climatic basis for selecting environmentally analogous sites for comparative studies. Moving beyond annual means, summer drought intensity is an important factor for distinguishing Mediterranean-climate environments (Lamont et al. 1985, Keeley 1992). Simple paradigms for how shrub species responded to seasonal moisture availability and summer drought in Mediterranean-climate regions were shown to be gross oversimplifications of what were varied and complex patterns of adaptations in plant water relations (Rundel 1995).

In early studies, rainfall reliability measured as interannual variation in annual and monthly rainfall and as the frequency of different sized rainfall events attracted surprisingly little interest as a factor that could explain convergence or divergence among Mediterranean-climate ecosystems. Summers are dry in all five regions, but have predictable small amounts of summer rainfall in some of the regions. With the recent recognition of the significance of rainfall reliability as a significant environmental and evolutionary factor in studies comparing the winter-rainfall Mojave Desert in California and Succulent Karoo in South Africa (Esler et al. 1999), there has been a renewed focus on this component of rainfall variability. Despite similar mean annual and winter rainfall levels across the Mediterranean-climate ecosystems, Cowling et al. (2005) showed that the rainfall regimes of the Cape Region and southwestern Australia are fundamentally different from those of the other Mediterranean-climate regions in having lower levels of environmental drought. Compared to California and Andalusia in the Mediterranean Basin, the Cape and southwestern Australian regions have: (a) greater amounts of summer rain; (b) substantially less interannual variability in winter rain; and (c) a higher frequency of rainfall events, especially small to moderate ones, in both winter and summer. Of these latter two regions, Australia has the more reliable rainfall regime, measured as the coefficient of variation of monthly rainfall and the frequency of small, low-intensity events. California has a highly variable regime and intense summer drought, whereas Andalusia has greater numbers of large and intense downpours, a notable feature of the Mediterranean Basin rainfall regimes (Blondel and Aronson 1999). These regional differences in rainfall occurrence and reliability have been hypothesized to have significant ecological and evolutionary implications for the floras of these regions.

Convergence and Divergence of Plant Life-History Traits

Today, there is an expanding interest in models for the evolution of plant life-history traits in Mediterranean-climate regions as an approach to better understanding the nature of past and present selective forces in each of these five regions. This approach has developed as the broader community-level vision of Mediterranean-climate "places" has been replaced by an evolutionary approach to understanding plant functional groups and their role in structuring communities. The occurrence and evolutionary consequences of plant life-history traits in these ecosystems is controlled by a variety of factors including plant growth form, rainfall reliability, fire regime, soil fertility, and phylogenetic history (Ackerly 2004, Pausas et al. 2006). The latter factor is of particular interest now that better cladistic data and more fossil records are available to look at species relationships. It has become increasing clear that many of our modern Mediterranean-climate shrub lineages first evolved in the late Tertiary in semiarid summer-rainfall regimes very different from today's Mediterranean-climate regime. Thus, it is not always easy to determine whether convergent plant traits represent adaptations to current selective forces or the legacy of genetically canalized or preadapted traits from a past selective regime (Cowling and Witkowski, 1994, Pausas and Verdú 2003, Verdú et al. 2003, Ackerly 2004).

Because selective regimes vary among Mediterranean-climate ecosystems, as does the phylogenetic composition of their floras, testing hypotheses based on differences in a single factor is a challenging and perhaps impossible task. It makes sense, as far as is feasible, to control for the other factors in selecting study systems. Cowling et al. (2004) hypothesized that reliable rainfall in many relatively small and low-intensity events after summer-autumn fires—summer and autumn being the normal fire season in Mediterranean-climate ecosystems—would select for plant traits that enhance germination in the immediate post-fire period. Thus, reliable soil moisture levels in the post-fire environment would select for such traits as synchronous post-fire germination, rapid post-fire growth of seedling populations, and early reproduction and rapid buildup of seed banks prior to the next fire. These traits would be less advantageous where rain falls in few large and unpredictable events, as prolonged drought periods would increase seedling mortality and therefore present a high risk of extinction of seedling cohorts.

Overall, comparative patterns of rainfall reliability viewed in the context of other components of the selective regime, most importantly fire, may provide a novel perspective on the distribution of plant life-history traits in Mediterranean-climate regions, and perhaps also in other ecosystems. Differences in rainfall reliability may well explain convergence and divergence in many ecological patterns and processes, with an approach of utilizing correlative relationships in regional and local conditions. Ultimately, this approach may contribute to explaining patterns of species richness in Mediterranean-climate ecosystems via differences and similarities in life-history traits as a step toward understanding causal relationships in a manner that Lawton (1999) might accept.

In both ecological and longer evolutionary time, there are significant issues related to the possible impacts of global change on the selective environments of Mediterranean-climate regions. While global-change models are still variable in their simulations for Mediterranean-climate regions, predicted changes include less seasonal rainfall, lower annual rainfall in some regions, and greater temperature extremes. If the seasonality of rainfall reliability has been a major selective force in Mediterranean ecosystems, then there is every reason to believe that we will be seeing a dramatic change in selective environment over coming decades with major impact on community structure and diversity (Midgely et al. 2002, Loarie et al. 2008).

Conclusions

Mediterranean-type ecosystems continue to provide many opportunities for comparative studies of the controlling factors in community structure and biodiversity, and on macroecological patterns in convergent community evolution. Does the predictive pattern of dominance by evergreen sclerophyllous shrubs in all five "places" rise to an example of a correlative law of ecology? Here again, this may be a semantic question that is subject to opinion.

Scientific research on these regions has developed from an early focus on convergence in general community traits and plant architecture, in a search for correlative "laws," to an approach of identifying differences in climatic regimes or other factors that may causally determine life history traits. Subtle differences in climatic conditions, topographic diversity, evolutionary history, and human impact have strongly influenced the differential patterns of biodiversity that are present. This system then fits Weber's (1999) conclusion that the exceptions help to understand the applicability of a broad ecological rule. Moreover, Mediterranean ecologists will frequently say that the more they visit other Mediterranean-climate "places," the better they understand their own region.

The convergent evolution model has had a strong influence in promoting knowledge of community structure and function in the Mediterranean-climate regions, building on early descriptive studies to develop and implement coordinated focused studies with international collaboration between Mediterranean-climate regions. The exceptions that exist to the general patterns of ecosystem convergence have been a strong motivating factor for addressing the specific causal factors involved with convergence, but perhaps not causal "laws" in the formal sense. Science has progressed historically from broad generalizations about these regions to comparative regional studies, and to ongoing regional research programs that focus on process and life history themes. All of these intensive regional programs, however, feed back to the broader macroecological questions of ecosystem convergence.

The remarkable biodiversity of Mediterranean-climate regions together with the large numbers of rare and endangered species in these regions gives a special significance to expanding studies of these regions to better understand the evolution of diversity, particularly by vascular plants. Today's serious threats of habitat transformation and degradation as well as global change make it critical that there be a better understanding of the conservation biology and sustainable resource management in all five Mediterranean-climate regions. An interactive international community of researchers today continues to make comparative Mediterranean ecosystem studies a dynamic and productive field of study.

Acknowledgments

My thoughts on the comparative ecology of Mediterranean-climate regions have been shaped by four decades of interactions with colleagues and friends representing all five of these regions. International MEDECOS meetings have been held every three to four years since their beginning in 1971, bringing Mediterranean ecologists together and promoting collaborative work and faculty and student exchanges.

References

Ackerly, D. D. 2004. Adaptation, niche conservatism, and convergence: Comparative studies of leaf evolution in the California chaparral. *American Naturalist* 163:654–71.

Arroyo, M. T. K., P. H. Zedler, and M. D. Fox, eds. 1994. *Ecology and Biogeography of Mediterranean Ecosystems in Chile, California, and Australia.* Springer, Berlin.

Barbour, M. G., and R. A. Minnich. 1990. The myth of chaparral convergence. *Israel Journal of Botany* 39:453–63.

Blondel, J., and J. Aronson. 1999. *Biology and Wildlife of the Mediterranean Region.* Oxford University Press, Oxford.

Brown, J. H. 1995. *Macroecology.* University of Chicago Press, Chicago.

Brown, J. H., and B. A. Mauer. 1989. Macroecology: The division of food and space among species on continents. *Science* 243:1145–50.

Cody, M. L., and H. A. Mooney. 1978. Convergence versus nonconvergence in Mediterranean-climate ecosystems. *Annual Review of Ecology and Systematics* 9:265–321.

Cowling, R. M.., F. Ojeda, B. B. Lamont, and P. W. Rundel. 2004. Climate stability in Mediterranean-type ecosystems: Implications for evolution and conservation of biodiversity. *In* M. Arianoutsou and V.P. Papanistasis, eds., *Ecology, Conservation and Management of Mediterranean Climate Ecosystems. Proceedings of the 10th MEDECOS Conference.* Millpress, Rotterdam. ISBN 90 5955 0161.

Cowling, R. M., F. Ojeda, B. B. Lamont, P. W. Rundel, and R. Lechmere-Oertel. 2005. Rainfall reliability, a neglected factor in explaining convergence and divergence of plant traits in fire-prone Mediterranean-climate ecosystems. *Global Ecology and Biogeography* 14:509–19.

Cowling, R. M., and S. Proches. 2005. Patterns and evolution of plant diversity in the Cape Floristic Region. *Biologiske Skrifter* 55:273–88.

Cowling, R. M., P. W. Rundel, B. B. Lamont, M. T. K. Arroyo, and M. Arianoutsou. 1996. Plant diversity in Mediterranean-climate regions. *Trends in Ecology and Evolution* 11:352–60.

Cowling, R. M., and E. T. F. Witkowski. 1994. Convergence and non-convergence of plant traits on climatically and edaphically matched sites in Mediterranean Australia and South Africa. *Australian Journal of Ecology* 19:220–32.

Cowling, R. M., E. T. F. Witkowski, A. Milewski, and K. R. Newbey. 1994. Taxonomic, edaphic and biological aspects of narrow plant endemism on matched sites in Mediterranean South Africa and Australia. *Journal of Biogeography* 21:651–64.

Di Castri, F. 1981. Mediterranean-type shrublands of the world. Pages 1–52 *in* F. di Castri, D. W. Goodall, and R. L. Specht, eds., *Mediterranean-Type Shrublands.* Elsevier, Amsterdam.

Di Castri, F., and H. A. Mooney, eds. 1973. *Mediterranean-Type Ecosystems: Origin and Structure.* Springer, New York.

Drude, O. 1890. *Handbuch der Pflanzengeographie.* J. Engelhorn, Stuttgart.

Esler, K. J., P. W. Rundel, and R. H. Cowling. 1999. The Succulent Karoo in a global context: plant structural and functional comparison with North American winter rainfall deserts. Pages 303–13 *in* W. R. J. Dean and S. J. Milton, eds., *The Karoo: Ecological Patterns and Processes.* Cambridge University Press, Cambridge.

Golley, F. B. 1993. *A History of the Ecosystem Concept in Ecology: More than the Sum of the Parts.* Yale University Press, New Haven, CT.

Grisebach, A. 1872. *Die Vegetation der Erde nach ihrer Klimatischen Anordnung: Ein Abriss der Vergleichenden Geographie der Pflanzen.* W. Engelmann, Leipzig.

Keeley, J. E. 1992. A Californian's view of fynbos. Pages 372–88 *in* R. M. Cowling, ed., *The Ecology of Fynbos: Nutrients, Fire and Diversity*. Oxford University Press, Cape Town.

Kruger, F. J., D. T. Mitchell, and J. U. M Jarvis, eds., *Mediterranean-Type Ecosystems: The Role of Nutrients*. Springer, Berlin.

Lamont, B. B., B. G. Collins, and R. M. Cowling. 1985. Reproductive biology of the Proteaceae in Australia and South Africa. *Proceedings of the Ecological Society of Australia* 14:213–24.

Lawton, J. H. 1999. Are there general laws in ecology? *Oikos* 84:177–92.

Loarie, S.R., B.E. Carter, K. Hayhoe, S. McMahon, R. Moe, C. M. Knight, and D. D. Ackerly. 2008 Climate change and the future of California's endemic flora. *PLoS ONE* 3: e2502 doi:10.1371/journal.pone.0002502.

Midgley, G. F., L. Hannah, D. Millar, D., M. C. Rutherford, and L. W. Powrie. 2002. Assessing the vulnerability of species richness to anthropogenic climate change in a biodiversity hotspot. *Global Ecology and Biogeography* 11: 445–51.

Miller, P. C., ed. 1981. *Resource Use by Chaparral and Matorral: A Comparison of Vegetation Function in Two Mediterranean Type Ecosystems*. Springer, Berlin.

Mooney, H. A., ed. 1977. *Convergent Evolution in Chile and California: Mediterranean-Climate Ecosystems*. Dowden, Hutchinson and Ross, Stroudsburg, PA.

Mooney, H. A., and E. L. Dunn. 1970a. Convergent evolution of Mediterranean-climate evergreen sclerophyll shrubs. *Evolution* 24:292–303.

———. 1970b. Photosynthetic systems of Mediterranean-climate shrubs and trees of California and Chile. *American Naturalist* 104:447–53.

Murray, B. G. 1999. Universal laws and predictive theory in ecology and evolution. *Oikos* 89: 403–8.

Naveh, Z. 1967. Mediterranean ecosystems and vegetation types in California and Israel. *Ecology* 48:445–59.

O'Hara, R. B. 2005. The anarchist guide to ecological theory; or, We don't need no stinkin' laws. *Oikos* 110:390–93.

Orians, G. H., and O. T. Solbrig, eds. 1977. *Convergent Evolution in Warm Deserts*. Dowden, Hutchinson and Ross, Stroudsburg, PA.

Pausas, J. G., J. E. Keeley, and M. Verdú. 2006. Inferring differential evolutionary processes of plant persistence traits in Northern Hemisphere Mediterranean fire-prone ecosystems. *Journal of Ecology* 94:31–39.

Pausas, J. G., and M. Verdú. 2005. Plant persistence traits in fire-prone ecosystems of the Mediterranean basin: A phylogenetic approach. *Oikos* 109:196–202.

Rundel, P. W. 1995. Adaptive significance of some morphological and physiological characteristics in Mediterranean plants: Facts and fallacies. Pages 119–39 *in* J. Roy, J. Aronson, and F. di Castri, eds., *Time-Scales of Water Stress Response of Mediterranean Biota*. SPB Academic Publishing, Amsterdam.

———. 1999. Disturbance in Mediterranean-climate shrublands and woodlands. Pages 271–85 *in* L. Walker, ed., *Ecosystems of Disturbed Ground*. Elsevier, Amsterdam.

———. 2000. Mediterranean-climate ecosystems. *Encyclopedia of Biodiversity*. Academic Press, San Diego.

———. 2004. Mediterranean-climate ecosystems: defining their extent and community dominance. *In* M. Arianoutsou and V. P. Papanistasis, eds., *Ecology, Conservation and Management of Mediterranean Climate Ecosystems. Proceedings of the 10th MEDECOS Conference*. Millpress, Rotterdam. ISBN 90 5955 0161.

Rundel, P. W., G. Montenegro, and F. Jaksic, eds. 1998. *Landscape Disturbance and Biodiversity in Mediterranean-Type Ecosystems*. Springer, Berlin.

Schimper. A. H. W. 1903. *Plant-Geography upon a Physiological Basis*. H. R. Engelmann (J. Cramer), Weinheim.

Specht, R. L. 1969a. A comparison of sclerophyllous vegetation characteristic of Mediterranean type climates in France, California, and southern Australia. 1. Structure, morphology, and succession. *Australian Journal of Botany* 17:277–92.

———. 1969b. A comparison of sclerophyllous vegetation characteristic of Mediterranean type climates in France, California, and southern Australia. 2. Dry matter, energy, and nutrient accumulation. *Australian Journal of Botany* 17:293–308.

Turchin, P. 2001. Does population ecology have general laws? *Oikos* 94: 17–26.

Verdú, M., P. Dávila, P. García-Fayos, N. Flores-Hernández, and A. Valiente-Banuet. 2003. Convergent traits of Mediterranean woody plants belong to pre-Mediterranean lineages. *Biological Journal of the Linnaean Society* 78:415–27.

Weber, M. 1999. The aim and structure of ecological theory. *Philosophy of Science* 66:71–93.

6 Ecological Insights into the Causes of an Adaptive Radiation from Long-Term Field Studies of Darwin's Finches

Peter R. Grant and B. Rosemary Grant

Figure 6.0. Rosemary (a) and Peter (b) Grant working mistnets on the Galápagos Islands. Photographs by Thalia Grant, 1999.

Abstract

Darwin's finches are a model system for investigating the causes of speciation and adaptive radiation because the finches are a classical example of diversification of many species from a single ancestor in an environment that has been relatively little disturbed by humans. In this chapter we draw upon our experience of studying intertwined problems of ecology, behavior, and evolution to identify the important insights we gained, how they were generated, and where they led. The chapter provides more than one example of unexpected events, and opportunities to take advantage of them, when organisms are studied in one place, every year, and for a long time.

Populations of Darwin's ground finches (*Geospiza* spp.) have been studied on the small island of Daphne Major for more than three decades. The most important results of the study include repeated natural selection as a result of droughts and the evolutionary consequences of selection in the next generation, identification of morphological and song components of the barrier to interbreeding, genetic effects of rare hybridization, and establishment of a new population on the island.

The study produced five major insights into finch evolution and ecology. First, populations are evolutionarily labile when subjected to environmental change. Second, continuous monitoring of the environment is as important for an understanding of long-term evolution as is the continuous monitoring of the finch populations themselves. Third, genetic disadvantages from inbreeding can be overridden by the ecological advantages of abundant food and absence of density-dependent effects on survival. Fourth, paternal song is the primary cue for both males and females in hybridizing and backcrossing pairs, whereas beak traits become the primary cue when potential mates differ greatly in size. This key insight into how song and beak traits are used in mate choice emerged from analysis of pedigrees. Fifth, the evolutionary fate of one species can be genetically dependent upon another through occasional interbreeding.

Continuous, annually repeated study can yield more insight into the causes of change as well as more precise measures of change itself. We have gained benefits that would have eluded us if the study had been intermittent as opposed to continuous on a scale of years. Perhaps the most important finding is that long-term effects of rare but strong events can have long-lasting consequences for the environment and for the organisms that exploit it. This insight emerged from a study of the effects of the El Niño event of 1983. Short-term or intermittent studies are likely to miss such events and not be able to interpret the consequences. Moreover, ecological conditions at the beginning of droughts that follow El Niño events are not all the same, but depend upon preceding conditions. This important insight emerged from a comparison of such events in 1983, 1987, and 1991. Since other long-term studies have yielded insights similar to ours, we conclude the chapter with some generalizations.

Introduction

Some biological processes are beyond the reach of direct study because they unfold over centuries or millennia. Insights into these hidden processes are gained in a variety of ways. A powerful claim from ecologists is that long-term and continuous field studies of a single system provide a rich source of insights, and through the extrapolation of results they help to bridge the gap between different temporal scales (Strayer et al. 1986, Likens 1989, Edmondson 1991, Cody and Smallwood 1996).

We subscribe to this view, adding that evolutionary insights are as rich a reward as ecological insights in well-chosen systems, because evolution occurs on ecological time scales (Hairston et al. 2005). For example, insights into the causes of variation in individual fitness and evolution have come from a variety of recent studies of plants (Campbell et al. 1996, Carney et al. 2000), invertebrates (Majerus 1998, Saccheri et al. 1998, Watt et al. 2003), and vertebrates (Reznick et al. 1996, Clutton-Brock and Pemberton 2003, Garant et al. 2004, Reid et al. 2005), to mention just a few. Annually collected and archived material, such as wheat sampled in the ultra-long-term study at Rothamsted, UK, for more than 150 years (Bearchell et al. 2005), and copepods sampled from the bottom of lakes (Hairston 1996, Weider et al. 1997), is another source of information about contemporary evolutionary change. Such samples have documented unexpected evolutionary trends caused by anthropogenic effects upon the environment (Bearchell et al. 2005, Hairston et al. 2005), which short-term studies would have failed to detect.

In this chapter we draw upon our experience of studying intertwined problems of ecology, behavior, and evolution of Darwin's finches in the Galápagos archipelago for more than three decades, to identify the important insights we gained, how they were generated, and where they led. By simultaneously taking account of ecological, behavioral, and genetic aspects of evolution, and being alert to their interconnectedness, we have hoped to reach a greater understanding of finch evolution than would be possible from a narrower focus on just one of these things. This chapter thus illustrates two aspects of the ecology of place theme discussed by Price and Billick (chapter 1): that a knowledge of natural history helps to frame initial questions and guides observation, and that it is indispensable for a comprehensive interpretation of the results. We also provide more than one example of unexpected events and the opportunities to take advantage of them when organisms are studied in one place, every year, and for a long time.

Background

We chose Darwin's finches as a model system to investigate the causes of speciation and adaptive radiation because the finches are a classical example of diversification of many species from a single ancestor in an environment that has been relatively little disturbed by humans (Lack 1947, Bowman 1961,

Grant and Grant 2008). Three interrelated questions preoccupied us from the outset: how, and how often, does adaptation to the ecological environment occur; how is speciation to be explained; and how is quantitative genetic variation in small populations maintained in the face of depleting forces of selection and drift? To address these questions we developed the strategy of combining archipelago-wide comparative studies of different combinations of finch species with intensive long-term studies of population processes on the islands of Genovesa (11 years) and Daphne Major (33 years).

Adaptation

Regardless of how species are formed and whether they diversify, a fundamental evolutionary process is adaptation to the environment through natural selection on heritable phenotypic variation. Therefore we begin with this topic. Usually adaptation is inferred; rarely is it studied directly (Schluter 2000). Our studies have been able to accomplish both.

Adaptation to exploiting different foods was inferred by Lack (1947) and Bowman (1961) from the association between beak size and shape on the one hand and diet characteristics on the other. Our studies have strengthened the inference through quantification of the physical characteristics of foods available in the environment and foods actually eaten (Abbott et al. 1977, Grant and Grant 1980, Boag and Grant 1984; see also Herrel et al. 2005).

Two key insights of David Lack into the evolutionary history of the finches were (a) competitive interactions between species explain some mutually exclusive distribution patterns and enhanced beak size differences between sympatric finches (character displacement), and (b) competitive interactions are expected to be most severe in the dry season when food is likely to be in short supply. The evolutionary hypothesis of competition was challenged by Bowman (1961); nevertheless our comparative analyses found support for it (Abbott et al. 1977, Smith et al. 1978), as did a modeling exercise that produced a better fit between observed and predicted beak sizes when competitive interactions were incorporated than when they were not (Schluter and Grant 1984, Schluter et al. 1985). Dry season food supplies are indeed lower than wet season supplies, with some exceptions (Smith et al. 1978); and as food supply declines, so do population sizes (Grant 1985). Both of Lack's insights were valuable and correct.

We gained additional insights into adaptation through long-term studies of ground finch (*Geospiza*) populations on both Genovesa and Daphne Major. By capturing and measuring a large number of finches and marking them with unique combinations of colored leg bands, we were able to follow their fates. This simple and well-used technique of study revealed the tempo of ecological and evolutionary change. It showed that ground finches could live as long as 16 years, which was a surprise given the occasionally harsh climatic environment of lowland Galápagos habitats and the shorter maximum lives of northern temperate relatives (Grant and Grant 1989, Newton 1989). At roughly four to five years, generation lengths are moderately long, even though most Darwin's finch individuals live for a much shorter time and many die in their first year (Grant and Grant 1992a, Grant 1999). More remarkable was the discovery that significant adaptive change can take place in less than a year and is therefore, potentially at least, a rapid process and not immeasurably slow as previously had been believed.

Natural selection on morphological variation in medium ground finches (*Geospiza fortis*) was clearly demonstrated during the drought on Daphne Major in 1977 (Boag and Grant 1981, 1984). We were fortunately able to interpret it in terms of quantified changes in food supply. As seed supply diminished through consumption, finches turned increasingly towards the large and hard seeds, which were now in relatively high abundance. Only large finches with relatively large beaks could crack the seeds and extract the kernels; hence, they survived at a relatively high frequency. Natural selection had occurred. Over the 33 years we discovered that natural selection is not restricted to one trait or one species; it also occurred in the population of cactus finches (*Geospiza scandens*), and on both size and shape traits. Natural selection occurs frequently, and varies in direction and strength according to the particular set of environmental conditions (Gibbs and Grant 1987a, Grant and Grant 2002a).

We were also able to show that the morphological traits subject to natural selection (beak and body size) were highly heritable, and that genetic correlations among them were generally very strong and uniformly positive (Boag and Grant 1978, Boag 1983, Grant and Grant 1994, 2000, Keller et al. 2001). These results were used to predict evolutionary responses to repeated natural selection, and predictions were then compared to actual changes from generation to generation. Matches between observations and predictions were close, both after the 1977 drought (Grant and Grant 1995a) and following later selection events (Grant and Grant 2002a, 2006), thus demonstrating that natural selection drives frequent evolutionary changes in nature. This is an important insight into the evolutionary lability of populations when subjected to environmental change. For example, it helps to make sense of large changes in several organisms occurring in an apparently short period of time in the paleontological record (Gould 2003).

An additional insight was the realization that continuous monitoring of the environment is as important for an understanding of long-term evolution as is the continuous monitoring of the finch populations themselves. This insight emerged from a study of the effects of the El Niño event of 1983. Unprecedented in the century in its length and severity, it was followed two

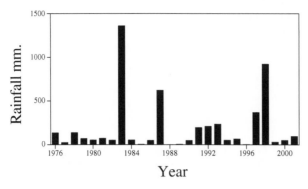

Figure 6.1. Annual rainfall on Daphne Major Island, Galápagos.

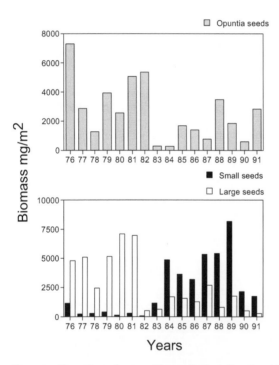

Figure 6.2. Change in seed composition on Daphne. *Opuntia echios* seeds (top) are intermediate in size and hardness between small seeds (bottom) and those in the woody components (mericarps) of *Tribulus cistoides* fruits. Both *Opuntia* and *Tribulus* were adversely affected by the extensive rains of the 1982–83 El Niño event; they were smothered by *Merremia aegyptica* and *Ipomoea linearifolia* vines, whereas many plants producing small and soft seeds grew prolifically. Each year seeds were counted in 50 randomly selected 1-m² quadrats in a permanent grid (Grant and Grant 1996b).

years later by a drought (figure 6.1). The contrast could scarcely have been more extreme. A simple expectation would have been that the selection event of 1977 would be repeated in 1985. That expectation would have been entirely wrong, because the vegetation and hence food environment had changed profoundly during 1983, from dominance of large-seed producing plants to dominance of small-seed producing ones (figure 6.2). This change has lasted for a long time. In the last six years, which have been relatively dry, the vegetation has changed both in composition and quantity back toward the pre-1983 condition, but has not reached it yet.

The important point is that conditions at the beginning of droughts are not the same. Droughts have selective or non-selective effects on the finches, depending on whether large seeds or small ones (and which ones) are plentiful at the start of a drought, and also depending on the finches' preceding evolutionary history. The 1977 drought had strong selective effects on the beak traits of two species of finches, the 1985–86 drought had weaker effects in an opposite direction on beak traits of one of the species, and the 1988–89

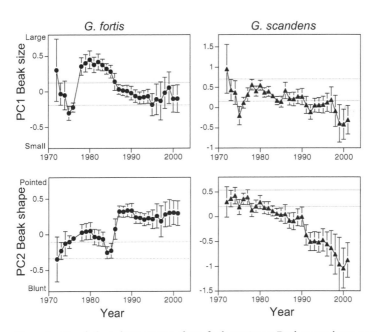

Figure 6.3. Morphological trajectories of two finch species on Daphne are shown as annual means with 95% confidence limits (vertical bars). The 95% confidence bands for the first large samples (1973) are shown as a pair of broken horizontal lines. Note the departure of all means from the initial confidence regions. The body size index was derived from an analysis of weight, wing length, and tarsus length measurements; beak size and shape indices were obtained from an analysis of length, depth, and width. From Grant and Grant (2002a). Beak size and shape are functionally related to different properties of the exploited seeds (Gibbs and Grant 1987a).

drought had no selective effects on either. Thus, by simply adding up the number of El Niño events and the number of droughts in a century, one would obtain a misleading estimate of the frequency of stresses on populations and the potential for evolutionary change. It would be an unreliable basis for extrapolating measured evolutionary change into the future.

The study provides an empirical foundation for general theories concerning adaptation to food resources with or without the diversification of species. Additionally it has yielded the surprising fact that as a result of natural selection, the finches on Daphne Major are not morphologically the same as their ancestors 30 years ago (Grant and Grant 2002a). Their environment has changed, and so have they (figure 6.3).

Speciation

Speciation involves ecological divergence and the establishment of barriers to gene exchange between populations descended from a common ancestor (Dobzhansky 1940, Mayr 1940, Lack 1947, Coyne and Orr 2004). The question is: how do these processes occur (Grant and Grant 2008)?

Coyne and Orr (2004) have made the most comprehensive attempt to address this question. While recognizing that speciation could occur sympatrically (in a single location) or parapatrically (in nearby locations with some gene exchange), they conclude from a critical survey of mechanisms and evidence that some form of speciation in allopatry (geographically distinct locations) is more likely for animal species. Their survey includes the Cocos finch, *Pinaroloxias inornatas*. This finch has been isolated long enough to give rise to more species, and the environment is heterogeneous enough for dietary diversification and specialization (Werner and Sherry 1987). These are circumstances in which sympatric speciation might be expected, yet the Cocos finch remains a single species (see also Coyne and Price 2002). Evidence for sympatric speciation on the Galápagos Islands is likewise equivocal (Ford et al. 1973, Grant et al. 2005, Huber et al. 2007). We focus on different forms of allopatric speciation.

Founder effects

Speciation begins when a new population is established. One possibility is that the crucial steps in the process are taken in the early stages of the founding of a new population by a few individuals. Rapid evolution might occur as a result of inbreeding, drift, the loss of genetic variation, and selection on sets of genes whose interactions change as a result of substantially altered allele frequencies. If the changes are profound enough they may yield a new species, reproductively isolated from the population that gave rise to it. This

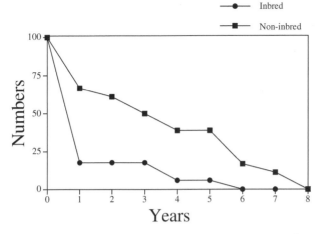

Figure 6.4. Inbreeding depression of *Geospiza magnirostris* on Daphne. Survival of inbred ($f \geq 0.125$, $N = 17$) and non-inbred ($f < 0.125$, $N = 18$) members of the 1991 cohort are contrasted. Numbers are scaled to an initial 100 of each (from Grant et al. 2001).

is the classical founder effect model of speciation of Mayr (1954, 1992). The closest we have come to a direct assessment of this model happened, appropriately, by chance. We were fortunate to witness the founding of a new population on Daphne Major (Gibbs and Grant 1987a, 1987b); it was an example of being in the right place at the right time. Two female and three male large ground finches, *Geospiza magnirostris*, established a breeding population at the end of 1982 when a major El Niño event was just beginning. We followed their fates and the fates of their descendants for the next 20 years (Grant and Grant 1995b, Grant et al. 2001).

As expected from the founder-effects model, inbreeding depression occurred in the first 10 years (figure 6.4), and was actually more severe than had ever been recorded in the two resident populations of *G. fortis* and *G. scandens* (Keller et al. 2002). Not all inbred individuals suffered, however, and survival of the first inbred individuals was unexpectedly high (Grant et al. 2001). Often more can be learned from exceptions like this than from observations that agree with expectations. The insight gained from this surprising observation was that genetic disadvantages from inbreeding can be overridden by the ecological advantages of abundant food and the absence of density-dependent effects on survival. Another unexpected observation was recurrent immigration, from more than one island source and not just from the island closest to Daphne (Grant et al. 2001). Recurrent immigration may have contributed to the elimination of any long-lasting effects of inbreeding, and also to the fact that no substantial change in either phenotypic traits or in microsatellite allele frequencies occurred, as should have happened according to the founder effects model. Insofar as the observations can be generalized, they provide no support for the model of founder effect speciation, and actually show two reasons why it can fail. Although this is a negative conclusion, the study is valuable because founder events and evolutionary effects are hypothesized frequently and studied extremely rarely, especially in nature (Clegg et al. 2002). Even if the specific model is not applicable, speciation may nonetheless be more likely to proceed in small insular populations than in large ones.

Byproduct of adaptation

An alternative to speciation through founder effects is Dobzhansky's (1937) adaptive byproduct hypothesis of speciation. If the same morphological traits that undergo adaptive change also function in a reproductive context when individuals choose mates, then speciation might proceed as a simple consequence of the adaptive change. No special genetic compatibility factors need be invoked to explain the sundering of ancestral and derived populations into separate species, since ecological factors are the primary driving force of differentiation, and pre-mating isolation is a passive secondary effect. This has been aptly described as ecological speciation (Schluter 2000). The two species could be, but need not be, genetically incapable of producing fertile offspring.

Our early studies on Daphne, Genovesa, and several other islands showed the importance of ecological factors—mainly the composition of the food supply, which varies from island to island. With this in mind, it is easy to see how finches that dispersed from one island and established a new breeding population on another would encounter a different constellation of foods and undergo evolutionary adaptation in beak and body size traits to the new food supply. This would by itself raise barriers to gene exchange between the original and derived populations, if beak and body size were also the key traits used by finches in choosing a mate. Species discrimination experiments with stuffed specimens of birds supported the hypothesis. Responding birds of different ground finch species tested on Daphne, Genovesa, and other islands discriminated between local members of their own species and either congeneric residents (Ratcliffe and Grant 1983a) or phenotypically different conspecifics from another island (Ratcliffe and Grant 1983b). They did so more on the basis of beak size and shape than by body size, and not at all by plumage color or pattern.

One awkward fact does not fit the byproduct hypothesis: In some cases the largest members of a small species (e.g., the small ground finch, *G. fuliginosa*) are morphologically more similar to the smallest members of a larger species (*G. fortis*) than they are to most members of their own population, on Daphne and elsewhere. If size traits are the only factor in mate choice, the species should hybridize. Moreover, interbreeding pairs should be similar in beak size. Long-term studies were needed to establish the fact that some species of Darwin's finches do indeed hybridize, though rarely, all previous evidence being ambiguous. Gene exchange between species was confirmed with microsatellite DNA-based assessment of parentage (Grant et al. 2004). Nevertheless the study on Daphne yielded the unexpected result that when *G. fortis* and *G. fuliginosa* hybridize they do so without apparent regard to their respective sizes (Boag and Grant 1984, Grant and Grant 1997a). This contradicts the main prediction, and shows that the byproduct hypothesis is not so much wrong as insufficient. Something else is involved in mate choice that usually leads to conspecific pairing, but occasionally gives rise to hybridization.

That something else is song. The initial argument for the importance of song was developed by Bowman (1979, 1983), with rich support from analyses of song and sound transmission profiles. He showed that some features of song (e.g., pitch) varied among species as a function of their body size, and others varied as a function of the sound transmitting properties of the habitats they occupied. That birds pay attention to particular features of each other's songs was demonstrated in a series of species-discrimination experiments with different pairs of ground finch species on several islands (Ratcliffe and Grant 1985). Thus two sets of experiments show that song and beak morphology are used as cues in mate choice. Why, then, do species sometimes hybridize? The answer lies partly in how finches acquire their songs and the other information they use later in choosing a mate.

A small number of laboratory experiments with finches exposed to tape-recorded song early in life showed that songs are learned in an imprinting-like process between the ages of 10 and 30 or more days (Bowman 1983). Only one song is sung, by males and not females, and once learned it remains constant for life (Grant and Grant 1996a). These results, combined with pedigree analyses on both Daphne and Genovesa, demonstrated that song is typically transmitted from father to son. Nevertheless the normal cultural transmission process can be perturbed. Usually this is not witnessed, and we learn of it only a year later by its result: misimprinting of young birds on heterospecific song. Insights into the causes of perturbation have been gleaned over a very long time on Daphne—with difficulty, both because it is rare and because the circumstances are idiosyncratic. For example, in one case an aggressive pair of *G. scandens* displaced the *G. fortis* owners of a nest and raised the offspring that hatched from one *G. fortis* egg together with their own offspring. The fostered *G. fortis* male later sang a *G.*

scandens song. In another case, a pair of *G. fortis* nested near an unpaired *G. magnirostris* male who sang vigorously and frequently and chased the male *G. fortis* whenever it sang. The son of the *G. fortis* pair later sang a *G. magnirostris* song.

A key insight into how song and beak traits are used in mate choice emerged from the pedigrees: paternal song is the primary cue for both males and females in hybridizing and backcrossing pairs, whereas beak traits become the primary cue when there is a large difference in beak size between the hybrid or backcross individual and the potential mates of the species that sing its song. For example, *G. fortis* individuals that have misimprinted on the song of the much larger *G. magnirostris* on Daphne have not hybridized. A parsimonious explanation for the development of mate preferences is that song is learned from parents and other conspecific individuals in association with morphological traits, primarily beak size and shape (Grant et al. 2000).

A model of adaptation and chance

Speciation begins in allopatry as a result of local adaptation in genetically transmitted beak traits that also function in mate choice. Culturally transmitted song traits also diverge. Some song features, such as frequency, vary in relation to body size (Bowman 1983), and others, such as note repetition rate, vary in relation to beak size and jaw musculature, possibly independent of body size (Podos 2001, Podos and Nowicki 2004, Podos et al. 2004). Thus, several facts about Darwin's finches are consistent with the adaptive byproduct hypothesis of Dobzhansky (1937). Yet there are three reasons for thinking that a more comprehensive model of speciation is needed for the finches.

First, there is a need to make explicit the role of learning in the development of mate preferences; mate preferences are not inherent or "genetically fixed." One consequence of learning is that species can sing each other's songs, within limits, despite beak shape differences, and base their choice of mates largely upon learned song. Thus a crucial question is just what features of song are used in mate choice (Podos and Nowicki 2004), and how these diverge (Grant and Grant 1996a)? Whatever causes songs to diverge in allopatry promotes speciation.

Second, there is a need to explicitly incorporate the role of chance, because it is an important factor in song divergence. Change in the composition of individual songs from one generation to the next probably occurs through copying errors, as we have seen manifested on Daphne in minor ways. The most egregious example was an unusual song variant in the *G. scandens* population, apparently caused by a cactus spine in the throat of one male. Its sons sang the same strange song, and at least two generations later there were five males on the island singing this song! Frequencies of song variants are subject to random change across generations, much as the frequencies of selectively neutral alleles drift at random. We have observed this also on Daphne, in both *G. fortis* and *G. scandens* populations, as well as in the *G. magnirostris* population during the first few years following colonization of the island. A substantial change in frequencies of *G. magnirostris* song types when numbers were low is more plausibly attributed to chance than to any hypothetical inherent variation in attractiveness (Grant and Grant 1995b).

The importance of chance in speciation is underlined by the fact that populations of the same species on adjacent islands with nearly identical habitats may sing different songs. The strongest example is the sharp-beaked ground finch, *Geospiza difficilis*, on the northern islands of Wolf and Darwin (Grant et al. 2000, Grant and Grant 2002b). There are no habitat or morphological differences between the populations that would help to explain the differences in song, and we are left with randomness in the colonization of the two islands, or random cultural drift in song characteristics, akin to random genetic drift, as explanations. *Geospiza difficilis* individuals on a third northern island, Genovesa, sing songs that sonographically resemble the songs on Wolf and Darwin. Responses to playback of tape-recorded songs on Genovesa show that these individuals do not distinguish between their own songs and songs from Darwin, but do discriminate between their own songs and songs from Wolf. Some subtle differences between Genovesa and Wolf songs, probably in temporal patterning, appear to set up a potential pre-mating barrier between these two populations. A stronger barrier is likely to exist between the Wolf and Darwin populations because their songs differ much more.

Third, natural selection and sexual selection may jointly result in a fit between characteristics of songs and sound-transmitting properties of the environment (Bowman 1983, Grant et al. 2000). Just as the association between beak size variation and ecological variation is adaptive, so too is the association between song and habitat structure adaptive—the difference being that in the former case, the trait (beak) is genetically heritable and in the second case the trait (song) is culturally heritable.

To conclude, a pre-mating barrier to gene flow arises in allopatric populations, partly as a result of adaptation to the environment and partly as a result of chance. The choice of mates is based on the learning of song in relation to beak morphology. Learning features of parents and other (conspecific) individuals that resemble them consolidates a barrier created by divergence. The barrier is leaky, and rare hybridization leads to gene exchange.

Mutation accumulation

Another possible explanation for speciation is that populations stay isolated on different islands long enough for barriers to arise simply as a result of the accumulation of different mutations. Then, at the extreme, when members of the two populations encounter each other they are unable to produce viable and fertile offspring, even if they choose to mate with each other, for reasons of genetic incompatibility arising from negative epistasis in the combined genomes (Dobzhansky 1937, Muller 1940). If so, time-dependent molecular genetic change adequately provides a metric of speciation.

Two sets of entirely allopatric populations of warbler finches (*Certhidea*) are prime candidates for testing this hypothesis because they have been isolated from each other for longer (1.5–2.0 MY) than any other group of populations, according to microsatellite (Petren et al. 1999) and mitochondrial DNA differences (Freeland and Boag 1999, Petren et al. 2005, Tonnis et al. 2005). It is possible that members of the two sets are genetically so different that whenever a warbler finch of one disperses to an island occupied by the other, it breeds but fails to produce fertile offspring for genetic reasons. There has been no test of this possibility. What is known is that there is no apparent barrier to mating; members of the two populations are very similar morphologically and they respond to playback of each other's tape-recorded song. Pre-mating isolation usually evolves earlier than post-mating isolation (Price and Bouvier 2002); therefore even in this example of long-isolated populations it is unlikely that genetic incompatibilities have evolved. Nevertheless, reflecting our uncertainty over whether they have reached a level of reproductive isolation sufficient to deserve being called species, we refer to the warbler finches as two genetic lineages, *olivacea* and *fusca* (Grant and Grant 2002c, Petren et al. 2005).

Genetic incompatibilities have not been detected between any of the four hybridizing pairs of ground finch species on Daphne and Genovesa. This statement applies to both survival and reproductive aspects of hybrid and backcross fitness. In following the fates of hybrids and backcrosses to determine where they might be at an expected disadvantage in relation to the parental species, we gained the important insight that their fitness is not fixed but is environment-dependent: just as environmental variation alters the fitness of inbred birds (Keller et al. 2002), so does it influence the fitness of hybrids and backcrosses (Grant and Grant 1992b, 2002a). In the first seven years the few hybrids that were produced by *G. fortis* breeding with either *G. fuliginosa* or *G. scandens* on Daphne did not survive to breed (nor did many of the non-hybrids). Genetically, hybridization led nowhere. The first hybrids bred in the El Niño (1983) year of abundance, and the breeding of hybrids and backcrosses has continued in most years since then. Before

1983 it would have been easy to draw the conclusion that the hybrids were incapable of breeding, whereas later events have shown this conclusion to be wrong. Moreover, since 1983 the survival of hybrids and backcrosses has been equal to or even greater than survival of the pure species. We interpret both results—the difference between fitness before and after 1983, and the relative survival fitness after 1983—in terms of an altered environment: a more favorable constellation of foods available to hybrids after 1983 than before. Thus, speciation can occur through pre-mating barriers to gene exchange well before genetic incompatibilities arise (Grant and Grant 1997b, Price and Bouvier 2002). Species formation is not a simple linear function of time, at least within the approximate time frame of two to three million years for this radiation. With the possible exception of the warbler finches, the mutation accumulation hypothesis does not apply to the finches.

Maintaining the Potential for Evolutionary Change

In the last million years or more, oscillations between glacial and interglacial conditions at temperate latitudes have caused sea level to rise and fall. Small islands have been repeatedly created and submerged in the Galápagos archipelago. Their populations of finches could have been the starting point for species formation through selection and hybridization, but were also prone to become extinct for ecological and genetic reasons, and not just because the islands disappeared! Contemporary study of finch populations on a small island such as Daphne can throw light on these dynamics, and in particular on the question of how quantitative genetic variation is maintained.

Genetic variation is lost through drift and selection and is regenerated by mutation. In principle rates of losses and gains could be equal, yielding an equilibrial level of genetic variation that is maintained for a long time. At the outset this simple scheme seemed insufficient to explain the high levels of phenotypic and presumed genetic variation in quantitative traits in Darwin's finch populations that were first quantified by Lack (1945, 1947) and then by Bowman (1961). Movement of birds from one island to another, resulting in a high rate of gene exchange, is a possible resolution of this problem. A combination of direct observations and indirect information from a phylogeographic study (Petren et al. 2005) indicates gene flow at a generally low, not high, rate. But even if gene flow does occur at a high rate, it raises another question: why, in the face of extensive gene flow, do the means of adaptive traits of different populations differ as much as they do?

An alternative to intraspecific gene flow is gene flow between species. This hypothesis was theoretically interesting when first considered, but empirically doubtful because Lack had looked specifically for evidence of hybridization in a field study in 1939 and failed to find it, and Orr (1945) had failed to produce crosses in captivity beyond the early nestling stage. Results of the long-term studies on Daphne and Genovesa showed that hybridization and ensuing gene exchange could indeed restore depleted genetic variation after all.

Continuation of the long-term study on Daphne Major yielded four additional insights into the process of gene exchange and its significance for the maintenance of genetic variation. First, the evolutionary fate of one species can be genetically dependent upon that of another because of occasional interbreeding. Second, the interdependence fluctuates as the direction of backcrossing changes. Here it has changed from a predominant flow of genes from *G. scandens* to *G. fortis* to a predominant flow in the opposite direction (Grant et al. 2004).

Third, fluctuating levels of genetic variation are as probable as a constant equilibrial level. The frequency of hybrids and backcrosses has not risen to a point at which rates of genetic gains and losses are balanced; instead, genetic input into the *G. scandens* population has increased in the last two decades. Most individuals now have the genetic signature of backcrosses, and there is no indication yet that the rate of losses has come into balance with the rate of gains through introgression. Losses through directional selection must have been smaller in recent years than in the earlier years when selection coefficients were at their largest, and losses through drift must have been small in recent years because the population of *G. scandens* has not been particularly small. This remarkable state of affairs has occurred apparently through high hybrid and backcross survival rather than through an increase in hybridization, although we lack the studies of pairing since 1998 that are needed to be sure this is correct.

Geospiza fortis and *G. scandens* are currently converging phenotypically and genetically (figure 6.5) (Grant et al. 2004). The outcome, if the process continues, is difficult to predict. Either these will remain separate species as a result of the distinctness of their songs, or they will eventually fuse into a single panmictic population as a result of an increase in the frequency of hybridization in each generation due to the species' morphological similarity. However, before fusion is completed the environment may revert to pre-1983 conditions, in which case gene exchange could diminish, the fitness of hybrids might decline under the changed ecological conditions, and with a selective advantage to divergent members of the two populations we would be back to where we were at the beginning of our study in 1973.

The fourth insight is that introgressive hybridization can weaken genetic correlations between traits (Grant and Grant 1994). From this we have argued that hybridization of species with different allometries is evolution-

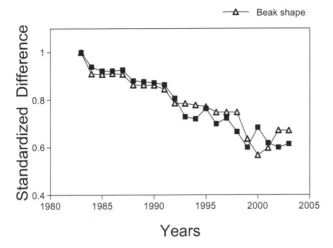

Figure 6.5. Recent morphological and genetic convergence of *Geospiza fortis* and *G. scandens* on Daphne in relation to standardized differences in 1982. Morphology is bill shape as indexed by PC 2 in a principal components analysis of three bill dimensions (see figure 6.3). Genetic difference is Nei's D applied to allele frequencies at 16 microsatellite loci (from Grant et al. 2005).

arily important because it reduces genetic constraints on future directions of evolution, as well as increasing the variation on which selection can act. Small, transitory islands with unique ecologies might have been arenas for introgression that facilitated novel evolutionary trajectories and contributed to the adaptive radiation (Grant et al. 2005). At the least, the evolutionary fate of one species is genetically dependent upon another through occasional interbreeding.

Discussion

The value of continuous long-term study of a single system

Repeated sampling of the same population at intervals of several years has contributed importantly to an understanding of just how widespread and rapid contemporary processes of evolution are in nature (Hendry and Kinnison 1999, Kingsolver et al. 2001). For example, field studies spanning periods of about 30 to 50 years have established change in the frequency of a chromosomal inversion in *Drosophila pseudoobscura* populations against a background of stability (Anderson et al. 1991), shifts in the dynamics of hybridizing populations of sunflowers (Carney et al. 2000) and spadefoot toads (Pfennig 2003), and geographical changes in the positions of hybrid zones between some species (Brumfield et al. 2001, McDonald et al. 2001, Rowher et al. 2001) but not others (Mallet 1986, Rising 1996).

Continuous, annually repeated study can potentially yield more insight into the causes of change, as well as more precise measures of change itself. We have gained benefits that would have eluded us if the study had been intermittent as opposed to continuous at a scale of years. Perhaps the most important finding is that long-term effects of rare but strong events can have long-lasting consequences for the environment and for the organisms that exploit it. Short-term or intermittent studies are likely to miss such events and not be able to interpret the consequences. For example, we do not know why the population of the large cactus finch, *G. conirostris*, on Genovesa, was subdivided into two ecological and morphological groups prior to 1978, because our long-term study on that island only began in 1978 (Grant and Grant 1989). By then the population had apparently undergone partial fission, and we studied it in a phase of fusion without gaining much insight into its fission.

The best example of the same phenomenon on Daphne is the critical El Niño event of 1982–83. If we had stopped before that year, we could have concluded—incorrectly—that droughts invariably select for large body and beak size, hybrids do not breed, and the island cannot support another species (e.g., *G. magnirostris*). If we had started after 1983, we would have been mystified about the origins of the current interbreeding and convergence of *G. fortis* and *G. scandens;* we would have missed the colonization by *G. magnirostris* and would have no reason to believe that it had not always been a member of the community of breeding finches. If we had stopped 20 years later, we would have missed a dramatic demonstration (in 2005) of ecological character displacement caused by *G. magnirostris* (Grant and Grant 2006). In short, our appreciation of the ecological and evolutionary importance of annual environmental variation would have been greatly diminished.

The rewards from continuous long-term study of a single system are not just reaped in the Galápagos. For another example in a strikingly different environment, see the study of Serengeti lions in east Africa by Packer et al. (2005), in which repeated sharp changes in lion numbers after periods of stasis were each associated with ecological factors: increasing numbers of wildebeest and buffalo, increase in removal of dead grasses, decrease in grass fires, and increase in regeneration of woodland. As another example, persistent studies in nature are revealing the frequency of hybridization and the genetic and ecological consequences of introgression. At least 10% of terrestrial animal species and 25% of plant species hybridize and potentially exchange genes, according to a recent survey by Mallet (2005). Measurable effects of introgression in both animals and plants are generally small, although there are exceptions (Arnold 1997, Rieseberg et al. 2003, Seehausen 2004). Whether they are evolutionarily important or trivial can only be answered with long-term studies, such as the study of hybridizing sticklebacks (*Gasterosteus*) in British Columbia (Taylor et al. 2006).

Generalizations about long-term studies

A limnologist, Tommy Edmondson, had some pithy things to say about continuous long-term studies (Edmonson 1991) that are germane to ours. First, most such studies start out as short-term projects (Strayer et al. 1986). Second, most are conceived by one or at most two people, and not by teams. Third, their continuity is easily broken by funding agencies that do not value the third renewal as being as great or exciting as the first proposal, and which may have the impression that cutting-edge hypothesis testing has been replaced by simple monitoring without much purpose. Fourth, their optimal length varies according to the nature of the problems addressed. Each long-term study has its own time scale for completion, and some may have no definite completion; the time scale is often unpredictable. Fifth, a major justification is that they not only yield unpredictable events, but that the preceding work and discoveries help to interpret those events; and contrariwise, a break in continuity reduces or eliminates that power of interpretation. Sixth, there are many similarities between long-term and ordinary research. "The main difference is that with a prolonged study of a natural system, the scientist is not in charge of the schedule. He must continue to work until the system has completed the change that is being studied." (Edmonson 1991, p. 235).

All of these justifications apply to our study, from the beginning (points 1 and 2) onwards. A break in continuity at certain times, especially 1983, would have severely constrained our ability to interpret selection and hybridization on Daphne and the colonization by a new species (point 5). Hays et al. (2005) have described some dismal examples of a break in the continuity of marine studies that were only restarted when it was realized that the results could shed light on global warming (Beaugrand and Reid 2003). With regard to point 3, by curious coincidence or by serendipity, the continuity of the Daphne program was placed in jeopardy most severely at the third renewal. Unpredictability has been the hallmark of the environment in which we have worked (points 4 and 6).

After 30 years of study, we realize that there is no logical ending point and that there will still be important questions at the end. This differs in a minor way from point 6. Ecological and evolutionary changes may tend to be cyclic, but in our experience the system does not return to its starting point.

Therefore, in a strict sense change is not completed, although it may be complete after 100 years. Moreover, as we noted in reviewing ecological findings after two decades of research, there is an inherent uncertainty about when rare events will occur and how long their effects will last. As a consequence of uncertainty, "some long-term studies are virtually unplannable, which is an argument for conducting such studies in an open-ended way" (Grant and Grant 1996b, p. 375). Yet another argument is that not all environmental conditions can ever be experienced during a study if, as is suggested by Bell (1992), temporal environmental variance increases indefinitely.

Laws of long-term studies

These considerations lead us to propose two laws of long-term studies. The first is they start too late. Observations made at the beginning cannot be fully interpreted, because the necessary information in previous years has not been obtained. The second law of these studies is that they stop too soon. This follows from our argument that there is no logical stopping point and not all questions are ever answered—if anything, more questions remain to be answered at the end than were present at the beginning. In the face of these laws, the best strategy is to start a study as early as possible, postpone its end for as long as possible, and hope that the remaining questions are few and small!

Conclusions

We conclude by emphasizing that our study, like other long-term ecological studies in one specific place, did not progress with a steady accumulation of knowledge and understanding on an undeviating linear trajectory. Expectations at the outset were conditioned by our understanding from having read the literature, and almost inevitably surprises were encountered on the way. El Niños and droughts were far more intense than we expected, and evolutionary change was far more rapid. We gradually became aware of the importance of song and the interaction of cultural and genetic evolution, change in the fate of hybrids (and the implications for speciation), the fact that changes in vegetation as a result of changes in climate could have evolutionary consequences as well as the more obvious ecological ones, and the fact that a reversal of weather did not result in a reversal of vegetation characteristics because there was an inertia to the system. The importance of temporal scale shifted from annual to decadal. Some observations forced us to rethink theory: for example, the success of some of the inbred *G. magnirostris*. Finally, we had no basis for expecting that the populations would not be the same phenotypically at the end as they were at the beginning of the study. Ramifications of this last result are still being explored.

References

Abbott, I., L. K. Abbott, and P. R. Grant. 1977. Comparative ecology of Galápagos ground finches (*Geospiza* Gould): Evaluation of the importance of floristic diversity and interspecific competition. *Ecological Monographs* 47:151–84.

Anderson, W. W., J. Arnold, D. G. Baldwin, A. T. Beckenbach, C. J. Brown, S. H. Bryant, J. A. Coyne, L. G. Harshman, W. B. Heed, D. E. Jeffrey, L. B. Klaczko, B. C. Moore, J. M. Porter, J. R. Powell, T. Prout, S. W. Schaeffer, J. C. Stephens, C. E. Taylor, M. E. Turner, G. O. Williams, and J. A. Moore. 1991. Four decades of inversion polymorphisms in *Drosophila pseudoobscura*. *Proceedings of the National Academy of Sciences (USA)* 88:10367–71.

Arnold, M. 1997. *Natural Hybridization and Evolution*. Oxford University Press, Oxford.

Bearchell, S. J., B. A. Fraaije, M. W. Shaw, and B. D. L. Fitt. 2005. Wheat archive links long-term fungal pathogen population dynamics to air pollution. *Proceedings of the National Academy of Sciences (USA)* 102:5438–42.

Beaugrand, G., and P. C. Reid. 2003. Long-term changes in phytoplankton, zooplankton, and salmon related to climate. *Global Change Biology* 9:801–17.

Bell, G. 1992. Five properties of environments. Pages 33–56 *in* P. R. Grant and H. S. Horn, eds., *Molds, Molecules and Metazoa*. Princeton University Press, Princeton, NJ.

Boag, P. T. 1983. The heritability and external morphology in Darwin's ground finches (*Geospiza*) on Isla Daphne Major, Galápagos. *Evolution* 37:877–94.

Boag, P. T., and P. R. Grant. 1978. Heritability of external morphology in Darwin's finches. *Nature* 274:793–94.

———. 1981. Intense natural selection in a population of Darwin's finches (Geospizinae) in the Galápagos. *Science* 214:82–85.

———. 1984. The classical case of character release: Darwin's finches (*Geospiza*) on Isla Daphne Major, Galápagos. *Biological Journal of the Linnean Society* 22:243–87.

Bowman, R. I. 1961. Morphological differentiation and adaptation in Galápagos finches. *University of California Publications in Zoology* 58:1–302.

———. 1979. Adaptive morphology of song dialects in Darwin's finches. *Journal für Ornithologie* 120:353–89.

———. 1983. The evolution of song in Darwin's finches. Pages 237–537 *in* R. I. Bowman, M. Berson, and A. E. Leviton, eds., *Patterns of Evolution in Galápagos Organisms*. American Association for the Advancement of Science, Pacific Division, San Francisco.

Brumfield, R. T., R. W. Jernigan, D. B. McDonald, and M. J. Braun. 2001. Evolutionary implications of divergent clines in an avian (*Manacus*: Aves) hybrid zone. *Evolution* 55:2070–87.

Campbell, D. R., N. M. Waser, and M. V. Price. 1996. Mechanisms of hummingbird-mediated selection for flower width in *Ipomopsis aggregata*. *Ecology* 77:1463–71.

Carney, S. E., K. A. Gardner, and L. H. Rieseberg. 2000. Evolutionary changes over the last fifty-year history of a hybrid population of sunflowers (*Helianthus*). *Evolution* 54:462–74.

Clegg, S. M., S. M. Degnan, J. Kikkawa, C. Moritz, A. Estoup and I. P. F. Owens. 2002. Genetic consequences of successive founder events by an island colonizing bird. *Proceedings of the National Academy of Sciences (USA)* 99:8127–32.

Clutton-Brock, T., and J. M. Pemberton. 2003. *Soay Sheep: Population Dynamics and Selection in an Island Population*. Cambridge University Press, Cambridge.

Cody, M. L., and J. A. Smallwood, eds. 1996. *Long-Term Studies of Vertebrate Communities*. Academic Press, New York.

Coyne, J. A., and H. A. Orr. 2004. *Speciation*. Sinauer Associates, Sunderland, MA.

Coyne, J. A., and T. D. Price. 2000. Little evidence for sympatric speciation in island birds. *Evolution* 54:2166–71.

Dobzhansky, T. 1937. *Genetics and the Origin of Species*. Columbia University Press, New York.

———. 1940. Speciation as a stage in evolutionary divergence. *American Naturalist* 74:312–21.

Edmondson, W. T. 1991. *The Uses of Ecology: Lake Washington and Beyond*. University of Washington Press, Seattle.

Ford, H. A., D. T. Parkin, and A. W. Ewing. 1973. Divergence and evolution in Darwin's finches. *Biological Journal of the Linnean Society* 5:289–95.

Freeland, J. R., and P. T. Boag. 1999. The mitochondrial and nuclear genetic homogeneity of the phenotypically diverse Darwin's ground finches. *Evolution* 53:1553–63.

Garant, D., L. E. B. Kruuk, R. H. McCleery, and B. C. Sheldon. 2004. Evolution in a changing environment: A case study with great tit fledging mass. *American Naturalist* 164:E115–29.

Gibbs, H. L., and P. R. Grant. 1987a. Oscillating selection on Darwin's finches. *Nature* 327: 511–513.

———. 1987b. Adult survivorship in Darwin's ground finch (*Geospiza*) populations in a variable environment. *Journal of Animal Ecology* 56:797–813.

Gould, S. J. 2003. *Evolution*. Harvard University Press, Cambridge, MA.

Grant, B. R., and P. R. Grant. 1989. *Evolutionary Dynamics of a Natural Population*. University of Chicago Press, Chicago.

———. 1996. Cultural inheritance of song and its role in the evolution of Darwin's finches. *Evolution* 50:2471–87.

———. 2002a. Simulating secondary contact in allopatric speciation: An empirical test of premating isolation. *Biological Journal of the Linnean Society* 76:545–56.

———. 2002b. Lack of reproductive isolation at the base of a phylogenetic tree. *American Naturalist* 160:1–19.

Grant, P. R. 1985. Interspecific competition in fluctuating environments. Pages 173–91 *in* J. M. Diamond and T. J. Case, eds., *Community Ecology*. Harper and Row, New York.

———. 1999. *Ecology and evolution of Darwin's Finches*. Princeton University Press, Princeton, NJ.

Grant, P. R., and B. R. Grant. 1980. The breeding and feeding characteristics of Darwin's finches on Isla Genovesa, Galápagos. *Ecological Monographs* 50:381–410.

———. 1992a. Demography and the genetically effective sizes of two populations of Darwin's finches. *Ecology* 73:766–84.

———. 1992b. Hybridization in bird species. *Science* 256:193–97.

———. 1994. Phenotypic and genetic effects of hybridization in Darwin's finches. *Evolution* 48:297–316.

———. 1995a. Predicting microevolutionary responses to directional selection on heritable variation. *Evolution* 49:241–51.

———. 1995b. The founding of a new population of Darwin's finches. *Evolution* 49:229–40.

———. 1996. Finch communities in a fluctuating environment. Pages 343–90 *in* M. L. Cody and J. A. Smallwood, eds., *Long-term Studies of Vertebrate Communities*. Academic Press, New York.

————. 1997a. Hybridization, sexual imprinting, and mate choice. *American Naturalist* 149: 1–28.

————. 1997b. Genetics and the origin of bird species. *Proceedings of the National Academy of Sciences USA* 94:7768–75.

————. 2000. Quantitative genetic variation in populations of Darwin's finches. Pages 3–40 *in* T. A. Mousseau, B. Sinervo, and J. A. Endler, eds., *Adaptive Genetic Variation in the Wild*. Oxford University Press, New York.

————. 2002. Unpredictable evolution in a 30-year study of Darwin's finches. *Science* 296: 707–11.

————. 2006. Evolution of character displacement in Darwin's finches. *Science* 313:224–26.

————. 2008. *How and Why Species Multiply: The Radiation of Darwin's Finches*. Princeton University Press, Princeton, NJ.

Grant, P. R., B. R. Grant, and K. Petren. 2000. The allopatric phase of speciation: the sharp-beaked ground finch (*Geospiza difficilis*) on the Galápagos Islands. *Biological Journal of the Linnean Society* 69:287–317.

————. 2001. A population founded by a single pair of individuals: Establishment, expansion, and evolution. *Genetica* 112/113:359–82.

————. 2005. Hybridization in the recent past. *American Naturalist* 166:56–67.

Grant, P. R., B. R. Grant, J. A. Markert, L. F. Keller, and K. Petren. 2004. Convergent evolution of Darwin's finches caused by introgressive hybridzation and selection. *Evolution* 58:1588–99.

Hairston, N. G., Jr. 1996. Zooplankton egg banks as biotic reservoirs in changing environments. *Limnology and Oceanography* 41:1087–92.

Hairston, N. G., Jr., S. P. Ellner, M. A. Geber, T. Yoshida, and J. A. Fox. 2005. Rapid evolution and the convergence of ecological and evolutionary time. *Ecology Letters* 8:1114–27.

Hairston, N. G., Jr., C. M. Kearns, L. P. Demma, and S. W. Effler. 2005. Species-specific *Daphnia* phenotypes: a history of industrial pollution and pelagic ecosystem response. *Ecology* 86:1669–78.

Hays, G. C., A. J. Richardson, and C. Robinson. 2005. Climate change and marine plankton. *Trends in Ecology and Evolution* 20:337–44.

Hendry, A. P., and M. T. Kinnison. 1999. The pace of modern life: measuring rates of micro-evolution. *Evolution* 53:1637–53.

Herrel, A., J. Podos, S. K. Huber, and A. P. Hendry. 2005. Evolution of bite force in Darwin's finches: A key role for head width. *Journal of Evolutionary Biology* 18:669–75.

Huber, S. K., L. F. De León, A. P. Hendry, E. Bermingham, and J. Podod. 2007. Reproductive isolation of sympatric morphs in a population of Darwin's finches. *Proceedings of the Royal Society of London B* 274:1709–14.

Keller, L. F., P. R. Grant, B. R. Grant, and K. Petren. 2001. Heritability of morphological traits in Darwin's finches: Misidentified paternity and maternal effects. *Heredity* 87:325–36.

————. 2002. Environmental conditions affect the magnitude of inbreeding depression in survival of Darwin's finches. *Evolution* 56:1229–39.

Kingsolver, J. G., H. E. Hoekstra, J. M. Hoekstra, D. Berrigan, S. N. Vignieri, C. E. Hill, A. Hoang, P. Gilbert, and P. Beerli. 2001. The strength of phenotypic selection in natural populations. *American Naturalist* 157:245–61.

Lack, D. 1945. The Galápagos finches (*Geospizinae*): A study in variation. *Occasional Papers of the California Academy of Sciences* 21:1–159.

————. 1947. *Darwin's Finches*. Cambridge University Press, Cambridge.

Likens, G. E. 1989. *Long-Term Studies in Ecology: Approaches and Alternatives.* Springer, New York.

Majerus, M. E. N. 1998. *Melanisms: Evolution in Action.* Oxford University Press, Oxford.

Mallet, J. 1986. Hybrid zones of *Heliconius* butterflies in Panama and the stability and movement of warning colour clines. *Heredity* 56:191–202.

———. 2005. Hybridization as an invasion of the genome. *Trends in Ecology and Evolution* 20:229–37.

Mayr, E. 1940. Speciation phenomena in birds. *American Naturalist* 74:249–78.

Mayr, E. 1954. Change of genetic environment and evolution. Pages 157–80 *in* J. S. Huxley, A. Hardy, and E. B. Ford, eds., *Evolution as a Process.* Allen and Unwin, London.

———. 1992. Controversies in retrospect. *Oxford Surveys in Evolutionary Biology* 8:1–34.

McDonald, D. B., R. P. Clay, R. T. Brumfield, and M. J. Braun. 2001. Sexual selection on plumage and behavior in an avian hybrid zone: Experimental tests of male-male interactions. *Evolution* 55:1443–51.

Muller, H. J. 1940. Bearings of the "Drosophila" work on systematics. Pages 185–268 in J. Huxley, ed., *The New Systematics.* Clarendon Press, Oxford.

Newton, I., ed. 1989. *Lifetime Reproduction in Birds.* Academic Press, New York.

Orr, R. T. 1945. A study of captive Galápagos finches of the genus *Geospiza. Condor* 47:177–201.

Packer, C., R. Hillborn, A. Mosser, B. Kissui, M. Borner, G. Hopcraft, J. Wilmshurst, S. Mduma, and A. R. E. Sinclair. 2005. Ecological change, group territoriality, and population dynamics in Serengeti lions. *Science* 307:390–93.

Petren, K., B. R. Grant, and P. R. Grant. 1999. A phylogeny of Darwin's finches based on microsatellite DNA length variation. *Proceedings of the Royal Society of London B* 266:321–29.

Petren, K., P. R. Grant, B. R. Grant, and L. F. Keller. 2005. Comparative landscape genetics and the adaptive radiation of Darwin's finches: The role of peripheral isolation. *Molecular Ecology* 14:2943–57.

Pfennig, K. 2003. A test of alternative hypotheses for the evolution of reproductive isolation between spadefoot toads: Support for the reinforcement hypothesis. *Evolution* 57:2842–51.

Podos, J. 2001. Correlated evolution of morphology and vocal signal structure in Darwin's finches. *Nature* 409:185–88.

Podos, J., and S. Nowicki. 2004. Beaks, adaptation and vocal evolution in Darwin's finches. *BioScience* 54:501–10.

Podos, J., J. A. Southall, and M. R. Rossi-Santos. 2004. Vocal mechanics in Darwin's finches: Correlation of beak gape and song frequency. *Journal of Experimental Biology* 207:607–19.

Price, T. D., and M. M. Bouvier. 2002. The evolution of F_1 postzygotic incompatibilities in birds. *Evolution* 56:2083–89.

Ratcliffe, L. M., and P. R. Grant. 1983a. Species recognition in Darwin's finches (*Geospiza* Gould). I. Discrimination by morphological cues. *Animal Behaviour* 31:1139–53.

———. 1983b. Species recognition in Darwin's finches (*Geospiza* Gould). II. Geographic variation in mate preference. *Animal Behaviour* 31:1154–65.

———. 1985. Species recognition in Darwin's finches (*Geospiza* Gould). III. Male responses to playback of different song types, dialects and heterospecific song. *Animal Behaviour* 33:290–307.

Reid, J. M., P. Arcese, A. L. E. V. Cassidy, S. M. Hiebert, J. N. M. Smith, P. K. Stoddard, A. B. Marr, and L. F. Keller. 2005. Fitness correlates of song repertoire size in free-living song sparrows (*Melospiza melodia*). *American Naturalist* 165:299–310.

Reznick, D. N., M. J. Butler IV, H. Rodd, and P. Ross. 1996. Life-history evolution in guppies (*Poecilia reticulata*) 6. Differential mortality as a mechanism for natural selection. *Evolution* 50:1651–60.

Rieseberg, L. H., A. Widmer, A. M. Arntz, and J. M. Burke. 2003. The genetic architecture necessary for transgressive segregation is common in both natural and domesticated populations. *Philosophical Transactions of the Royal Society of London B* 325:1141–47.

Rising, J. D. 1996. The stability of the oriole hybrid zone in western Kansas. *Condor* 98:658–63.

Rowher, S., E. Bermingham, and C. Wood. 2001. Plumage and mitochondrial DNA haplotype variation across a moving hybrid zone. *Evolution* 55:405–22.

Saccheri, I. J., M. Kuussaari. M. Kankare, P. Vikman, W. Fortelius and I. Hanski. 1998. Inbreeding and extinction in a butterfly metapopulation. *Nature* 392:491–94.

Schluter, D. 2000. *The Ecological Theory of Adaptive Radiation.* Oxford University Press, Oxford.

Schluter, D., and P. R. Grant. 1984. Determinants of morphological patterns in communities of Darwin's finches. *American Naturalist* 123:175–96.

Schluter, D., T. D. Price, and P. R. Grant. 1985. Ecological character displacement in Darwin's finches. *Science* 227:1056–59.

Seehausen, O. 2004. Hybridization and adaptive radiation. *Trends in Ecology and Evolution* 19:198–207.

Smith J. N. M., P. R. Grant, B. R. Grant, I. Abbott, and L. K. Abbott. 1978. Seasonal variation in feeding habits of Darwin's ground finches. *Ecology* 59:1137–50.

Strayer, D., J. S. Glitzenstein, C. G. Jones, J. Kolasa, G. E. Likens, M. J. McDonnell, G. G. Parker, and S. T. A. Pickett. 1986. Long-term ecological studies: An illustrated account of the design, operation, and importance to ecology. *Occasional Publication of the Institute of Ecosystem Studies, no. 2.* New York Botanical Garden, Millbrook, NY.

Taylor, E. B., J. W. Boughman, M. Groenenboom, M. Sniatynski, D. Schluter, and J. L. Gow. 2006. Speciation in reverse: Morphological and genetic evidence of the collapse of a three-spined stickleback (*Gasterosteus aculeatus*) species pair. *Molecular Ecology* 15:343–55.

Tonnis, B., P. R. Grant, B. R. Grant, and K. Petren. 2005. Habitat selection and ecological speciation in Galápagos warbler finches (*Certhidea olivacea* and *Certhidea fusca*). *Proceedings of the Royal Society of London B* 272:819–26.

Watt, W. B., C. W. Wheat, E. H. Meyer, and J. F. Martin. 2003. Adaptation at specific loci. VII. Natural selection, dispersal and the diversity of molecular-functional variation patterns among butterfly species complexes (*Colias*: Lepidoptera, Pieridae). *Molecular Ecology* 12:1265–75.

Weider, L. J., W. Lampert, M. Wessels, J. K. Colbourne, and P. Limburg. 1997. Long-term genetic shifts in a micro-crustacean egg bank associated with anthropogenic changes in the Lake Constance ecosystem. *Proceedings of the Royal Society of London B* 264:1613–18.

Werner, T., and T. W. Sherry. 1987. Behavioral feeding specialization in *Pinaroloxias inornata*, the "Darwin's finch" of Cocos Island, Costa Rica. *Proceedings of the National Academy of Sciences (USA)* 84:5506–10.

7 Individual Fitness, Social Behavior, and Population Dynamics of Yellow-Bellied Marmots

Kenneth B. Armitage

Figure 7.0. Ken Armitage weighing a marmot in a bag at the Rocky Mountain Biological Laboratory. Photograph by Dennis Johns, 1981.

Abstract

Some fundamental biological questions can be answered only by taking an "ecology of place" approach—by studying a focal system over long periods of time. One of these fundamental questions is the degree to which population dynamics are a function of social behavior. Long-term studies of yellow-bellied marmot populations of known pedigree have examined the role of kinship in social behavior and organization. Additional studies of physiological energetics, feeding behavior and nutrition, and hibernation revealed how marmots adapt to life in a seasonal environment, while studies of genetics, dispersal, and demography uncovered the demographic variables that underpin local and regional population dynamics. My research team and I determined that these montane marmots live in mother-daughter-sister groups, or matrilines, that are defended by an unrelated male. The groups are localized near rocky outcrops that provide two limiting resources—suitable burrows for hibernation and predator escape, and access to meadow foraging areas. Population dynamics are largely a function of fertility and age of first reproduction. Both parameters are strongly affected by the social behavior of dominant females, who can delay reproductive onset in younger subordinates. Although by delaying reproduction subordinate females lose direct fitness that is not compensated by increased indirect fitness (fitness of relatives), the alternative of dispersing is worse: dispersal increases the probability of mortality and does not increase the likelihood of early reproduction. The fundamental reproductive strategy of a female is to join or form a matriline even if she faces possible reproductive suppression. Consequently, the survival and fecundity variables that drive the dynamics of local yellow-bellied marmot populations result largely from the fitness-maximizing strategies of individuals. These same strategies determine the dynamics and persistence of the regional metapopulation because they cause large colonies to export individuals to vacant patches, which then become occupied by the new matriline. This case study, which has been rooted in a single location for more than 40 years, illustrates how population dynamics can be understood from the perspective of individuals acting to maximize their individual fitness in a patchy, seasonal environment.

Introduction

Behavior and population dynamics

More than 50 years ago, I became familiar with the then-revolutionary idea that population dynamics might be treated from a social behavior perspective rather than from a demographic perspective (Calhoun 1952). This idea was also expressed by Errington (1956), who argued that "we must not ignore the role of social intolerance" as a factor limiting higher vertebrate populations. While there had long been an understanding that factors such as food resources, weather, disease, or predation could affect population dynamics, this new line of thinking stressed that natality, mortality, and dispersal could be influenced by interactions among individuals. Socially mediated events, such as whether individuals were accepted by groups or decided to leave them, might reasonably be viewed as having a profound influence on the number of individuals in a population.

Calhoun (1952) marshaled several lines of evidence to support the importance of social interactions to population dynamics. Perhaps most striking was his work with a laboratory population of Norway rats given as much food as they wanted. He noted that the population stabilized at a density substantially lower than what could be achieved if rats were individually housed in cages stacked into the same overall space. Removal of the cage walls allowed social interactions, and population densities dropped dramatically as a consequence. It was a powerful illustration, even if the context was artificial, that social dynamics could be critical to population dynamics.

Consideration of such work led me to realize that the role of interactions among individuals in population dynamics could be readily examined under natural conditions. What was required was a social species whose behavior could be observed and its consequences for population dynamics measured. Additionally, because social interactions might be only one among multiple factors (e.g., severe weather; Armitage 1994) such a study must be carried out over several generations. And, because of the potential for geographic variation in predation or climate, it was necessary to track populations in a single place through time rather than make inferences based upon comparisons among populations. Finally, I wanted to study several local populations simultaneously so that possible effects of habitat variation could be evaluated. From such place-based research, it seemed to me that general principles relating social behavior to population dynamics could be identified despite the variability inherent in biological systems.

Based in part on these initial ideas, I initiated a study of the yellow-bellied marmot, *Marmota flaviventris*, in the Upper East River Valley in the vicinity of Gothic, Colorado, in the United States (38° 58' N, 106° 59' W; elevation 2900 m) that has continued for 41 years. A wealth of detailed information has accumulated during this long-term study on the basic ecology and dynamics of these high-elevation populations; on the life histories, kin relationships, and social behavior of numerous individual marmots; and on the population consequences of rare extreme climatic or predation events. These data have contributed to a general understanding not only of how social behavior affects population dynamics, but also of how social behavior and group structure reflect individual strategies for maximizing inclusive fitness. In this chapter I review how a long-term study, conducted in a single location, has provided general insights into relationships between individual fitness, social behavior, and population dynamics.

The social species

The yellow-bellied marmot is a diurnal, ground-dwelling squirrel that occupies meadow habitat in a forested landscape. The animals live in permanent burrows dug in patchily-distributed rocky outcrops or talus. Because they are diurnal, are localized near their burrows, and are readily trapped and marked, the behavior and life histories of individuals were readily characterized. In the Upper East River Valley, marmots hibernate for about eight months each year in a permanent burrow that is suitable for over-wintering. Not all their burrows can function as hibernacula; thus this special burrow is a critical resource. The animals emerge from hibernation in late April or early May, mate soon after, and give birth after about 30 days of gestation. Young-of-the-year are weaned after three to four weeks of lactation and emerge above ground between June and mid-July. During the remaining active season, all age classes forage in the meadow, grow, accumulate fat, and prepare to hibernate in late August or early September (Armitage 1998). Because young must reach a minimal size and reproductive females must accumulate fat for successful hibernation, the short growing season restricts reproduction to once annually. This greatly simplifies characterizing age-specific reproduction and survival.

Initial results

We initiated work on yellow-bellied marmots in 1962, recording social behavior and population numbers of five discrete clusters of individuals ("colonies") for 10 years. Over this time period, changes in population numbers were not correlated among the five colonies; numbers declined at some sites and increased at others. Because there were no obvious environmental factors that caused this variation in population dynamics, we saw a potential role for social behavior. We therefore asked whether variation in social behavior could explain the variation in population dynamics.

We found considerable variation in overall rates of social interaction as well as in rates of amicable versus agonistic behaviors both within and among colonies (Armitage 1977). Although overall rates of social behavior were unrelated to changes in colony numbers, agonistic behavior (social intolerance) was associated with population decline and failure to recruit new residents whereas amicable behavior (social tolerance) was associated with recruitment of new females into groups that shared home ranges (Armitage 1975). Further detailed analysis revealed that patterns of social behavior were affected not just by colony numbers, but also by the age-sex structure of the population, the individual behavioral phenotypes of the residents and potential recruits, the number of years residents had lived together, and the way in which space was shared (Svendsen and Armitage 1973, Armitage 1977).

Thus, at the end of this initial study I had evidence for the link between social behavior and population dynamics postulated by Calhoun and

Errington. But the links were much more complex than a simple feedback loop in which social behavior adjusted reproduction, recruitment, and dispersal in a way that regulated colony size. Instead of responding only to colony numbers, social behavior seemed to depend on idiosyncrasies of colony structure and history. It is important to note that I could discern the variability and complexity of the relationship between behavior and population dynamics because I followed the same populations in the same places for several generations. If I had observed one population for one or two years, I might have reached different conclusions depending on which population I observed in which year.

The book *Sociobiology* (Wilson 1975) stimulated me to rethink my research program. Prior to Wilson's book, social behavior was thought to be a group-selected trait that prevented overexploitation of resources by regulating population at an optimal density (Wynne-Edwards 1962, 1965). Wilson and other evolutionary biologists like G. C. Williams (1966) suggested instead that social behavior reflected the fitness-maximizing strategies of individuals, and that population dynamics reflected the demographic consequences of those individual strategies. I found these ideas exciting; they suggested a new way of thinking about the relationship between social behavior and population biology. Consequently I asked a new research question: Can marmot social behavior be understood as the result of selection to maximize individual fitness? Because individuals gain fitness not only through their own reproduction but also through that of kin (Hamilton 1964), understanding marmot group structure was critical for answering this new question. We therefore spent considerable effort characterizing marmot social organization.

Social Organization

In the Upper East River Valley yellow-bellied marmots occupy habitat patches that range in size from 0.15 to 7.24 ha (Armitage 2004a). In the early years of our study we concentrated on large patches that supported large aggregations, called colonies. On these sites adult males maintained non-overlapping territories that included home ranges of 2.3 adult females on average (Armitage 1986), and we concluded that the fundamental social group in yellow-bellied marmots was a harem: an adult male plus one or more females. But there was more to group structure than that. As we followed the turnover of individuals in the study populations, we learned that all adult males were immigrants to the sites, whereas most of the females were born there. Females formed mother-daughter-sister groups that persisted through time with extensive home-range overlap. We called these groups "matrilines" (Armitage 1984). Unrelated immigrant males attached themselves to, and defended, one or more matrilines. Because all juvenile males disperse but only about half of the females do, and because adult male residents at a site rarely exceed two, population dynamics essentially are a consequence of what females do (Armitage 1991).

In contrast, a group of yellow-bellied marmots living on small patches, formerly designated as satellite sites (Svendsen 1974), often consisted of a male and an adult female together with young-of-the-year or yearlings. Initially we thought that the social organization on small patches differed from that on large patches, because the small patches contained a greater proportion of immigrant females. However, by observing populations on satellite sites for many years we determined that they, too, consisted of matrilines—just fewer, smaller ones than on colonial sites (Armitage and Schwartz 2000). We therefore concluded that the social organization of yellow-bellied marmots was matrilineal, and that patch area influenced the size and number of matrilines but not the basic social structure.

Could the matrilineal structure account for the patterns of social behavior? Analysis of correlations between relatedness and social behavior suggested that it is critical. Adult males are antagonistic toward other males regardless of relatedness, which can account for the lack of male home-range overlap and the fact that all juvenile males disperse (Armitage 1974). Amicable behavior predominated among members of a matriline, whose coefficient of relatedness (r, the probability that homologous alleles in two individuals are identical by descent from a common ancestor) equals 0.5 on average, whereas agonistic behavior predominated among adult females of different matrilines ($r \leq 0.25$; Armitage and Johns 1982), which explains the greater space-use overlap (space-use is a measure of the sharing of the critical resources of food and burrows) among close kin ($r = 0.5$) than among more distantly related individuals ($r < 0.5$; Armitage 1996a). A recent analysis of alarm calling in yellow-bellied marmots further suggests that kinship matters for behavioral interactions: 42% of the variation among individuals in the rate of calling was attributable to whether the caller was an adult female with newly-weaned young (Blumstein et al. 1997).

Costs and Benefits of Living in a Matriline

What are the fitness consequences of this differential treatment of kin and the resulting matriline social structure? Individuals of social species may gain fitness, measured as the number of gene copies contributed to future generations, either directly or indirectly. Direct fitness accrues to individuals who produce offspring, whereas indirect fitness accrues to individuals who assist in the reproduction of non-descendant kin. Total or inclusive fitness is the sum of direct and indirect fitness (Hamilton 1964). We therefore evaluated consequences of matriline life for these two routes to fitness, and for inclusive fitness.

There is considerable variation in reproductive success among females within matrilines. On average, only 48% of the adult female yellow-bellied marmots reproduce in any given year (Blumstein and Armitage 1998). Much of this reproductive skew occurs because females vary considerably in age of first reproduction. Although yellow-bellied marmots are reproductively mature at the age of two years (in their third summer), many delay breeding; only 34% of females observed throughout their lives reproduced as two-year-olds. The average age of first reproduction was 3.15 years, and some females had their first litter at age six years (Armitage 2003a). The average age of first reproduction does not capture the considerable reproductive skew; some females may reproduce only once during as many as 10 years of residency, whereas other females may reproduce as many as five times (Armitage 1991, 2002). Age of first reproduction does not affect the subsequent per-year probability of breeding, litter size, or survival (Armitage 2003a). Consequently, females that start reproducing earlier pay no cost, and individual fitness decreases as age of first reproduction increases (Oli and Armitage 2003).

Why do females delay reproduction? Because forgoing reproduction in a given year represents a loss of direct fitness, females who do so will have lower inclusive or total fitness unless there is a compensatory increase in indirect fitness. Our extended tracking of 428 females from weaning until their disappearance from our study sites provided the data we required to estimate the direct and indirect components of inclusive fitness using Oli's (2003) method, and thus to assess whether delayed reproduction could be explained by kin selection (Hamilton 1964). Delayed breeders did not survive better or produce larger litters, and they had lower direct fitness (Oli and Armitage 2003). However, nonreproductive females had higher indirect fitness than reproductive females (Oli and Armitage 2008) because all of their fitness came from the offspring produced by their reproductive kin, and their presence enhanced the reproductive output of older females (Armitage 2003a). By contrast, reproductive females gained no indirect fitness from their nonreproductive kin; all of their fitness was direct. Overall, the inclusive fitness of reproductive females was 2.3 times that of nonreproductive females (Oli and Armitage 2008).

Before we concluded that delayed reproduction is not a fitness-maximizing strategy, we considered the possibility that delayed breeders could compensate to some extent by overproducing sons; a son who survives to reproduce will sire an average of 12.7 young during his lifetime, compared to 7.1 for a successful female (Armitage 2004a). However, to be successful a son (or daughter) must survive to reproductive age, and males suffer higher mortality than females. Yellow-bellied marmots are long-lived; female generation time is 4.5 years and some females live beyond age 10 years, but no known-aged male lived beyond nine years (Schwartz et al. 1998). Male and female mortality are similar in the first year of life; but from age one year onward, male mortality exceeds that of females. This difference in mortality is related to the different reproductive strategies of males and females. Male yearlings avoid adult males; any social contact between yearling and adult males is almost entirely agonistic (Armitage 1974). As a consequence, yearling males typically disperse, whereas only about half of the female yearlings do so. Males continue to move widely at age two years, apparently seeking females. Even after settling down, adult males are more active than females; they spend long days patrolling their territory and expand it to include widely dispersed females (Salsbury and Armitage 1994). Marmots are vulnerable to predators when they are moving above ground, particularly if they are dispersing through unfamiliar areas or areas far from burrows; hence, males suffer higher mortality than females (Van Vuren and Armitage 1994). Only about 16% of one-year-old males live to age three, and only about 6% of these successful males account for the difference in average reproductive output of males and females who survive to breeding age. Thus about 1% of yearling males become highly successful. This calculation suggests that females are unlikely to gain fitness by overproducing sons. Indeed, females who delay reproduction produce the same 1:1 sex ratio of weaned young as other females. Thus, we can exclude the likelihood that females compensate for a loss of direct fitness by producing highly successful sons.

If forgoing reproduction reduces inclusive fitness, why do females delay reproduction? Could many two-year-olds simply be too immature or lack enough food to breed? We tested these possibilities by comparing the body masses of reproductive and nonreproductive two-year-olds, and by providing supplemental food to two colonies. Nonreproductive females did not have lower body masses than reproductive two-year-olds; the key difference was whether older females were present (Armitage 2003a). Similarly, a slightly lower fraction of two- and three-year-old females reproduced in the supplemented colonies, whereas all older females in both colonies reproduced in every year of the five-year experiment (Woods and Armitage 2003). In both colonies, young females reproduced only when an older female died, or after moving to the edge of the matrilineal home range (Woods and Armitage 2003).

The association of early reproduction with absence of older females suggested that delayed reproduction could result from reproductive suppression. Further analysis indicated that indeed, a young adult female is unlikely to wean a litter if an older female is present, even if the older female is her mother. About 20% of two-year-old females reproduce in the presence of older females, whereas about 48% reproduce when older females are absent. Similarly, about 33% reproduce for the first time as three-year-olds when older adults are present, whereas 58% do so if older adults are absent. Some of the two- and three-year-olds who succeed in reproducing when older adults are present do so by moving away and establishing a home range that overlaps only slightly with that of the older adults. Reproductive suppression was not density-dependent (Armitage 2003a); it occurred at the same frequency whether one or two older females were present. Analysis of behavioral data suggested that dominant, older females actively suppress litter production in subordinate, younger females via agonistic behavior (Armitage 1965).

Why should older females suppress reproduction by younger kin? We suspect that there is a limit to the number of individuals that a matrilineal home range can support, because average per-capita fitness, measured as net reproductive rate (average number of offspring produced per individual in her lifetime), initially increases with matriline size, only to decrease in the largest matrilines (Armitage and Schwartz 2000). The decrease probably reflects greater reproductive suppression in large matrilines (Armitage 2007), but it also could reflect resource constraint. If there is a limit to matriline size, females would maximize their direct fitness by filling those limited spaces with daughters ($r = 0.5$) rather than with more distantly-related nieces or grand-daughters ($r = 0.25$). Hence, we expect reproductive competition to occur within matrilines, as well as between matrilines.

Reproductive competition is mediated through agonistic behaviors that determine whether individuals are allowed to share a matrilineal home range, and, if so, whether they are allowed to reproduce. Adult females may force yearling females—even their full sisters—to disperse from their natal site (Armitage 1986). They suppress the reproduction of other females (whether related or not) through infanticide or aggressive behavior that either reduces litter production or forces females to move to a new, and probably inferior, home range (Armitage 2003a). One such pattern of conflict between unrelated females continued for seven years. The dominant female caused the subordinate female to change home ranges in the fourth year. The abandoned space was colonized by descendants of the dominant female, who consequently recruited into her matriline six reproductive daughters who produced 63 grand-offspring, whereas the subordinate female recruited only two reproductive daughters who produced seven grand-offspring (Armitage 1992). The aggressive behavior of the dominant female clearly enhanced her direct fitness.

Because females in all of our study areas formed matrilines, we hypothesized that group living brings fitness benefits. We found several benefits. First, recruiting daughters into the natal group eliminates dispersal and its attendant mortality. Furthermore, average survival of matriline residents increases as group size increases (Armitage and Schwartz 2000), perhaps because larger groups are more likely to detect predators. Living in a matriline also enhances reproductive success. Fitness, measured as net reproductive rate (average number of lifetime production of offspring per individual), increased with matriline size except in the largest matrilines (Armitage and Schwartz 2000). Reproduction was especially high for females living with younger or same-age kin in the absence of older females (Armitage 2003a). As yet we do not understand the mechanisms behind these reproductive effects of group living.

Our long-term longitudinal studies of individual yellow-bellied marmots suggest that the major components of the reproductive strategy of females enhance direct fitness. They involve selective tolerance of home range sharing by kin and suppression of reproduction by younger, subordinate females (Armitage 1999, 2003a, 2007). Recruitment of daughters into the matriline increases survivorship, aids in defense of the matrilineal home range against incursions of neighbors or immigrants (Armitage 2003a, 2003c), and enhances reproduction. Recruitment of daughters increases the likelihood that a female will produce reproductive descendants (Armitage 2002). Reproductive suppression of younger females increases the probability of recruiting additional daughters that are more closely related than granddaughters. Populating the matriline with reproductively suppressed daughters is an effective strategy for maximizing a female's genetic legacy to future generations because it will be daughters, rather than more distantly related kin, who will reproduce to continue the genetic lineage (Armitage 1991, 1992, 2003a).

An Alternative Strategy

But one major fitness problem remains: why do young females remain in the natal matriline if it means they will lose direct fitness that is not compensated for by increased indirect fitness? Is there an alternative strategy that would increase inclusive fitness?

The obvious alternative is to disperse and attempt to breed in another habitat patch. Most females who disperse do so as yearlings (Armitage 1991); they experience higher mortality than yearlings who stay home (Van Vuren and Armitage 1994). Even if they survive dispersal, they have a low probability of becoming established and breeding in the non-natal site. Only six of 32 female yearling dispersers trapped at a colony site were successful immigrants, and none was successful if philopatric yearlings (marmots born into the colony) were present (Armitage 2003c). Furthermore, successful immigrant females are older when they reproduce than are philopatric females. A philopatric female can reproduce at age two years, but an immigrant is unlikely to reproduce before age three years (Armitage 2003c): only 20 (26%) of 77 immigrants of breeding age reproduced in their year of immigration (Armitage and Schwartz 2000). By contrast, 39.6% of philopatric females reproduced as two-year-olds (Armitage 2003a).

Thus, dispersal and subsequent immigration into a new site does not appear to be a viable strategy. Compared to philopatric females, dispersers face greater mortality while dispersing, have a lower chance of reproducing even if they do survive, and are older when they reproduce. Furthermore, the philopatric female has a finite chance of avoiding reproductive suppression; in some highly successful matrilines nearly all females weaned litters (Armitage 2007). Even if the philopatric female does suffer suppression, it may be short-lived if the older female loses her competitive edge or dies young; and she may have the opportunity to move to a vacant space in the habitat patch, and bud off a new matriline of her own (Armitage 2003a, Armitage and Schwartz 2000). Indeed, yearling females appear to prefer to remain in their natal colony: experimental removal of adult females greatly decreased dispersal rates (Brody and Armitage 1985), indicating that much dispersal is forced by adult aggression.

We conclude that joining a matriline is the best strategy for female yellow-bellied marmots. Matrilines provide direct fitness benefits, through enhanced survival and reproduction of successful females, that outweigh the fitness costs of reproductive suppression. In effect, we can view matrilines as social groups characterized by cooperation (reproductive enhancement) and competition (reproductive suppression).

Reproductive Strategies and Population Dynamics

Having established how individual behavior affects the demographic parameters of survivorship, reproduction, and recruitment, we can now examine how individual behavior affects population dynamics. But first we need to examine some characteristics of changes in population numbers. As was true for the first 10 years of study, population numbers and growth rate at the study sites continued to fluctuate widely over 40 years of observation and were uncorrelated among sites (Armitage 1977, 2003b; Oli and Armitage 2004).

Equivalent temporal and spatial variation in demographic parameters accompanied this population-level variation. Survival rates varied both spatially and temporally; juvenile survival rates varied among sites and over time, whereas adult survival rates varied only among sites (Ozgul et al. 2006a). Similarly, the breeding probability of reproductive and nonreproductive adult females and litter size varied over space, whereas the breeding probability of reproductive adults varied only over time (Ozgul et al. 2007).

We could attribute some of the fluctuation in demographic variables to stochastic events. For example, we observed instances where a badger (*Taxidea taxus*) predator heavily impacted juvenile survival at one colony in one year, but did not affect any other colony (Armitage 2004b). Similarly, we observed cases where late summer drought or unusually late snow melt reduced both juvenile and adult overwinter survival and fertility in some sites, whereas sites only 5 km away were characterized by typical demographic patterns (Armitage 1994, 2003b). Although weather variation explains some of the variation in demography (Schwartz and Armitage 2005), much of the variation among colonies and through time appeared to be due to variation in demographic parameters that were affected by social behavior, such as age of first reproduction and fertility.

We could assess the relative contributions of these demographic variables by analyzing life tables that we constructed from our long-term longitudinal life history data. We first pooled our data into a single age-structured projection matrix and used elasticity analysis to explore how much a proportional change in each demographic variable affects the projected finite rate of population growth (Oli and Armitage 2004). This elasticity analysis indicated that juvenile survival would have the largest relative influence on population growth rate, and that the importance of the demographic factors did not differ between years characterized by negative or positive population growth. We then conducted a retrospective analysis (a life table response experiment, or LTRE; Caswell 2001) that documents the extent to which observed changes in population growth rate is a function of individual demographic variables. The LTRE indicated that changes in fertility, followed by age of first reproduction, made the largest contributions to observed annual changes in the population growth rate (Oli and Armitage 2004). Increased fertility, increased juvenile survival, and decreased age of first reproduction made almost equal contributions to increases in that rate, whereas decreased fertility followed by increased age of first reproduction made the largest contributions to decreases in that rate (Oli and Armitage 2004). Because this study focused on the same populations over many years, we had sufficient statistical power to demonstrate that the demographic factors driving population increases might differ from those driving population decreases. This study was one of the first to demonstrate such differences.

We can now ask the question: How do individual fitness strategies translate into population growth and decline? Change in the number of adult residents is closely related to change in the size and number of matrilines; immigration is of minor importance (Armitage 2003c). Matriline size increases when females recruit their daughters, and the number of matrilines increases primarily by the fission of large matrilines (Armitage 1996b, 2002).

Recovery following population decline is particularly instructive. In one colony that was reduced to one adult female, the number of resident adults increased to four in two years because three two-year-old daughters were recruited and two of them weaned litters. Population growth rate was high because of the early age of first reproduction and the high fertility of the surviving female and her daughters. Population growth rate subsequently declined when a two-year-old recruit failed to reproduce and fertility of the older adults declined (Armitage 1973). In another colony, population recovery following a weather-induced decline also occurred through the recruitment of daughters. The number of adult females was reduced to three. One failed to reproduce, one probably became coyote (*Canis latrans*) prey, and the third founded a matriline. Over a seven-year period, nine adult females were recruited in the multigenerational matriline. This kin group spread out over the habitat patch to occupy burrow sites and foraging areas, and then it split into three matrilines of closely related individuals (Armitage 2003b). Interestingly, only one of the nine recruits reproduced at the age of two years, and she was one of two two-year-olds who split off to form one of the new matrilines.

These specific examples, combined with quantitative analyses of demography, make clear that the fitness-maximizing decisions of females drive the size and number of matrilines at a site and hence have large effects on the dynamics of local populations.

Because we monitored five major sites throughout our long-term study and up to 18 sites in some years, we could use our data to test new ideas in population biology. Our studies of genetics and dispersal (Schwartz and Armitage 1980, Van Vuren and Armitage 1994) demonstrated that the local marmot populations were not isolated; gene flow occurred through dispersal and immigration. We realized then that we were working with a metapopulation—a regional cluster of local populations whose dynamics were linked to some extent by movement of individuals—and that we could use our data to test metapopulation models (Ozgul et al. 2006a, 2006b, 2007). A presence/absence analysis (a stochastic patch occupancy model, or SPOM) of marmots at 21 sites (Ozgul et al. 2006a, 2006b, 2007) revealed that persistence of the yellow-bellied marmot metapopulation depended strongly on large, highly productive colony sites and that smaller, less productive satellite sites made a significant contribution to long-term persistence when regional stochasticity was high (Ozgul et al. 2006b). These results are in large part a consequence of female reproductive strategies: recruitment behavior of adult females at large colony sites produces large matrilines with high reproductive success (Armitage 1984, Armitage and Schwartz 2000). When yearlings are not recruited, they disperse and may colonize available habitat patches—either satellite sites or available space in colony sites (Armitage

2003c). In this way the regional production of young is distributed among all suitable sites, which stabilizes the metapopulation in the face of spatial and temporal environmental stochasticity.

Environment, Body Size, Social Organization, and Life History

Our long-term, place-based studies suggested that a seasonal environment with short summers, long, cold winters, and a patchy distribution of suitable habitat leads to decisions by individuals about when to disperse. These decisions shape the social organization of the marmots in a way that has consequences for *M. flaviventris* life history and population dynamics. Is this model of the relationship between environmental constraints, individual decision-making, life history, and population consequences consistent with what we see in other species?

Analysis of different marmot species makes it clear that variation in dispersal patterns can help explain variation in survivorship. Patterns of age-specific survival differ between the yellow-bellied marmot and three other marmot species (*M. caudata aurea*, the golden marmot; *M. olympus*, the Olympic marmot; and *M. vancouverensis*, the Vancouver Island marmot). These three species have greater survivorship than do yellow-bellied marmots beyond the age of one year, and survivorship does not differ between the sexes (Blumstein et al. 2002). Why do these differences occur among such similar species? Our long-term research suggested that age of dispersal is the key. Yellow-bellied marmots disperse at least one year earlier than do the other three species. We lack long-term studies of the other species, but the available data indicate that males disperse at an older age than yellow-bellied males and move less widely. Because dispersal and movement are risky, survivorship is higher than in yellow-bellied marmots. Cross-species comparisons support the notion that examining the age of dispersal can help to explain life history parameters such as survivorship.

The differences in dispersal patterns also help explain differences in social structure and patterns of reproduction among marmot species. Yellow-bellied marmots have a relatively simple group structure in which single males defend groups of females. The other three species have more complex group structure, and hence can be considered more social. Males of these other marmot species do not always disperse; they form social groups each consisting of a dominant monogamous reproductive pair and a variable number of adult nonreproductive subordinates of both sexes (Armitage 2000). The dominant male mates only with the dominant female (Cohas et al. 2006) and sires on average only six young (Lenti Boero 1999). A greater proportion of males mate, but the reproductive success of those individuals appears to be much lower than that of successful yellow-bellied marmot males.

Females of the three more social marmot species also delay dispersal to age two or older and disperse at about the same rate as males. Because many yellow-bellied marmot females disperse as yearlings, their survivorship is lower than that of the more social marmots, and this difference persists at older ages. The age of first reproduction for these more social species is at least one year older than that of yellow-bellied marmots. We do not know why yellow-bellied marmots differ from these other marmot species in their dispersal, but we can speculate that environmental differences affecting the cost/benefit of dispersal are responsible.

The role of environmental factors in explaining dispersal is supported by broadening our cross-species comparisons to include ground-dwelling squirrels. All ground-dwelling squirrels (*Cynomys*, the prairie dogs; *Spermophilus*, the ground squirrels; and *Marmota*, the marmots) live in a seasonal environment and in social groups of varying complexity (Armitage 1981, Blumstein and Armitage 1998). The level of social complexity, as defined by the extent of cohabitation of overlapping generations, appears to be a function of body size and the length of the growing season: larger species, such as marmots, with a short growing season are socially more complex than either smaller ground squirrels or large species with a long growing season. The major factor underlying this pattern of sociality is that it takes longer for the young of larger species to reach a body size that allows them to be independent. In areas with a short growing season, it may take several years for young to get big enough to disperse. Sociality is the result of the young deciding to remain in their natal areas for multiple years (Armitage 1981, 2007). We can predict that further examination of species of marmot other than the yellow-bellied marmot will reveal that differences in dispersal patterns and sociality will ultimately be explained by environmental differences such as size and dispersion of suitable habitat patches, forage quality, and temporal uncertainty that affect the costs and benefits of dispersal.

This comparison of yellow-bellied marmots with other marmot species and ground squirrels suggests that the approach of examining the relationship between behavior, social structure, and population dynamics has explanatory power that extends beyond a single species. Calhoun and Errington posited the importance of social interactions to population dynamics. Wilson's work provided a basis, by explaining behavioral decisions in terms of fitness, for understanding social interactions. Our detailed studies of yellow-bellied marmots demonstrated how decisions by individuals lead to social structure, which in turn has consequences for life history and population dynamics. By linking individual decisions to environmental constraints, our studies of yellow-bellied marmots also provide a framework to explain observed patterns in other, less well-studied species.

Why Have General Insights Emerged from the Study of Marmots in One Place?

This study began with a relatively narrow focus on testing general ideas about social behavior and population regulation. At the beginning I did not anticipate that it would continue for 40 years (it still continues under the direction of Dan Blumstein), or that it would expand into a broad research program with relevance to such new areas as sociobiology, metapopulation ecology, life-history evolution, or global change biology.

Several factors contributed to the ability of our research to contribute so broadly. Of primary importance was the obvious need to collect basic life-history data consistently in the same populations and over long periods of time. If we missed a year, we lost critical information on the life histories and kin relationships of these long-lived animals. Furthermore, depending on which year was skipped, we could have missed the effects of drought or late spring snow cover on survival and reproduction, or the effects of a major predation event (see also Grant and Grant, chapter 4). These rare, stochastic events often provide unexpected insights into ecological problems.

Because funding agencies did not support long-term research, the marmot research program was actually a collection of many short-term projects, on various fundable topics, that were carried out by graduate students (table 7.1). Although these projects could have been conducted with yellow-bellied marmot populations anywhere, they benefited from and contributed to the accumulated knowledge of the Upper East River Valley and its inhabitants. Graduate students assisted in the core activities of trapping and marking marmots and recording life-history data. But perhaps more importantly, their work provided an integrated picture of how this species is adapted for living in an environment with a short summer and a long, cold winter.

For example, physiological and energetics studies revealed that yellow-bellied marmots are heat intolerant, highly conservative of energy and water, and have high growth efficiency. These characteristics seem to be related to the need to conserve energy during the long period of hibernation and to achieve reproduction and growth during the short summer. Genetic studies revealed that marmot social groups are outbreeding primarily because all young males disperse and resident, territorial males are immigrants. Female yearlings did not disperse when adult females were absent or experimentally removed (Brody and Armitage 1985), and the timing of yearling dispersal was strongly affected by social interactions between yearlings and adults (Downhower and Armitage 1981).

Phenological studies indicated that marmots adjust the timing of their emergence from hibernation to local cues associated with impending snowmelt (Armitage 2005), and that late snowmelt at higher sites is associated

TABLE 7.1. Topics of yellow-bellied marmot research

Topic	Principal Investigators
Social behavior and organization	K. B. Armitage, D. W. Johns
Habitat characteristics	G. E. Svendsen
Energy flow	D. L. Kilgore, Jr.
Food selectivity and foraging behavior	B. A. Frase, K. B. Armitage
Food supplementation	B. C. Woods
Individuality	K. B. Armitage, G. E. Svendsen, D. H. Van Vuren
Genetics	O. A. Schwartz
Metabolic rate, water balance and environmental temperature	J. M. Ward, Jr.
Thermal energy exchange	J. C. Melcher
Energy conservation and hibernation physiology	K. B. Armitage, D. T. Blumstein, B. C. Woods
Reproductive energetics	K. B. Armitage, C. M. Salsbury
Body size and growth rates	C. M. Salsbury, J. F. Downhower, G. E. Svendsen
Scent marking	K. Brady
Play behavior	S. Nowicki, S. H. Jamieson
Time budgets	K. B. Armitage, C. M. Salsbury, E. L. Barthelmess
Dispersal	A. K. Brody, J. F. Downhower, D. H. Van Vuren

References in Armitage 1991, 1998; Armitage and Van Vuren 2003

with reduced litter size and a smaller proportion of females breeding (Van Vuren and Armitage 1991), perhaps because growing season length is determined by snowmelt date. Recent changes in the association of emergence cues with actual snowmelt may explain the recent finding that yellow-bellied marmots are emerging from hibernation about 23 days earlier than they were in 1976 and are emerging through a deeper snow cover (Inouye et al. 2000). Because the beginning of the growing season has not changed over the same time period, marmots face longer periods of activity before food becomes available. How do they cope with this increased demand on their fat reserves, and how does that affect reproductive success and competition? What cues initiate the termination of hibernation, and are these cues now maladaptive? Answering these questions may be critical; an analysis of marmot climate space and phylogeny indicates that marmots may be more vulnerable than many mammals to global climate change (Davis 2005).

Energy-flow studies indicated that yellow-bellied marmots use a small percentage of available plant production. This finding stimulated research on foraging behavior that revealed that foraging areas are influenced by plant distribution and predation risk (marmots dig new escape burrows in foraging areas; Armitage 2003e); that sharing of foraging areas is strongly influenced by kinship, and that marmots are selective feeders (Armitage 2003d). These studies suggest that marmots are not food limited. Food supplementation experiments support this interpretation: food addition did not increase survival of young (Woods and Armitage 2003a), survival rates of adult females, female recruitment, or litter size. Nor did it decrease mean age of first reproduction (Woods and Armitage 2003b).

Finally, predation and mortality during hibernation were implicated as major causes of population decline (Armitage 1996b); these factors plus female competition for reproductive success (which limits immigration and recruitment) appear to maintain yellow-bellied marmot population density below the limits set by food availability.

The diverse insights into determinants of marmot social organization, behavior, life history, fitness, and population biology have certainly been key to our ability to predict and understand these same attributes in other species, and hence to generalize beyond one species in one place. But perhaps of equal importance has been the quality of the detailed demographic data. These proved amenable to analysis by new methods that were unavailable when the data were collected (e.g., life table response experiments) for the purpose of answering completely new questions (for example, the dynamics of metapopulations or the potential consequences of climate change). As yet, only a fraction of all the data collected over 40 years of study has been analyzed. Imagine what can be learned by capturing more from long-term, place-based studies like ours!

References

Armitage, K. B. 1965. Vernal behaviour of the yellow-bellied marmot (*Marmota flaviventris*). *Animal Behaviour* 10:319–31.

———. 1973. Population changes and social behavior following colonization by the yellow-bellied marmot. *Journal of Mammalogy* 54:842–54.

———. 1974. Male behaviour and territoriality in the yellow-bellied marmot. *Journal of Zoology, London* 172:233–65.

———. 1975. Social behavior and population dynamics of marmots. *Oikos* 26:341–54.

———. 1977. Social variety in the yellow-bellied marmot: A population-behavioural system. *Animal Behaviour* 25:585–93.

———. 1981. Sociality as a life history tactic of ground squirrels. *Oecologia* 48:36–49.

———. 1984. Recruitment in yellow-bellied marmot populations: Kinship, philopatry, and individual variability. Pages 377–403 *in* J. O. Murie and G. R. Michener, eds., *Biology of Ground-dwelling Squirrels*. University of Nebraska Press, Lincoln.

———. 1986. Marmot polygyny revisited: Determinants of male and female reproductive strategies. Pages 303–31 *in* D. S. Rubenstein and R. W. Wrangham, eds., *Ecological Aspects of Social Evolution*. Princeton University Press, Princeton, NJ.

———. 1991. Social and population dynamics of yellow-bellied marmots: Results from long-term research. *Annual Review of Ecology and Systematics* 22:379–407.

———. 1992. Social organization and fitness strategies of marmots. Pages 89–94 *in* B. Bassano, P. Durio, V. Gallo Orsi, and E. Macchi, eds., *First International Symposium on Alpine Marmot (Marmota marmota) and on genus Marmota*, Turin.

———. 1994. Unusual mortality in a yellow-bellied marmot population. Pages 5–13 *in* V. Rumiantsev, ed., *Actual Problems of Marmots Investigation*. ABF Publishing House, Moscow.

———. 1996a. Resource sharing and kinship in yellow-bellied marmots. Pages 129–34 *in* M. Le Berre, R. Ramousse, and L. Le Guelte, eds., *Biodiversity in Marmots*. International Network on Marmots, Lyon.

———. 1996b. Social dynamics, kinship, and population dynamics of marmots. Pages 113–28 *in* M. Le Berre, R. Ramousse, and L. Le Guelte, eds., *Biodiversity in Marmots*. International Marmot Network, Lyon.

———. 1998. Reproductive strategies of yellow-bellied marmots: Energy conservation and differences between the sexes. *Journal of Mammalogy* 79:385–93.

———. 1999. Evolution of sociality in marmots. *Journal of Mammalogy* 80:1–10.

———. 2000. The evolution, ecology, and systematics of marmots. *Oecologia Montana* 9:1–18.

———. 2002. Social dynamics of yellow-bellied marmots: Strategies for evolutionary success. Pages 9–16 *in* K. B. Armitage and V. Y. Rumiantsev, eds., *Holarctic Marmots as a Factor of Biodiversity*. ABF Publishing House, Moscow.

———. 2003a. Reproductive competition in female yellow-bellied marmots. Pages 133–42 *in* R. Ramousse, D. Allainé, and M. Le Berre, eds., *Adaptive Strategies and Diversity in Marmots*. International Marmot Network, Lyon.

———. 2003b. Recovery of a yellow-bellied marmot population following a weather-induced decline. Pages 217–24 *in* R. Ramousse, D. Allainé, and M. Le Berre, eds. *Adaptive Strategies and Diversity in Marmots*. International Marmot Network, Lyon.

———. 2003c. Dynamics of immigration into yellow-bellied marmot colonies. *Oecologia Montana* 12:15–18.

———. 2003d. Observations on plant choice by foraging yellow-bellied marmots. *Oecologia Montana* 12:21–24.

———. 2003e. Nesting activities of yellow-bellied marmots. Pages 27–32 *in* R. Ramousse, D. Allaine, and M. Le Berre, eds., *Adaptive Strategies and Diversity in Marmots*. International Marmot Network, Lyon.

———. 2004a. Life time reproductive success of territorial male yellow-bellied marmots. *Oecologia Montana* 13:28–34.

———. 2004b. Badger predation on yellow-bellied marmots. *American Midland Naturalist* 151:378–87.

———. 2005. Intraspecific variation in marmots. Pages 39–48 *in* V. Sánchez-Cordero and R. A. Medellín, eds. *Contribuciones Mastozoológicas en Homenje a Bernardo Villa*. Instituto de Biología, UNAM; Instituto de Ecología, UNAM; CONABIO, México.

———. 2007. Evolution of sociality in marmots: It begins with hibernation. Pages 356–67 *in* J. O. Wolff and P. W. Sherman, eds., *Rodent Societies: An Ecological and Evolutionary Perspective*. University of Chicago Press, Chicago.

Armitage, K. B., and D. W. Johns. 1982. Kinship, reproductive strategies and social dynamics of yellow-bellied marmots. *Behavioral Ecology and Sociobiology* 11:55–63.

Armitage, K. B., and O. A. Schwartz. 2000. Social enhancement of fitness in yellow-bellied marmots. *Proceedings of the National Academy of Science (USA)* 97:12149–52.

Armitage, K. B., and D. H. Van Vuren. 2003. Individual differences and reproductive success in yellow-bellied marmots. *Ethology Ecology & Evolution* 15:207–33.

Blumstein, D. T., and K. B. Armitage. 1998. Life history consequences of social complexity: A comparative study of ground-dwelling sciurids. *Behavioral Ecology* 9:8–19.

Blumstein, D. T., J. C. Daniel, and W. Arnold. 2002. Survivorship of golden marmots (*Marmota caudata aurea*) in Pakistan. Pages 82–85 *in* K. B. Armitage and V. Y. Rumiantsev, eds. *Holarctic Marmots as a Factor of Biodiversity*. ABF Publishing House, Moscow.

Blumstein, D. T., J. Steinmetz, K. B. Armitage, and J. C. Daniel. 1997. Alarm calling in yellow-bellied marmots: II. The importance of direct fitness. *Animal Behaviour* 53:173–84.

Brody, A. K., and K. B. Armitage. 1985. The effects of adult removal on dispersal of yearling yellow-bellied marmots. *Canadian Journal of Zoology* 63:2560–64.

Calhoun, J.B. 1952. The social aspects of population dynamics. *Journal of Mammalogy* 33: 139–59.

Caswell, H. 2001. *Matrix Population Models: Construction, Analysis, and Interpretation*. Sinauer Associates, Sunderland, MA.

Cohas, A., N. G. Yoccoz, A. Da Silva, B. Goossens, and D. Allainé. 2006. Extra-pair paternity in the monogamous alpine marmot (*Marmota marmota*): The roles of social setting and female mate choice. *Behavioral Ecology and Sociobiology* 59:597–605.

Davis, E. B. 2005. Comparison of climate space and phylogeny of *Marmota* (Mammalia: Rodentia) indicates a connection between evolutionary history and climate preference. *Proceedings of the Royal Society B* 272:519–26.

Dobson, F. S., and M. K. Oli. 2001. The demographic basis of population regulation in Columbian ground squirrels. *American Naturalist* 158:236–47.

Downhower, J. F., and K. B. Armitage. 1981. Dispersal of yearling yellow-bellied marmots. *Animal Behaviour* 29:1064–69.

Errington, P. L. 1956. Factors limiting higher vertebrate populations. *Science* 124:304–7.

Hamilton, W. D. 1964. The genetical evolution of social behaviour. *Journal of Theoretical Biology* 7:1–52.

Inouye, D. W., B. Barr, K. B. Armitage, and B. D. Inouye. 2000. Climate change is affecting altitudinal migrants and hibernating species. *Proceedings of the National Academy of Sciences (USA)* 97:1630–33.

Lenti Boero, D. 1999. Population dynamics, mating system and philopatry in a high altitude colony of alpine marmots (*Marmota marmota* L.). *Ethology Ecology & Evolution* 11:105–22.

Oli, M. K. 2003. Hamilton goes empirical: Estimation of inclusive fitness from life-history data. *Proceedings of the Royal Society of London B* 270:307–311.

Oli, M. K., and K. B. Armitage. 2003. Sociality and individual fitness in yellow-bellied marmots: Insights from a long-term study (1962–2001). *Oecologia* 136:543–50.

———. 2004. Yellow-bellied marmot population dynamics: Demographic mechanisms of growth and decline. *Ecology* 85:2446–55.

———. 2008. Indirect fitness benefits do not compensate for loss of direct fitness in yellow-bellied marmots. *Journal of Mammalogy* 89:874–81.

Oli, M. K., N. A. Slade, and F. S. Dobson. 2001. Effect of density reduction on Uinta ground squirrels: Analysis of life table response experiments. *Ecology* 82:1921–29.

Ozgul, A., K. B. Armitage, D. T. Blumstein, and M. K. Oli. 2006a. Spatiotemporal variation in survival rates: Implications for population dynamics of yellow-bellied marmots. *Ecology* 10:1027–37.

Ozgul, A., K. B. Armitage, D. T. Blumstein, D. H. Van Vuren, and M. K. Oli. 2006b. Effects of patch quality and network structure on patch occupancy dynamics of a yellow-bellied marmot metapopulation. *Journal of Animal Ecology* 75:191–202.

Ozgul, A., M. K. Oli, L. E. Olson, D. T. Blumstein, and K. B. Armitage. 2007. Spatiotemporal variation in reproductive parameters of yellow-bellied marmots. *Oecologia* 154:95–106.

Salsbury, C. M., and K. B. Armitage. 1994. Home-range size and exploratory excursions of adult, male yellow-bellied marmots. *Journal of Mammalogy* 75:648–56.

Schwartz, O. A., and K. B. Armitage. 1980. Genetic variation in social mammals: The marmot model. *Science* 207:665–67.

———. 2005. Weather influences on demography of the yellow-bellied marmot (*Marmota flaviventris*). *Journal of Zoology, London* 265:73–79.

Schwatrz, O. A., K. B. Armitage, and D. Van Vuren. 1998. A 32-year demography of yellow-bellied marmots (*Marmota flaviventris*). *Journal of Zoology, London* 246:337–46.

Svendsen, G. E. 1974. Behavioral and environmental factors in the spatial distribution and population dynamics of a yellow-bellied marmot population. *Ecology* 55:760–71.

Svendsen, G. E., and K. B. Armitage. 1973. An application of mirror-image stimulation to field behavioral studies. *Ecology* 54:623–27.

Van Vuren, D., and K. B. Armitage. 1991. Duration of snow cover and its influence on life-history variation in yellow-bellied marmots. *Canadian Journal of Zoology* 69:1755–58.

———. 1994. Survival of dispersing and philopatric yellow-bellied marmots: What is the cost of dispersal? *Oikos* 69:179–81.

Williams, G. C. 1966. *Adaptation and Natural Selection*. Princeton University Press, Princeton, NJ.

Wilson, E. O. 1975. *Sociobiology*. Harvard University Press, Cambridge, MA.

Woods, B. C., and K. B. Armitage. 2003a. Effect of food supplementation on juvenile growth and survival in *Marmota flaviventris*. *Journal of Mammalogy* 84:903–14.

———. 2003b. Effects of food addition on life history of yellow-bellied marmots. *Oecologia Montana* 12:1–8.

Wynne-Edwards, V. C. 1962. *Animal Dispersion in Relation to Social Behaviour*. Oliver and Boyd, Edinburgh.

———. 1965. Self-regulating systems in populations of animals. *Science* 147:1543–48.

8 *The Aleutian Archipelago*

ADDRESSING THE FUNCTIONAL IMPORTANCE OF CARNIVORY THROUGH VARIATION IN SPACE AND TIME

James A. Estes

Figure 8.0. Research vessel in the Aleutian Islands. Photograph by James A. Estes, 2003.

Abstract

The ecology of place mind-set promotes field studies designed to explore questions for which a given place is particularly well suited. Islands, because of their isolation and unique individual histories, can often provide powerful insights into the pattern and process of ecology and evolution. My colleagues and I have used these simple properties of islands to explore the ecological importance of mammalian carnivory. This chapter is a synoptic account of 36 years of research in the Aleutian archipelago that addresses the direct and indirect effects of three different predators and their prey: sea otters and sea urchins, killer whales and coastal-living marine mammals, and arctic foxes and seabirds. In each case the predators reduce prey abundance by one to two orders of magnitude, in turn setting off a diverse array of indirect effects that permeate their respective interaction webs. These indirect effects involve both top-down and bottom-up forcing processes, in some cases establishing important linkages across ecosystems. Striking landscape-level effects are an ultimate manifestation of mammalian carnivory.

Introduction

In the quest to understand how nature works, there are at least three good reasons for taking the time that is necessary to know a place well. One simply is that learning the essentials of natural history—what a species eats and by whom it is eaten, when and where it reproduces, how individuals move through space and time, and the numerous factors that influence mortality and reproduction—itself takes time. Another is that the behavior of every ecosystem is dictated by interactions among species on the one hand and between these species and their physical environment on the other. Relatively long periods of time are required for even the most clever and perceptive of human minds to begin to grasp the dynamic nature of these immensely complex interaction webs. Finally, time inevitably brings change, and observing change through time can provide insights into interaction-web dynamics that are difficult to obtain in any other way. Time spent in one place is not a prescription for understanding, of course; but such place-based studies, when conducted purposefully and thoughtfully, can be used to answer ecological questions that are difficult to resolve in more traditional ways. One such question concerns the functional roles of large carnivores.

There are a number of fairly obvious obstacles to understanding the functional roles of large carnivores. For one, although large carnivores once were common nearly everywhere, they have been among the first species to disappear under the growing spectrum of human influence (Ray et al. 2004), and thus are now rare or absent in most places. The process of predation is hard to observe and even more difficult to understand under such circumstances. Another difficulty derives from the emotional connections that people have for large animals, especially mammals. These emotions have led to ethical and legal prohibitions against many kinds of research. A final obstacle emerges from the culture of ecology, where the prevailing mind-set has long been that top-down forcing processes are relatively unimportant compared with bottom-up forcing and physical change. As a result, large carnivores are often viewed more as passengers than as drivers in ecosystem dynamics. The intense debate that has recently emerged over the relative importance of ocean change versus killer whale predation in the collapse of Steller sea lions in southwest Alaska, which I will discuss below, is testimony to these different cultures. Accordingly, our lack of understanding of large carnivores is due in part to a lack of looking.

This chapter provides an overview of the work my colleagues I have done in the remote Aleutian archipelago over the past 36 years. In order to emphasize the value of time spent in this place, I have chosen to write the chapter as a personal chronology of some of the things I have seen and learned. I will begin by describing the region, emphasizing various physical and biological features that make the islands of this archipelago so well suited for large-scale spatial analyses. This introduction is followed by overviews of three case studies, one of which concerns the functional role of sea otters (*Enhydra lutris*) in the coastal marine system, the second of which concerns the influence of killer whales (*Orcinus orca*) on other species of coastal-living marine mammals, and the last of which concerns the functional role of arctic foxes (*Alopex lagopus*) in the adjacent terrestrial realm. Understanding in each case is based on observing and interpreting patterns of variation in space and time. Over the course of the nearly four decades I have spent in the Aleutian archipelago, opportunities have arisen to view this system in a dynamic state through both purposeful retrospection and serendipity. I will end with an overview of what we have learned about carnivory and interaction-web dynamics from this long-term place-based study, and a brief discussion of the broader relevance of our findings to ecology and conservation.

The Aleutian Archipelago

The Aleutian archipelago extends westward from the Alaska Peninsula to the Commander Islands, thereby loosely joining Asia with North America and defining the boundary between the North Pacific Ocean and the Bering Sea (figure 8.1). The islands were formed during the late Eocene or early Oligocene by tectonic uplift and volcanism (Gard 1978, Lallemant and Oldow 2000). The easternmost Fox Islands lie on the North American continental shelf, and as such are land bridge islands; the remaining islands in the Aleutian archipelago are oceanic.

Climate has changed appreciably over the geological history of the Aleutians. Subtropical conditions prevailed at the time of their genesis, but substantial cooling began at about the end of the Miocene with the onset of the most recent Glacial Age (Addicott 1969, Cervato and Burckle 2003, Barron 2003). The modern climate is subarctic maritime (Armstrong 1977). Surface sea temperatures range between a summer high of about 7°C and a winter low of about 1° C. Air temperatures commonly range from winter lows of about −5° C to summer highs of about 20° C. Prevailing ocean currents are dominated by the westward-flowing Alaskan Stream to the south and the eastward-flowing Aleutian North Slope Current to the north (Stabeno et al. 1999). Nutrient-rich seawater and high summer solar radiation result in

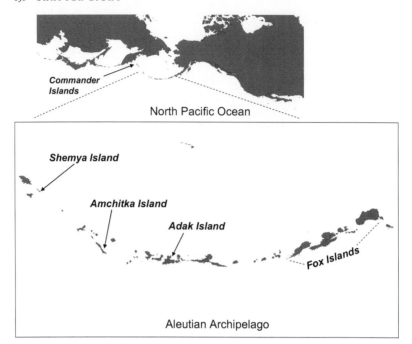

Figure 8.1. The North Pacific Ocean, showing the Aleutian archipelago and places referred to in the text.

elevated primary productivity (Walsh et al. 1989), which in turn supports spectacular numbers of marine birds, mammals, and fishes. Like most other tundra ecosystems, the land is intrinsically nutrient-impoverished and thus relatively unproductive. In contrast with most tropical islands and archipelagos, climate (temperature and precipitation) is remarkably uniform within and among the Aleutian Islands, especially at lower elevations.

Also in contrast with most tropical oceanic islands (e.g., Galápagos and Hawaiian), terrestrial and coastal marine biotas of the Aleutian archipelago are highly impoverished and contain few endemic species—qualities that likely resulted from glacially induced extinctions during the Pleistocene and reinvasions from Asia or North America following the last glacial recessions in the early Holocene. Exotic species are also rare, with several important exceptions, especially in the terrestrial realm. Arctic foxes were introduced to most of the Aleutian Islands to augment the Pacific maritime fur trade (Bailey 1993). The history of the introduction and extinction of this exotic species is central to the design of our terrestrial ecosystem studies, as is explained later in the chapter. Norway rats (*Rattus norweigicus*) have become established on some of the islands following shipwrecks and World War II military activity. Our terrestrial studies have avoided these rat-infested islands. The species composition of a few wide-ranging vertebrates, like sea birds, also varies considerably among islands depending on local oceanographic conditions (Springer et al. 1996). Otherwise, species composition on land and in the sea is remarkably consistent across the archipelago.

One might fairly ask why I chose the Aleutian archipelago over some less remote and more hospitable location as a place to work. One reason is that the islands are intrinsically well suited for comparative studies of exogenous disturbances. There is an abundance of islands and thus almost limitless opportunity for replication. Furthermore, the physical and biological homogeneity of these islands, the lack of human activity (at least in recent years and in the immediate vicinity of the islands), and the general absence of exotic species minimize the confounding effects that commonly plague studies based on inter-island comparisons. Another reason for working in the Aleutians is that historical accidents have caused the distribution and abundance of several mammalian carnivores to vary among the islands. This useful feature, coupled with the availability of a large number of otherwise physically and biologically similar islands, provides simple yet powerful opportunities to infer the influences of these predators on their associated ecosystems. Another personal reason for venturing into the Aleutians is simply a love for the place. The region provides at least a semblance of true wilderness. There are no people, no cell phones, no roads, no houses, no airplanes—just rock, tundra, sea, and the wild things that live there. I have been drawn back to this place not only to learn, but to nourish my spirit and my soul.

Sea Otters and Kelp Forests

Sea otters were extraordinarily abundant in the Aleutian archipelago when European explorers first visited this region. Reports of these riches following the Bering Expedition's return to Russia in 1742 set off the Pacific maritime fur trade, thus establishing an important historical context for our research. Together with northern fur seals (*Callorhinus ursinus*), sea otters were intensively hunted during the eighteenth and nineteenth centuries. By the beginning of the twentieth century both species had become so rare that an international treaty was established to protect those that remained.

More than a hundred thousand sea otters lived in the Aleutian archipelago before the fur trade (Burn et al. 2003). By the end of the fur trade only a few hundred remained (Kenyon 1969). However, the survivors somehow coalesced into two or three remnant colonies, and with protection from human exploitation their numbers grew. Although the protected populations increased at about 20% yr^{-1}, the species' theoretical r_{max} (Estes 1990), range expansion to the many unoccupied islands was inhibited by deep, broad ocean passes, and the sea otters' combined needs of diving to the sea floor to obtain their prey and feeding frequently. The result was a "natural experiment"—populations recovered to high densities at some islands while remaining absent from others.

Early naturalists such as Olaus Murie and Karl Kenyon recognized these patterns in the abundance of sea otters and understood their causes (Murie 1959, Kenyon 1969). Kenyon also recognized that sea otters had a significant limiting influence on their benthic invertebrate prey. However, little thought apparently had been given to the broader effects of this predator-prey interaction or to the question of how the coastal system might serve to inform the larger issues of top-down influences on food web dynamics.

In 1970, when I began working in the Aleutian Islands, I was interested in community studies but had little sense of how to proceed beyond a vague notion that the important questions somehow must center on the effects of habitat quality and bottom-up forcing processes—that is, how otters were influenced by their ecosystem. Although the now-classic papers by Hairston et al. (1960) and Paine (1966), purporting significant predator-induced effects on ecosystems, were already in the literature, I did not grasp their importance at the time. My thinking and early research efforts thus focused on a single place—Amchitka Island—because sea otters were abundant there, and I saw little reason to look elsewhere. Then, in summer of 1971 I met Bob Paine, who turned my thinking toward the question of how sea otters might be influencing this ecosystem. This was a revelation. I knew that the distribution of sea otters in the Aleutian Islands had been fragmented by the fur trade. Hence, contrasting islands with and without otters suddenly became an important and exciting thing to do. Shortly thereafter, several colleagues and I visited Shemya Island (from which otters were absent), intent on making that contrast. We knew beforehand that sea otters ate sea urchins and that sea urchins ate kelp. We also knew from a recently published experimental study (Paine and Vadas 1969) that kelp stands could be reduced or eliminated by sea urchin grazing, so it was easy to imagine that the loss of sea otters might cause a reduction or loss of kelp.

The sea otter's role in maintaining the kelp forest was immediately obvious to us as we first glimpsed Shemya Island's shoreline (Estes and Palmisano 1974). Whereas the shallow coastal habitats at Amchitka (where otters were present) were characterized by dense kelp stands, similar habitats at Shemya were characterized by large and abundant sea urchins and a dearth of kelp. We thus concluded that by limiting sea urchins, sea otters indirectly maintained a system in which kelps were abundant. This functional role provided early empirical support for the "green world" hypothesis of

Hairston et al. (1960), an idea that subsequently morphed into the modern concepts of top-down forcing (Hunter and Price 1991) and trophic cascades (Paine 1980, Carpenter and Kitchell 1993).

One might ask whether it was really necessary to view the kelp forest ecosystem in the perturbed state (i.e., with sea otters removed) to understand the species' ecological role. With information on the demography, nutritional requirements, digestive efficiencies, and nutritional value of kelps, sea urchins, and sea otters, might not it have been possible to deduce this role solely from the study of a place like Amchitka Island where sea otters were abundant? The answer is no. It is possible to calculate the size-specific loss rates of sea urchins to sea otter predation based on the abundance and foraging behavior of sea otters, although the recruitment dynamics of the sea urchins are more complex, as these apparently involve both settlement from the plankton and the movements of adults from deep to shallow water. Predicting the indirect effect of sea otter predation on urchin-kelp dynamics presents an even more fundamental problem, because of density-dependent variation in herbivore behavior. When kelp is abundant in the system (as it was at Amchitka in the early 1970s) the avenue of nutritional flux from kelps to sea urchins is via detritus (Duggins et al. 1989). The highly productive and rapidly growing kelps produce large quantities of particulate organic carbon through tissue sloughing distal to their intercalary meristems. This material, visually evident as mounds of beach wrack, falls to the seafloor as "drift algae" where it is consumed by herbivores and detritivores, including the sea urchins. Under these circumstances, sea urchins are sessile, "sit-and-wait" consumers whose nutritional needs are provided by the kelps at little or no cost to the living plants. However, when sea otters are removed from the system and sea urchins become larger and more abundant, the sea urchins exhaust this detrital food resource and therefore switch to a strategy of mobile searching and consumption of living plant tissue. It is this behavioral switch that results in overgrazing and the deforested community state that characterizes systems lacking sea otters (Konar and Estes 2003). These dynamics would have been difficult or impossible to predict by simply applying demographic/energetic analyses to a static system with abundant sea otters. Only by contrasting islands with and without sea otters did the species' ecological importance become clear.

Further study in the 1980s revealed that the aforementioned dichotomy between kelp-dominated and deforested ecosystems was a broadly recurrent and highly predictable feature of habitats with and without sea otters in the Aleutian archipelago (Estes and Duggins 1995). The dynamic interactions among predators, herbivores, and plants provided one of the earliest examples of top-down forcing and trophic cascades—processes now known to occur in diverse ecosystems (Pace et al. 1999). Trophic cascades, however,

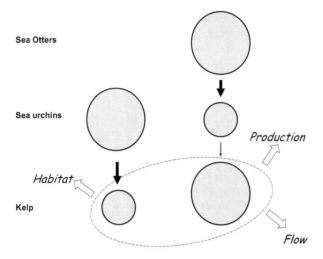

Figure 8.2. Schematic diagram of the sea otter/urchin/kelp trophic cascade. The food chain on the left is for islands with otters absent, and on the right for those with otters abundant. Circle size indicates relative biomass. Large dark arrows indicate strong interactions; smaller grey arrows indicate weak interactions. These interactions result in three classes of indirect effects on other interaction-web processes: altered habitat, production, and water flow.

occur amidst numerous other species that are linked together through an almost unfathomable complexity of interaction-web pathways. We have long been interested in how these linkages influence other species and ecosystem processes. Demonstrating or even inferring these processes in systems that are static, as Amchitka Island seemed to be in the early 1970s, is nearly impossible because of the immense biological and physical complexity. Here again, contrasts of islands with and without sea otters have provided a reasonably clear view of some of the ways in which sea otter predation impinges upon complex interaction-web pathways.

My colleagues and I have approached the task of understanding these complex interactions by first recognizing that indirect effects of the sea otter–induced cascade can be separated into three associated processes: increased production, altered physical habitat, and reduced water flow (figure 8.2). The underlying logic for this categorical scheme is fairly straightforward. Kelps and other macroalgae in high latitude oceans grow at extremely high rates (Steneck et al. 2003) because none of the four essential ingredients for photosynthesis—inorganic carbon, water, light, and nutrients—is limiting, at least during summer. It was thus easy to imagine that systems with and without abundant stands of kelp would differ considerably in total production. In a manner analogous to trees in terrestrial ecosystems, kelps provide three-dimensional structure to seafloor ecosystems, in turn potentially providing requisite habitats for other species. Finally, kelps project large surface areas into a viscous water mass and the resulting surface tension attenuates water movement from waves and currents in a manner again analogous to the attenuation of wind on a forest floor (Duggins 1988).

These three processes—altered production, habitat structure, and flow—have strong influences on the distribution, abundance, and behavior of various species. For example, the increased production in otter-dominated systems generates a threefold to fivefold increase in the growth rate of filter-feeding barnacles and mussels (Duggins et al. 1989). Loss of kelp forest habitat causes roughly an order of magnitude reduction in the abundance of rock greenling (*Hexagrammus lagocephalus*), a dominant kelp forest fish in the Aleutian archipelago (Reisewitz et al. 2006). The diet and foraging behavior of glaucous-winged gulls (*Larus glaucescens*) differ fundamentally between sea otter-dominated kelp forests and otter-free deforested habitats (Irons et al. 1986). Gulls in otter-dominated habitats are largely piscivorous, whereas in otter-free habitats they feed predominantly on intertidal invertebrates. Even the diets of bald eagles (*Haliaeetus leucocephalus*), another apex predator, appear to be influenced by sea otters (Anthony et al. 2002). Newborn sea otter pups and rock greenling are important prey of nesting bald eagles at sea otter–dominated islands, whereas eagles consume fewer of these prey and more seabirds at otter-free islands. Many other species and interaction-web pathways are known or believed to be influenced by the otter-urchin-kelp trophic cascade (Estes 1996, 2005; Estes et al. 2004).

Killer Whales and Coastal Marine Mammals

All of the aforementioned patterns and processes were explored and discovered from prospective visions and tests of a priori hypotheses. Our questions in each case grew from a rudimentary knowledge of place-based natural history, process-related information from work done by others in different places, and expectations of how the interaction-web dynamics should vary between otter-dominated and otter-free ecosystems. Our perspectives and questions were also based on the implicit assumption that kelp-forest ecosystems were closed to important interaction-web linkages with either the land or the open sea, although in truth that view was founded more on convenience than on knowledge. We understood that dispersive life stages of many plants, invertebrates, and fishes were planktonic, and thus capable of long-distance movements. We understood that nutrients, sediments, and other materials flowed from land to the sea. We even had concrete evidence that oceanic production fueled the coastal ecosystem—as, for example, when an episodic spawning migration of smooth lumpsuckers (*Aptocyclus ventricosus*) subsidized the sea otter population at Amchitka Island in the early

1990s by suddenly providing a majority of the caloric input to their winter diet (Watt et al. 2000). For our broader purposes, however, it was convenient to think of each island as a closed system, since the methodological foundation of our "experiment" was based on differences between islands with and without sea otters, or on changes through time as otter-free islands were recolonized. Although these views and approaches served us well through the 1970s and 1980s, they began to fail us in the 1990s. Events that have since occurred cannot be understood if island kelp forests are closed systems, or by employing a prospective vision of pattern and change.

In 1991 several colleagues and I set about to characterize the behavior and demography of sea otters at Amchitka Island. Technology by then allowed us to surgically implant radio transmitters in sea otters and continuously follow the tagged individuals for several years. The abundant population at Amchitka Island was believed at the time to be at carrying capacity, limited by nutritional resources. This assessment was based on information from earlier studies, including high winter-spring mortality (Kenyon 1969, Estes 1977), poor body condition, a diverse diet (Estes et al. 1981), a high proportion of time invested in foraging (Estes et al. 1982), and a generally high and stable population density during the preceding several decades (Kenyon 1969, Estes 1977). Our goal was to instrument a large number of animals and then describe demographic and behavioral features of the population through longitudinal records of individual diet, activity, movement, reproductive success, and mortality.

The results of these studies provided the first hints that the Aleutian system wasn't so simple and well behaved as I had imagined. An island-wide survey of sea otters at Amchitka in 1993 provided a somewhat lower count (about 4000) than expected (>6000; Estes 1977), although I discounted this because of the singular nature of the data point and the notorious uncertainty of wildlife population surveys. However, numerous other findings also were counter to my initial expectations. Perhaps the most striking was the almost total absence of beach-cast sea otter carcasses between late winter and early spring, a period in which starvation and stranding had been common earlier (Kenyon 1969, Estes 1974). We reconciled this and several other anomalies by believing that the episodic influx of lumpsuckers subsidized an otherwise food-limited sea otter population during this critical period (Watt et al. 2000). The time otters spent foraging was less than had been measured earlier using a scan-sampling approach (Estes et al. 1982). We reconciled this difference by imagining that the glut of lumpsuckers had reduced the pressure on sea otters to feed so intensively, and also by convincing ourselves (in the absence of any evidence) that the earlier scan-sample data were somehow flawed, even though other information strongly suggested that scan-sampling/telemetry comparisons should be biased in the opposite direction. Mortality rates of the instrumented sea otters were higher than expected from population models based on life table statistics and the assumption of population stability (Monson et al. 2000). We reconciled this finding by presuming that the radio-transmitter failure rates were higher than expected (a conclusion for which there was also empirical support), and thus that the calculated mortality rate was too high. Perhaps the most perplexing inconsistency was in the sea otters' diet. Earlier studies had shown that roughly 60% of the diet (by mass) was comprised of kelp forest fishes, whereas these fishes were rarely eaten in the early 1990s (Watt et al. 2000). I was unable to explain this stunning disparity, so I simply assumed that the diet had changed for some unknown reason. None of our imagined explanations for these inconsistencies was very compelling, and in retrospect they were all wrong.

It had become evident by the mid-1990s that sea otter numbers in the western Aleutian Islands were in rapid decline. The first clear indication was seen in a time series of population counts at Adak Island, about 250 km east of Amchitka. By 1996 the counts at Adak had declined to about 10% (871) of the pre-decline population estimate (about 8500). We returned to Amchitka in the spring of 1997 to discover a similar decline. By 1997 sea otters had become so depleted that the coastal ecosystem shifted from kelp forests to sea urchin barrens across large areas of the Aleutian archipelago. This shift provided additional support for the sea otter's keystone role (sensu Paine 1980) in the coastal ecosystem, demonstrating also that populations well in excess of 10% of carrying capacity were required to maintain that function (Soulé et al. 2003). The generally synchronous nature of the sea otter decline across the Aleutian archipelago (Doroff et al. 2003) indicated that it was somehow connected to a larger-scale process. Our tagging and telemetry data from Amchitka and Adak demonstrated that fertility and reproduction had not changed during the decline, that sea-otter movement patterns were so limited that redistribution could not account for the decline, and thus that elevated mortality must have been the culprit.

Our efforts shifted toward identifying the cause or causes of this elevated mortality. Increased sea urchin densities (Estes et al. 2004), improved sea otter body condition (Laidre et al. 2006), and the absence of stranded carcasses convinced us that nutritional limitation was not the explanation. Similarly, the search for diseases or environmental toxins proved fruitless (Estes et al. 1998, Hanni et al. 2003). At the same time we were perplexed by the persistence of abundant sea otters in Clam Lagoon, a shallow embayment on Adak Island with a narrow and shallow connection to the open sea that apparently was rarely, if ever, entered by killer whales. In addition, we learned that the number of killer whale sightings in coastal waters had increased markedly, and attacks on sea otters were first seen in 1991 (Hatfield et al. 1998). Although initially I was intrigued by the possibility that killer whale predation had caused the sea otter decline, I discounted this explanation because of the small number of observed attacks. However, when we calculated the expected number of observed deaths by killer whale attacks from the area of instantaneous sightability (about one km to either side of the observer) and the amount of time that had been spent in the field watching (about 4,000 hrs), we discovered that this value (5.05 attacks) and the observed number of attacks (six) were remarkably similar. We also doubted that the area supported enough killer whales to consume the tens of thousands of sea otters that would have to have been eaten to drive the decline. However, when we computed the caloric value of a sea otter and compared it with the energy requirements of a killer whale (see Williams et al. 2004 for details), we were astonished to discover that all of the missing sea otters could have been eaten by just 3.6 killer whales. Based on this information and the resulting weight of evidence, we concluded that the sea otter population decline probably had been driven largely or exclusively by increased killer whale predation (Estes et al. 1998).

Why did this happen? Here our understanding of the forcing processes becomes less certain, because they transcend the scale of coastal ecosystems and individual islands and require an understanding of history that has not been well chronicled. Nonetheless, a growing body of evidence provides some clues. The sea otter decline roughly coincides geographically with an earlier population collapse first of harbor seals (*Phoca vitulina*) and then Steller sea lions (*Eumetopias jubata*). Thus, an intriguing potential explanation is that killer whales expanded their diet to include sea otters as their preferred pinniped prey became rare. The similar trajectories, common geographical ranges, and sequential nature of these declines, coupled with the well-known fact that pinnipeds are eaten by killer whales (Jefferson et al. 1991, Baird and Dill 1995, Ford et al. 1998, Ford 2002), supports the switching hypothesis and suggests that the pinniped declines also may have been caused by increased killer whale predation (National Research Council 2003). This explanation is consistent with both energetic/demographic analyses (Williams et al. 2004) and an initial behavioral analysis (Mangel and Wolf 2006). A long-favored alternative hypothesis is that the pinniped declines were driven by food limitation, which in turn resulted from competition with fisheries or oceanographic change. I think the evidence for this is thin at best. Nonetheless, it would be unfair not to acknowledge that some uncertainty and a great deal of controversy currently surround the issue.

If the population collapses of sea otters and pinnipeds are interconnected—and if killer whale predation was responsible for the pinniped declines, as appears to be true for sea otters—what was the ultimate cause? Like nearly everyone else, I initially assumed that the answer somehow related to ocean climate change and overfishing. However, another fortuitous professional association, this time with Alan Springer, turned my thinking away from this widely held view. Alan had been pondering many of these same issues and had already recognized several important inconsistencies with the nutritional limitation hypothesis. He also knew that killer whales preyed on both pinnipeds and large whales, and that the abundance of large whales had been significantly reduced by industrial whaling. With help from several other colleagues, we grew to believe that whaling was indeed an important driver in this megafaunal collapse (Springer et al. 2003). This view is based on the facts that the historical biomass of great whales in the North Pacific Ocean and southern Bering Sea exceeded that of pinnipeds by almost ten times, and that the whaling-induced collapse of great whale populations predated the onset of the pinniped declines. A number of biologists who study large whales doubt this explanation because they believe there is little evidence that large whales were ever an important source of nutrition for killer whales. However, there is a large and growing body of evidence that killer whales commonly attack, kill, and consume various species of great whales (Springer et al. 2008), and our further analyses strongly suggest that the killer whale/marine mammal predator-prey system was unsustainable in the absence of an abundant large whale fauna (Estes et al. 2009). Whitehead and Reeves (2005) recently pointed out that killer whales may have been especially drawn to whale carcasses during the era of modern industrial whaling, because the exploding harpoons that were used to dispatch the whales advertised the time and location of kills, and the injection of gas to float the carcasses prevented them from sinking beyond the foraging range of a killer whale. Whitehead and Reeves's (2005) view was that 75 years of modern industrial whaling led killer whales to depend even more strongly on large whales—a relationship that ended abruptly with the cessation of commercial whaling.

Despite the various uncertainties, my time in the Aleutian Islands has led me to see the dynamics of predator-prey systems in different ways. I now view predators such as sea otters as intermediate players in the broader fabric of ecosystem dynamics, in the sense that they not only have diverse influences on their local systems but are influenced in important ways by their surrounding environment. The sea otter collapse also has led me to appreciate the vastly larger scales of interaction-web connectivity than I could have even imagined at the onset of my work in the Aleutians. I can now see that events in the open sea that began nearly a century ago must be included as significant parts of the sea otter/kelp forest interaction web.

Arctic Foxes and Terrestrial Ecosystems

As I look back over my years spent in the Aleutian Islands and reflect on the way my views of the system have changed, I recognize that serendipitous events in my own history started me down avenues of thought and discovery that otherwise might not have been explored. The earlier described friendships with Bob Paine and Alan Springer were two such events. Another was a discussion I chanced to have about 10 years ago with Vernon Byrd, a longtime friend and a biologist with the Alaska Maritime National Wildlife Refuge. As we shared stories and perspectives from our times in the Aleutians, I became intrigued by Vernon's ideas of how introduced arctic foxes may have influenced the terrestrial side of these island ecosystems.

As mentioned earlier in this essay, foxes were introduced to many islands in Alaska (including most of the Aleutian archipelago) in an effort to support the maritime fur trade (Bailey 1993). A few islands remained fox-free, either because they were too difficult to access or because the introductions failed. Here again, the fortuitous existence of islands with and without foxes created an opportunity for understanding their ecological roles. We had some ideas of what to look for at the onset. We knew that foxes were capable of greatly reducing the distribution and abundance of ground-nesting seabirds; we knew that marine ecosystems in this part of the world were nutrient-rich and highly productive, whereas the terrestrial ecosystems were intrinsically nutrient-impoverished (despite being volcanic-soil based); and we knew that seabirds, because of their mutual dependency on the land for nesting and the sea for feeding, were capable of vectoring nutrients between these habitats.

In order to explore the consequences of these potential interaction-web pathways, we selected 18 islands, 9 of which supported foxes and 9 of which were fox-free. We then established a uniform grid of points over each island, and at each point we characterized the plant communities as well as the nutrient profiles in plants and soils. We were particularly interested in the degree to which the nutrients were derived from the sea, so we measured both the concentrations of nitrogen and phosphorous and the stable isotopic composition of nitrogen. In addition, we measured the nitrogen isotopic composition of four consumer species: a fly, a spider, a slug, and a passerine bird.

The fox-free and fox-infested islands were strikingly different. Nesting seabird densities were more than two orders of magnitude greater on the fox-free islands. The flora on fox-free islands was dominated by graminoids (mostly *Lymus*, *Carex*, and *Festuca* spp.) and umbelliferous plants (especially cow parsnip, *Heracleum lanatum*), whereas fox-infested islands were characterized more by low-lying forbs and dwarf shrubs. These floral differences were immediately evident, as the fox-infested islands were easy to traverse by foot, whereas walking through the thigh- to waist-deep vegetation on the fox-free islands was much more difficult. Nutrient concentrations were substantially elevated in both soils and plants at fox-free islands. In addition, $\delta^{15}N$ values were elevated in plants and soils on fox-free islands, thus indicating that the nitrogen was coming from higher trophic levels (i.e., marine sources; Dawson et al. 2002, Schindler and Lubetkin 2004), while the $\delta^{15}N$ signatures from similar materials on fox-infested islands was indicative of internal derivation via nitrogen fixation (Croll et al. 2005, Maron et al. 2006). These different $\delta^{15}N$ signatures also occurred in each of the consumer species, thus indicating that the nutrient effects were being passed upward through the terrestrial food web (Croll et al. 2005).

We were not surprised to find that foxes dramatically reduced ground-nesting seabird populations. This effect had been known for decades and was the primary motivation behind the Maritime Refuge's program to rid the islands of introduced foxes (Byrd et al. 1994, Ebbert and Byrd 2002). We were more intrigued by the indirect effects of this predator-prey interaction, because of several interesting departures from the structure of Hairston, Smith, and Slobodkin's (1960) influential model (HSS), which has served for decades as the conceptual template for predator-herbivore-plant interactions. Two points are especially noteworthy. One is that in contrast with the HSS model and its subsequent elaborations by Fretwell (1987) and Power (1992), which involve a systematic cascade of top-down forcing processes across successively lower trophic levels, the fox-seabird system includes both top-down and bottom-up forcing. The other point is that this interaction web also includes significant linkages across what might otherwise be seen as discrete and very different ecosystems—the land and the coastal ocean. The late Gary Polis and his colleagues had identified similar linkages between ecosystems (Polis et al. 1997), although the role of predation was novel to our study.

Some General Conclusions

An ecologist's view of place can be small or large, depending upon his or her perspectives and needs. It seems to me that this view is typically dictated by three factors—characteristics of the environment, characteristics of the species of interest, and the philosophical approach taken in studying them. For instance, one's perspective of place should tend to be small in areas where regional diversity varies over short distances, but much larger in areas where it varies over greater distances. Similarly, one's perspective of place might be small if the biological focus is on sessile or weakly motile organisms, and relatively large for species that are highly mobile. Finally, people whose study methods are based on manipulative experiments may have a smaller view of place than those whose approach is more comparative or descriptive in nature, for the simple reason that it is difficult to manipulate organisms or resources over large areas.

My perception of place in the Aleutian archipelago is very large, in part because of the area's large-scale homogeneity, in part because my interests have involved highly mobile species (carnivorous mammals), and in part because the most interesting scale of biological variation has been among islands. Looking back, the decision to take this large-scale approach had both benefits and costs. The costs have been forfeiture in the understanding of detail. I simply couldn't afford to stay in one particular place long enough to know it well. And in the end, some of my perceptions of large-scale variation in space and time were still inadequate to encompass important interaction-web processes. The main benefit of this large-scale approach is that it has given me a view of interaction-web process that probably would have been difficult to obtain in most other places or from a methodological approach founded on a smaller-scale view of place. I will conclude by elaborating on these points.

I see space and time as being inextricably linked within the discipline of ecology, and thus contend that some knowledge of history is essential in order to make proper sense of spatial variation. This is certainly true for the three examples outlined above. For the case of sea otters and kelp forests, it would have been impossible to know where to look, or even how to interpret what we saw, without first knowing the consequences of the fur trade. The same general point can be made for foxes and seabirds. It would have been difficult to know where to look and how to interpret what we saw without first knowing something of the history of fox farming in Alaska. Our understanding of the ecological role of killer whale predation, such as it is, arose from a long-term presence in the region, and from paying attention to the history of species consumed by killer whales. For both sea otters/kelp forests and fox/seabirds, we might have attempted to ascribe some other environmental cause to the spatial variation if we had lacked an appreciation of context provided by an understanding of history.

It is difficult to overemphasize the power and importance of retrospection in our work. Although in this chapter I have not discussed our deeper and larger explorations into time and space, these too have been revealing. For example, by coupling a knowledge of modern ecological process with an assessment of the remains in Aleut kitchen middens, we were able to infer that sea otters were rare or absent from at least parts of the Aleutian archipelago during the majority of time that aboriginal peoples occupied the region (Simenstad et al. 1978). The most likely explanation for this pattern is that early humans overexploited sea otters—an idea that fits well with a growing body of evidence for prehistoric overkill (Barnosky et al. 2004, Martin 2005). An even broader and deeper look at space and time has shown us how the key elements of the sea otter–kelp forest system was assembled, providing interesting biogeographic and evolutionary perspectives on the modern-day system. In brief, these expanded views indicate that sea otters, urchins, and kelps arose and evolved together in the North Pacific Ocean, beginning at about the onset of the most recent glacial age. By contrasting the North Pacific system with its Southern Hemisphere analogues (which seem to have arisen independently, but without a predator of comparable influence to the sea otter), we have come to the view that intense predation in the North Pacific greatly lessened the intensity of herbivory on kelps, thus decoupling a potential evolutionary arms race for defense in the plants and resistance in the herbivores (Estes and Steinberg 1988, Steinberg et al. 1995). The poorly defended flora that consequently evolved in the North Pacific are highly vulnerable to destructive overgrazing, thus perhaps helping us to understand why the loss of sea otters has had such devastating ecological consequences.

A very long and intimate look at one place—the Aleutian archipelago—led to a view of interaction-web dynamics that may be of wider application. At one level the sea otter–kelp forest system operates like a classic HSS/Painean trophic cascade. As a two-trophic-level food chain (i.e., lacking both sea otters and killer whales), strong herbivore-plant interactions emerge. With the addition of sea otters as a third trophic level, herbivore-plant interactions diminish in strength, only to intensify again as a fourth trophic level (killer whales) is added to the system (figure 8.3). These findings lend credence to Fretwell's (1978) prediction that herbivore-plant interactions should be strong in even-numbered food chains and weak in odd-numbered food chains. Our studies further demonstrate that the influence of apex predators extends far beyond this traditional view of trophic cascades. The sea otter system has shown us that the interaction-web dynamics are substantially more complex, in that various other processes and species are connected with and influenced by the trophic cascade. The fox-seabird-terrestrial plant system has shown us further that apex predators can influence plant communities through pathways other than simple cascades of top-down forcing processes across successively lower trophic levels. Finally, we have seen that interaction-web pathways that control the workings of modern-day ecosystems can reach far back in time and extend between different ecosystems. The fox-seabird-plant work exemplifies a powerful predator-induced linkage between the land and the sea. The sea otter–kelp forest system has taught us that, while some of the details are unclear, the modern-day kelp forest

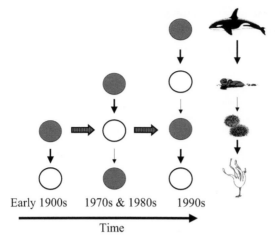

Figure 8.3. Diagram showing how plant-herbivore interaction strength in kelp forest ecosystems of the Aleutian archipelago changed over the course of the twentieth century from (1) early in this period when sea otters were rare or absent, to (2) the middle and late decades when otter populations were recovering, to (3) the 1990s, when killer whales entered the system and again reduced the number of sea otters. Bold and light arrows indicate strong and weak interactions, respectively. Filled and open circles indicate that species within that trophic level are limited by resources versus disturbance by their consumers, respectively. Plant-herbivore interaction strength shifts from strong to weak to strong as the growing food chain alternates between an even and odd number of trophic levels, as predicted by Fretwell (1978). From Estes (2005).

system is being influenced by activities that had their origin in the open sea many decades earlier. These findings are testimonies to the didactic value of place-based ecology.

Acknowledgments

I thank I. Billick, G. V. Byrd, K. Laidre, R. T. Paine, M. V. Price, A. M. Springer, M. T. Tinker, and G. van Vliet for comments on earlier drafts of this essay. To the various administrators who believed in my vision, to the various agencies who funded my work, and especially to the many friends and colleagues who have shared this exciting journey with me—I am eternally grateful.

References

Addicott, W. O. 1969. Tertiary climatic change in the marginal northeastern Pacific Ocean. *Science* 165:583–86.
Anthony, R. G., J. A. Estes, M. A. Ricca, A. K. Miles, and E. D. Forsman. 2008. Bald eagles and sea otters in the Aleutian archipelago: Indirect effects of trophic cascades. *Ecology* 89:2725–35.

Armstrong, R. H. 1977. Weather and climate. Pages 53–58 *in* M. L. Merritt and R. G. Fuller, eds., *The Environment of Amchitka Island, Alaska*. National Technical Information Service, United States Department of Commerce, TID-26712, Springfield, VA.

Bailey, E. P. 1993. Introduction of foxes to Alaskan islands:History, effects on avifauna, and eradication. *United States Department of the Interior, Fish and Wildlife Service, Resource Publication 193*, Washington, DC.

Baird, R. W., and L. M. Dill. 1995. Occurrence and behavior of transient killer whales: Seasonal and pod-specific variability, foraging behavior and prey handling. *Canadian Journal of Zoology* 73:1300–1311.

Barnosky, A. D., P. L. Koch, R. S. Feranec, S. L. Wing, and A. B. Shabel. 2004. Assessing the causes of Late Pleistocene extinctions on the continents. *Science* 306:70–75.

Barron, J. A. 2003. Planktonic marine diatom record of the past 18 MY: Appearances and extinctions in the Pacific and Southern Oceans. *Diatom Research* 18:203–24.

Burn, D. M., A. M. Doroff, and M. T. Tinker. 2003. Carrying capacity and pre-decline abundance of sea otters (*Enhydra nereis kenyoni*) in the Aleutian Islands. *Northwestern Naturalist* 81:145–48.

Byrd, G. V., J. L. Trapp, and C. F. Zeillemaker. 1994. Removal of introduced foxes: A case study in restoration of native birds. *Transactions of the North American Wildlife and Natural Resources Conference* 59:317–21.

Carpenter, S. R., and J. F. Kitchell, eds. 1993. *The Trophic Cascade in Lakes*. Cambridge University Press, New York.

Cervato, C. and L. Burckle. 2003. Pattern of first and last appearance in diatoms: Oceanic circulation and the position of polar fronts during the Cenozoic. *Paleoceanography* 18:1055, doi:10.1029/2002PA000805.

Croll, D. A., J. L. Maron, J. A. Estes, E. M. Danner, and G. V. Byrd. 2005. Introduced predators transform subarctic islands from grassland to tundra. *Science* 307:1959–61.

Dawson, T. E., S. Mambellli, A. H. Plamboeck, P. H. Templer, and K. P. Tu. 2002. Stable isotopes in plant ecology. *Annual Review of Ecology and Systematics* 33:507–59.

Doroff, A. M., J. A. Estes, M. T. Tinker, D. M. Burn, and T. J. Evans. 2003. Sea otter population declines in the Aleutian archipelago. *Journal of Mammalogy* 84:55–64.

Duggins, D. O. 1988. The effects of kelp forests on nearshore environments: Biomass, detritus, and altered flow. Pages 192–201 *in* G. R. VanBlaricom and J. A. Estes, eds., *The Community Ecology of Sea Otters*. Ecological Studies, vol. 65. Springer, New York.

Duggins, D. O., C. A. Simenstad, and J. A. Estes. 1989. Magnification of secondary production by kelp detritus in coastal marine ecosystems. *Science* 245:170–73.

Ebbert, S. E., and G. V. Byrd. 2002. Eradication of invasive species to restore natural biological diversity on Alaska Maritime National Wildlife Refuge. Pages 102–9 *in* C. R. Veitch and M. N. Clout, eds., *Turning the Tide: The Eradication of Invasive Species. Proceedings of the International Conference on Eradication of Island Invasives*. Occasional Paper of the IUCN Species Survival Commission, no. 27.

Estes, J. A. 1977. Population estimates and feeding behavior of sea otters. Pages 511–26 *in* M. L. Merritt and R. G. Fuller, eds., *The Environment of Amchitka Island, Alaska*. National Technical Information Service, United States Department of Commerce, TID-26712, Springfield, VA.

———. 1990. Growth and equilibrium in sea otter populations. *Journal of Animal Ecology* 59: 385–400.

———. 1996. The influence of large, mobile predators in aquatic food webs: Examples from sea otters and kelp forests. Pages 65–72 *in* S. P. R. Greenstreet and M. L. Tasker, eds., *Aquatic Predators and their Prey*. Fishing News Books, Oxford.

———. 2005. Carnivory and trophic connectivity in kelp forests. Pages 61–81 *in* J. C. Ray, K. H. Redford, R. S. Steneck, and J. Berger, eds., *Large Carnivores and the Conservation of Biodiversity*. Island Press, Washington, DC.

Estes, J. A., E. M. Danner, D. F. Doak, B. Konar, A. M. Springer, P. D. Steinberg, M. T. Tinker, and T. M. Williams. 2004. Complex trophic interactions in kelp forest ecosystems. *Bulletin of Marine Sciences* 74:621–38.

Estes, J. A., D. F. Doak, A. M. Springer, and T. M. Williams. 2009. Causes and consequences of marine mammal population declines in southwest Alaska: A food web perspective. *Philosophical Transactions of the Royal Society of London* 364:1647–58.

Estes, J. A., and D. O. Duggins. 1995. Sea otters and kelp forests in Alaska: Generality and variation in a community ecological paradigm. *Ecological Monographs* 65:75–100.

Estes, J. A., R. J. Jameson, and E. B. Rhode. 1982. Activity and prey selection in the sea otter: Influence of population status on community structure. *American Naturalist* 120:242–58.

Estes, J. A., and J. F. Palmisano. 1974. Sea otters: Their role in structuring nearshore communities. *Science* 185:1058–60.

Estes, J. A., and P. D. Steinberg. 1988. Predation, herbivory, and kelp evolution. *Paleobiology* 14:19–36.

Estes, J. A., M. T. Tinker, T. M. Williams, and D. F. Doak. 1998. Killer whale predation on sea otters linking coastal with oceanic ecosystems. *Science* 282:473–76.

Ford, J. K. B. 2002. Killer whale (*Orcinus orca*). Pages 669–76 *in* W. F. Perrin, B. Würsig, and J. G. M. Thewissen, eds., *Encyclopedia of Marine Mammals*. Academic Press, San Diego.

Ford, J. K. B., G. Ellis, L. Barrett-Lennard, A. B. Morton, R. S. Palm, and K. C. Balcomb. 1998. Dietary specialization in two sympatric populations of killer whales (*Orcinus orca*) in coastal British Columbia and adjacent waters. *Canadian Journal of Zoology* 76:1456–71.

Fretwell, S. D. 1987. Food chain dynamics: The central theory of ecology? *Oikos* 20:169–85.

Hairston, N. G., F. E. Smith, and L. B. Slobodkin. 1960. Community structure, population control, and competition. *American Naturalist* 94:421–25.

Hanni, K. D., J. A. K. Mazet, F. M. D. Gulland, J. A. Estes, M. Staedler, M. J. Murray, and D. A. Jessup. 2003. Clinical pathological values and assessment of pathogen exposure in southern and Alaskan sea otters. *Journal of Wildlife Disease* 39:837–50.

Hatfield, B. B., D. Marks, M. T. Tinker, and K. Nolan. 1998. Attacks on sea otters by killer whales. *Marine Mammal Science* 14:888–94.

Hunter, M. D. and P. W. Price. 1992. Playing chutes and ladders: Heterogeneity and the relative roles of bottom-up and top-down forces in natural communities. *Ecology* 73:724–32.

Irons, D. B., R.G . Anthony, and J. A. Estes. 1986. Foraging strategies of glaucous-winged gulls in rocky intertidal communities. *Ecology* 67:1460–74.

Jefferson, T. A., P. J. Stacey, and R. W. Baird. 1991. A review of killer whale interactions with other marine mammals: predation to co-existence. *Mammalian Review* 21:151–180.

Kenyon, K. W. 1969. The sea otter in the eastern North Pacific Ocean. *North American Fauna*, no. 68. 352 pages.

Konar, B., and J. A. Estes. 2003. The stability of boundary regions between kelp beds and deforested areas. *Ecology* 84:174–85.

Laidre, K. L., J. A. Estes, M. T. Tinker, J. Bodkin, D. Monson, and K. Schneider. 2006. Patterns of growth and body condition in sea otters from the Aleutian archipelago before and after the recent population decline. *Journal of Animal Ecology* 5:978–89.

Lallemant, H. G. A., and J. S. Oldow. 2000. Active displacement partitioning and arc-parallel extension of the Aleutian volcanic arc based on Global Positioning System geodesy and kinematic analysis. *Geology* 28: 739–42.

Mangel, M. and N. Wolf. 2006. Predator diet breadth and prey population dynamics: mechanism and modeling. Pages 279–85 *in* J. A. Estes, D. P. DeMaster, D. F. Doak, T. M. Williams, and R. L. Brownell, Jr., eds., *Whales, Whaling, and Ocean Ecosystems*. University of California Press, Berkeley.

Maron, J. L., J. A. Estes, D. A. Croll, E. M. Danner, S. C. Elmendorf, and S. L. Buckelew. 2006. An introduced predator alters Aleutian Island plant communities by thwarting nutrient subsidies. *Ecological Monographs* 76:3–24.

Martin, P. S. 2005. *Twilight of the Mammoths*. University of California Press, Berkeley.

Monson, D., J.A. Estes, D.B. Siniff, and J. B. Bodkin. 2000. Life history plasticity and population regulation in sea otters. *Oikos* 90:457–68.

Murie, O. J. 1959. Fauna of the Aleutian Islands. *North American Fauna*, no. 61. 406 pages.

National Research Council. 2003. Decline of the Steller sea lion in Alaskan waters: Untangling the food webs and fishing nets. National Academies Press, Washington, DC.

Pace, M. L., J. J. Cole, S. R. Carpenter, and J. F. Kitchell. 1999. Trophic cascades revealed in diverse ecosystems. *Trends in Ecology and Evolution* 14:483–88.

Paine, R. T. 1966. Food web complexity and species diversity. *American Naturalist* 100:65–75.

———. 1980. Food webs: linkage, interaction strength, and community infrastructure. *Journal of Animal Ecology* 49:667–685.

Paine, R. T., and R. L. Vadas. 1969. The effect of grazing by sea urchins *Strongylocentrotus* spp. on benthic algal populations. *Limnology and Oceanography* 14:710–19.

Polis, G. A., W. B. Anderson, and R. D. Holt. 1997. Toward an integration of landscape and food web ecology: The dynamics of spatially subsidized food webs. *Annual Review of Ecology and Systematics* 28:289–316.

Power, M. E. 1992. Top-down and bottom-up forces in food webs: Do plants have primacy? *Ecology* 73:733–46.

Ray, J. C., K. H. Redford, R. S. Steneck, and J. Berger, eds. 2005. *Large Carnivores and the Conservation of Biodiversity*. Island Press, Washington, DC.

Reisewitz, S. E., J. A.Estes, and S. A. Simenstad. 2006. Indirect food web interactions: Sea otters and kelp forest fishes in the Aleutian archipelago. *Oecologia* 146:623–31.

Schindler, D. E., and S. C. Lubetkin. 2004. Using stable isotopes to quantify material transport in food webs. Pages 25–42 *in* G. A. Polis, M. E. Power, and G. R. Huxel, eds., *Food Webs at the Landscape Level*. University of Chicago Press, Chicago.

Simenstad, C. A., J. A. Estes, and K. W. Kenyon. 1978. Aleuts, sea otters, and alternate stable state communities. *Science* 200:403–11.

Soulé, M. E., J. A. Estes, J. Berger, and C. Martínez del Rio. 2003. Recovery goals for ecologically effective numbers of endangered keystone species. *Conservation Biology* 17:1238–50.

Springer, A. M., J. A. Estes, G. B. van Vliet, T. M. Williams, D. F. Doak, E. M. Danner, K. A. Forney, and B. Pfister. 2003. Sequential megafaunal collapse in the North Pacific Ocean: an ongoing legacy of industrial whaling? *Proceedings of the National Academy of Science (USA)* 100:12223–228.

Springer, A. M., J. A. Estes, G. B. van Vliet, T. M. Williams, D. F. Doak, E. M. Danner, and B. Pfister. 2008. Mammal-eating killer whales, industrial whaling, and the sequential megafaunal collapse in the North Pacific: A reply to critics of Springer et al. 2003. *Marine Mammal Science* 24:414–42.

Springer, A. M., J. F. Piatt, and G. B. Van Vliet.1996. Seabirds as proxies of marine habitats and food webs in the western Aleutian Arc. *Fisheries Oceanography* 5:45–55.

Stabeno, P. J., J. D. Shumacher, and K. Ohtani. 1999. The physical oceanography of the Bering

Sea. Pages 1–28 *in* T. R. Loughlin and K. Ohtani, eds. *Dynamics of the Bering Sea: A Summary of Physical, Chemical, and Biological Characteristics, and a Synopsis of Research on the Bering Sea.* North Pacific Marine Science Organization (PICES), University of Alaska Sea Grant, AK-SG-99-03.

Steinberg, P. D., J. A. Estes, and F. C. Winter. 1995. Evolutionary consequences of food chain length in kelp forest communities. *Proceedings of the National Academy of Sciences (USA)* 92:8145–48.

Steneck, R. S., M. H. Graham, B. J. Bourque, D. Corbett, J. M. Erlandson, J. A. Estes, and M. J. Tegner. 2003. Kelp forest ecosystem: Biodiversity, stability, resilience and future. *Environmental Conservation* 29:436–59.

Walsh, J. J., C. P. McRoy, L. K. Coachman, J. J. Georing, J. J. Nihoul, T. E. Whitledge, T. H. Blackburn, P. L. Parker, C. D.Wirick, P. G. Shuert, J. M. Grebmeier, A. M. Springer, R. D. Tripp, D. A. Hansell, S. Djenidi, E. Deleersnijder, K. Henriksen, B. A. Lund, P. Andersen, F. E. Müller-Karger, and K. Dean. 1989. Carbon and nitrogen cycling with the Bering/Chukchi Seas: Source regions for organic matter affecting AOU demands of the Arctic Ocean. *Progress in Oceanography* 22:277–359.

Watt, J., D. B. Siniff, and J. A. Estes. 2000. Interdecadal change in diet and population of sea otters at Amchitka Island, Alaska. *Oecologia* 124:289–98.

Whitehead, H., and R. Reeves. 2005. Killer whales and whaling: The scavenging hypothesis. *Biology Letters* 1:415–18.

Williams, T. M., J. A. Estes, D. F. Doak, and A. M. Springer. 2004. Killer appetites: Assessing the role of predators in ecological communities. *Ecology* 85:3373–84.

Part 3

BUILDING AN UNDERSTANDING OF PLACE

Mary V. Price and Ian Billick

In this section we suggest that the understanding that derives from place-based research can be thought of as a conceptual model—a hypothetical description of a complex entity—that successfully accounts for the observed properties of the study system. We like this way of thinking about the understanding of place because it acknowledges the complexity of ecological systems and captures the idea that knowledge of them is constructed gradually out of interconnected pieces. This section considers the process by which ecological models are built and tested in place-based research.

We already know from the essays in parts 1 and 2 that place-based research programs gain insight from sustained study over long periods and comparison across sites. Long experience with a system allows one to distinguish exceptional from frequent states and to characterize slow processes—in short, to leverage knowledge out of temporal heterogeneity, and so to understand the fundamentally dynamic nature of ecological and evolutionary phenomena. Said in another way, long experience uncovers the temporal variance in ecological systems, which is arguably more important than their average properties. Similarly, exposure to multiple places allows one to know what is unique or representative about one's own place, and to leverage knowledge out of spatial heterogeneity. But the understanding of place does not automatically emerge from repeated observations over time or space. If it did, we would only have to tap the knowledge of our local octogenarian or globetrotter. No, there is more to it than that!

The four narratives in this part illustrate how the understanding of place develops. It isn't a simple by-product of accumulated observation; it is built via a deliberate process of progressive model refinement. This is more of a "Bayesian" approach to science—in which the model grows by accretion of new component hypotheses that emerge out of experimental results and natural history observations, so that confidence in the model grows in steps—than it is a "hypothetico-deductive" approach to science, in which predetermined hypotheses are accepted or rejected wholesale on the basis of, for example, "P values" (see also Pulliam and Waser, chapter 4; Feinsinger et al., chapter 18).

An example from the last section may clarify this distinction. Estes (chapter 8) wanted to understand the dynamics of sea otter populations in the Aleutian archipelago, which in the 1970s were recovering after being hunted almost to extinction. The initial model, which proposed that fecundity and mortality were determined primarily by food availability, was honed by several decades of study of otters' diets, behavior, and life history. During the 1990s, however, it became clear that the model could not account for a sudden change in the Amchitka Island population. At that time mortality increased and the population began to decline. If the decline were due to food limitation, fecundity should have been low, animals should have been in poor condition, they should have spent much time foraging, and beach-stranded carcasses—the result of starvation or disease—should have been common. Just the opposite was observed. These contrary results did not cause Estes to throw out the food-limitation model of otter dynamics, however. Instead, he modified the model to include another mortality factor—predation by killer whales—that had been suggested by a series of anecdotal observations. The model is still being refined, as Estes seeks to understand what causes prey-switching by killer whales and the resulting shift from food limitation to predator limitation in otter populations. This process of model development is not strictly hypothetico-deductive science, because the model, which represents a complex set of interconnected hypotheses, is not directly competed against alternative models in an "accept-versus-reject" dynamic. Nonetheless, hypothesis testing is certainly important for model development. Consider, for example, the apparent decline in otter populations that was suggested by survey samples in the early 1990s. Estes did not accept the "otter-population-decline" hypothesis right away, but instead competed it against an alternative: that the apparent decline was an artifact of survey methodology. He distinguished these hypotheses by looking at the rate of loss of otters fitted with radio transmitters. Consistent with the "real decline" hypothesis, marked individuals disappeared at a high rate. Eventually the weight of evidence confirmed the downward population trend which, together with other strands of investigation, prompted a change in the otter population model.

Each chapter in this section describes the evolution of a conceptual model that explains some attribute of a place. In the case of Peckarsky et al. (chapter 9), the model addresses how predatory fish affect the abundance of their insect prey in high-elevation streams. Peckarsky and her collaborators began with a model in which predators affect prey solely by eating them. From unanticipated discrepancies between the effects of natural and experimental manipulations of predator densities on prey densities, they realized that a better model would include various "nonconsumptive" effects of predators on prey behavior, life history, and dispersal.

Ostfeld and Jones (chapter 10) also began with a simple model in which the population dynamics of a tree-defoliating insect (the gypsy moth) are driven by its interaction with a major predator (the white-footed mouse). They soon realized that this pairwise interaction failed to capture important dynamics: the model needed to incorporate interactions between the mouse and its temporally-pulsed primary resource (acorns), as well as feedbacks between defoliation events caused by moth outbreaks and acorn production. This three-way interaction in turn was progressively linked to the spatial and temporal dynamics of other species—deer, ticks, Lyme disease, chipmunks, and birds—to arrive at a more complete model of the interaction web in their oak forest.

In the case of Paine et al. (chapter 11), the model is meant to account for the spatial distribution and diversity of species that attach themselves to the hard surfaces of the rocky intertidal zone of Tatoosh Island. The initial model grew out of basic natural history observations: that *Pisaster* starfish eat many barnacles and mussels and in the process open up space, and that mussel abundance decreases while species diversity increases in lower reaches of the intertidal zone where starfish are common. These observations suggested that *Pisaster* enhances diversity by preventing competitively dominant mussels from monopolizing space. After being tested, retested, augmented, and refined, the model has grown over the years into a rich account of how intertidal rocky-shore communities are molded by disturbance, direct and indirect food-web interactions, and successional processes that are mediated through patterns of individual survival, growth, and reproduction.

Curtin (chapter 12) describes two cases of models designed to guide sustainable resource management. The failure of the resulting management prescriptions highlights the practical importance of having correct models. In the western rangelands and Gulf of Maine fisheries of North America, existing management policies were failing: shrub encroachment continued to cause deterioration of grasslands, and ground fish stocks continued to decline. In both examples, a collaborative effort between scientists and local resource users has helped to improve the models upon which management was based. The informal natural history and economic knowledge of the resource users—ranchers and fishermen—helped to pinpoint key problems with the models, and the scientists provided the conceptual tools for using this local knowledge to modify the models and devise ways of testing them. Both case histories underscore the potential that managed systems hold for improving our understanding of particular ecosystems and extending that understanding to general ecological theory (see also Krebs, chapter 13).

These narratives, along with those in parts 1 and 2, illustrate four features of research strategy that facilitate the process of model development and refinement. First, they all point to the critical importance of deploying diverse types of natural history information about the study system in the design and interpretation of experimental or observational studies. In this context, "natural history information" refers to a set of reliable facts already known about the system. The observations on which the facts are based can be systematic or anecdotal. They can originate with trained scientists or lay people. They can include descriptive information about the environmental context of the system, the biology of study organisms, the history of disturbance events, or even observations of system responses to natural or manipulative experiments that were designed to answer some previous question.

Ostfeld and Jones (chapter 10) emphasize a second factor that contributes to the development of local understanding, which echoes a point made explicitly by Estes (chapter 8) and Waller and Flader (chapter 3). They point out how knowledge of historical events, such as the loss of the American chestnut from North American eastern deciduous forests, informs an understanding of how the system came to be the way it is, and allows testable predictions about how it might change in the future. A related subtheme that runs through all of these narratives is that attention to spatiotemporal heterogeneity and scale of ecological processes is critical to the understanding of how a system works and why it works in that way.

Third, the narratives highlight the value of a team approach. By this they do not mean teams of the "big science" sort, where each player is deployed to add to sample size or tackle a small piece of the research program. Instead they mean a loose consortium of independent investigators who bring diverse perspectives and skills and pursue different questions, who cooperate in designing and carrying out experiments, and who share information, insight, and experience. Why this "loose consortium" approach might be better than the "assembly line" approach is not immediately apparent. Both approaches can accommodate problems that are inherently multivariate, both can assure continuity of effort in maintaining long-term experiments, and both can bring complementary skills and perspectives to bear on a problem. Perhaps the difference lies in the assertion by Paine et al. (chapter 8) that "research independence rather than a primary group-based endeavor is the mother of innovation," and in Pulliam and Waser's assertion (chapter 4) that "excessive regimentation can stifle the creativity, akin to artistic creativity, that represents the most mysterious and productive aspect of the scientific endeavor." Could it be that the loose-consortium approach is better when the primary task is to build a model of a complex system such as an ecological system? Indeed, Curtin (chapter 12) makes it clear that even perspectives as disparate as those of laypersons versus scientists can be quite valuable in building understanding.

Finally, the chapters in this part and part 2 illustrate how model building is a dynamic process in which each planned result or unplanned observation leads to new interpretations, new questions, and new hypotheses—what Peckarsky et al. (chapter 9) call the "follow your nose" approach. How, one might wonder, can this be a recipe for efficiently building a coherent model? Isn't it instead a recipe for an intellectual random walk? We don't claim to have an answer to this question, because the process of scientific inquiry is not yet well understood. However, we suggest that a careful reading of these case histories points toward a "directed random walk" at worst. In each case, a single broad question guided the general course of investigation: what determines the population dynamics of sea otters; what determines the spatial distribution and species composition of intertidal communities; what determines relative abundances of prey that occupy different trophic levels; what determines the temporal dynamics of consumer populations in oak forests. And in each case, the relationship of "tangential" investigations of new hypotheses or new questions to the overarching question was kept firmly in mind.

Although the overarching question gives some coherence to the process, the narratives in this section paint a picture of scientific inquiry that seems considerably more disorganized than the "strong inference" process described by Platt (1964). In Platt's "strong inference" (figure p3.1a), investigators organize hypotheses into hierarchical sets of mutually exclusive and exhaustive groups, with the broadest sets at the first level and their more specific manifestations at each successive level.[1] This organization allows one to exclude whole groups of specific hypotheses with single experiments at each level, an efficient way to develop an ever more detailed answer to the original question. The process by which place-based conceptual models are built is far less linear. Instead of winnowing through a predetermined set of hypotheses to identify the single correct one, the outcomes of each experiment or observation feed back to add new components to the model and modify their relationships (figure p3.1b).

Why does model development in place-based ecological research follow this path? One possibility is that ecological systems are complex, with multiple processes influencing such things as population dynamics or species composition of a community. This complexity has several consequences. It means that system models may include alternative subhypotheses that are not mutually exclusive. For example, both predation risk and net rate of energy intake may affect foraging decisions. If hypotheses are not mutually exclusive, then more than one can be true simultaneously, and an experimental outcome that is consistent with one does not mean that the

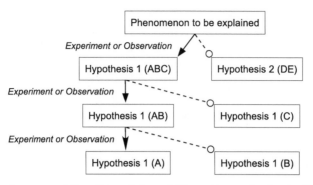

Figure p3.1a. A linear "strong inference" (Platt 1964) example of a scientific investigation. The first experiment or observation distinguishes group 1 from group 2 hypotheses: supported hypotheses are indicated by solid lines and arrows, discarded ones by dashed lines ending in open circles. The second experiment or observation distinguishes subsets of group 1 hypotheses, and the third experiment leaves only Hypothesis 1(A).

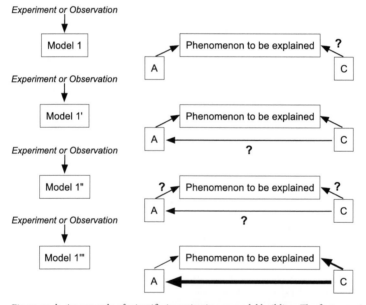

Figure p3.1b. An example of scientific investigation as model building. The first experiment or observation suggests an initial model 1 with a confirmed causal path from A. Successive observations or experiments test whether a new hypothesized causal process (c) should be added (model 1 to 1'), whether that process affects the phenomenon to be explained via indirect as well as direct pathways (model 1' to model 1"), and the relative strengths of the causal linkages (model 1" to 1'''), indicated by arrow thickness. The hypotheses that are entertained can be suggested either by general theories proposed to explain the phenomenon of interest, or by the specific natural history of the local system.

alternatives are false. As a consequence, each hypothesis must be evaluated separately. The complexity of ecological systems also means that we may not be able to identify all of the possible relevant factors at the outset of an ecological investigation, much less think about the variety of ways in which they might manifest themselves in individual systems. If one entertains alternative hypotheses that do not exhaust the possibilities, then all can be false, and one might interpret an experimental outcome as supporting a false hypothesis simply because the correct one has not yet been identified (e.g., Whitham 1980). The complexity of ecological systems therefore means that the process of model building is nonlinear: each component hypothesis of the model may need to be tested individually, and a supported hypothesis may need to be retested against a new alternative. Such ambiguities make the route to understanding complex systems a circuitous one that involves integration of disparate bits of information into reformulated hypotheses and repeated testing of the same hypothesis, not just a linear winnowing out of incorrect hypotheses.

The product of the Bayesian boiling of hypotheses described by the chapters in this part is a conceptual model of how a particular local system works that accounts for what is currently known about its properties. Such local understanding may be the primary goal of applied fields of ecology like range management or conservation biology, but it isn't the final goal of basic ecology—general understanding is. How does the local understanding generated by place-based research programs contribute to general understanding? And similarly, how does the arsenal of general concepts in a field relate to the development of a place-based local model? These issues are taken up in part 4.

References

Platt, J. R. 1964. Strong inference. *Science* 146:347–53.
Whitham, T. G. 1980. The theory of habitat selection examined and extended using *Pemphigus* aphids. *The American Naturalist* 115:449–66.

Notes

1. Although Platt (1964) did not explicitly point out this general-to-specific feature of logical tree structure, it is implicit in his remarks. For example, in step 1' of the scientific method, he talked about "making subhypotheses or sequential hypotheses to refine the possibilities that remain." A hierarchical structure is also implicit in his example of steps for chemical analysis of an unknown sample: "Add reagent A; if you get a red precipitate, it is subgroup alpha and you filter and add reagent B; if not, you add the other reagent, B'; and so on."

9 Understanding the Role of Predation in Open Systems

THE VALUE OF PLACE-BASED RESEARCH

Barbara L. Peckarsky, J. David Allan, Angus R. McIntosh, and Brad W. Taylor

Figure 9.0. (a) Experimental mesocosms on a slope above the East River. Mesocosms are being supplied with stream water that is circulated through a trout-occupied bin. (b) Experimental augmentation of trout odor in a natural trout stream with water from trout-occupied bins. Photographs by Angus R. McIntosh, 1995 (a) and 1996 (b).

Abstract

Studies of predation traditionally focus on how predators influence prey communities and prey population oscillations via mortality. Studies carried out for nearly 40 years in one place have enabled us to evaluate the generality of this focus. Early observations in one high-altitude, rocky-bottom stream in western Colorado fit the traditional model of negative correlation between predator and prey abundance, suggesting that predation by salmonid fishes may explain spatial variation in the abundance of stream insects. In contrast, a large-scale experiment showed that predation did not explain variation in abundance of primary consumers (mayflies) or large insect predators (stoneflies). We describe a series of studies conducted in more streams over many years that forced traditional thinking to give way to a new appreciation for predator-induced changes in prey behavior and life history. Our research used a "follow your nose" approach, which allowed the place to motivate the questions and resulted in the evolution of a conceptual model to explain how the place works. Natural history observations, combined with many years of phenomenological and mechanistic experiments, have contributed to the following generalizations about the importance of nonconsumptive effects of predators. First, mechanisms of trophic cascades may be predominantly nonconsumptive. Second, the effects of predation may be obscured or exaggerated by prey immigration or emigration. Third, if top and intermediate predators have opposite effects on emigration of shared prey, counterintuitive patterns of prey abundance may result when both kinds of predators are present. Finally, predator-induced changes in prey behavior and development can reduce prey fecundity, which may affect rates of prey population growth more than predator-induced mortality. Thus, intuition gained from studying one set of organisms in one place supports a general conceptual model according to which in open systems with high levels of prey dispersal, the predominant influence of predators may be nonconsumptive.

Introduction

The science of ecology, generally defined as the study of factors influencing the distribution and abundance of organisms, involves a variety of approaches (e.g., Dodson et al. 1998), including intuition from natural history (Power 1998). Natural history has been integral to our own approach, motivating our questions and hypotheses, informing the interpretation of experiments, and enabling our understanding of a complex system to evolve. In this chapter we describe how implementing our approach in one place gradually revealed a lack of fit to the traditional model in which predators affect prey population dynamics via mortality. Continuously fueled by new findings, new conceptual models have thus evolved which emphasize the importance of nonconsumptive effects of predators that induce risk-sensitive behavioral and life-history responses of prey. We further recognized that a key feature of our system is the mobility of organisms: streams are open systems (Reiners and Driese 2001) characterized by high levels of dispersal. Under those conditions prey mobility can swamp direct effects of predation (Allan 1982a, Allan 1983, Cooper et al. 1990, Englund 1997), resulting in patterns of predator and prey abundance that are inconsistent with traditional models of predator-prey dynamics.

Look Up, Look Down, Look All Around! Features of Our Place-Based Approach

Studying ecology in one place for a long time should not be equated with an overly narrow perspective or increasingly mechanistic approach, although those can be consequences. On the contrary, loyalty to one place has enabled us to develop a more comprehensive understanding of the system by extending our scope to larger spatial and temporal scales. Long-term observations over larger spatial scales have made it possible to identify unusual events (e.g., high-water years), generalize over generations of predators and prey, and incorporate spatial heterogeneity into our models. Our story thereby underscores the well-documented value of conducting long-term research (e.g., Strayer et al. 1986; Likens 1989; Magnuson 1990; Estes et al. 1998; Grant and Grant, chapter 6; Armitage, chapter 7; Estes, chapter 8).

Furthermore, our place-based research has fostered a team approach by attracting collaborators, graduate students, and undergraduate students. Those collaborators have contributed different perspectives and approaches, creating a multidisciplinary effort and diversifying the intellectual climate. To reap this benefit, pioneers of place-based research need to welcome new people to their places, not be overly possessive of ideas, embrace alternatives, and be willing to be wrong. Many other scholars have come to our place over the years attracted by the accumulated knowledge, and those people have enriched our story.

Our "place" includes high-altitude, rocky-bottom streams in western Colorado, near the Rocky Mountain Biological Laboratory (RMBL) and town of Mt. Crested Butte (figure 9.1). The smallest streams in this system arise from snowmelt, springs, or lake outlets. The confluence of small tributaries forms larger streams, many of which contain predatory fish that are excluded from some streams by waterfalls that prevent upstream dispersal. The structure of these drainage basins provides opportunities for studying replicate streams of varying size. Replication at small scale and in whole streams has strengthened our conclusions.

Our story follows a sequence of studies that implements a "follow your nose" approach and allows the place to motivate the research. We provide examples of patterns of distribution and abundance of organisms that were sometimes intuitive and sometimes not. We describe how our understanding of the mechanisms explaining those patterns evolved as we explored them at different temporal and spatial scales, diversified our approaches, and integrated across levels of organization. We present our work in the

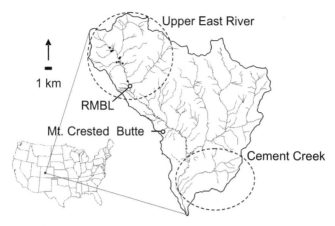

Figure 9.1. Map of the East River Drainage Basin—"the place"—indicating the Upper East River and Cement Creek sub-basins (circled) and specific sites (dots) in which we worked.

context of four generalizations that represent the current status ("the invisible present," Magnuson 1990) of our conceptual model, along with the progression of evidence for the importance of nonconsumptive effects of predators. We describe the chronology of observations and experiments that produced each generalization, and suggest where continued research in this place should reduce uncertainties.

Four Generalizations

Generalization 1: Mechanisms of trophic cascades may be predominantly nonconsumptive

The food webs of our study system have four trophic levels: top predators (e.g., brook trout, *Salvelinus fontinalis*), predatory insects (e.g., stoneflies, Plecoptera), grazing insects (e.g., mayflies, Ephemeroptera), and attached algae (periphyton). Our first observational study (Allan 1975) compared the distribution and abundance of intermediate consumers along an elevation gradient of one stream, Cement Creek (figure 9.1). This study showed that trout-free headwaters had densities of stream insects several times higher than those of downstream reaches containing trout, suggesting that predation by salmonid fishes could explain spatial variation in the abundance of stream invertebrates. Those observations were broadly consistent with the traditional idea of top-down control by predators (Allen 1951; Allan 1981, 1983).

To test the hypotheses that were generated by our original observations, one of us subsequently implemented a challenging, large-scale field experiment in Cement Creek over multiple years within the context of natural environmental variability. As an early indication of things to come, results of that experiment cast doubt on the traditional view. Experimental reduction of trout density in a 1.2 km reach of Cement Creek had no consistent effects on the abundance of primary consumers (primarily mayflies) or predatory stoneflies (Allan 1982a). Those results were unexpected, and they provided the first clue that a model of top-down control of prey abundance needed further scrutiny. Two hypotheses, highly speculative at the time, were that prey antipredator adaptations minimized predator effectiveness on intermediate consumers (the stoneflies), and that the continual renewal of mayfly prey due to downstream transport (by drift) obscured the effects of predation.

The inconsistency between that experiment and previous observations illustrates that underlying ecological processes may not be clearly detected by simply observing complex and dynamic ecological systems (Peckarsky et al. 1997). In those cases phenomenological experiments (mechanism-free experiments in which the system is modified to see what happens; Dunham and Beaupre 1998) may reveal cryptic processes that are obscured by other processes acting concurrently. We have repeatedly used phenomenological experiments as an effective means of generating explanatory hypotheses (Peckarsky 1998), despite initial obstacles.[1]

To more fully develop the new conceptual model suggested by studies in Cement Creek, we subsequently expanded the spatial and temporal scales of observations to include multiple years and streams within the Upper East River drainage basin (figure 9.1). Those observations showed that large predatory stoneflies achieved high densities in some trout streams and were rare in some fishless streams (Peckarsky et al. 2001). We thus realized that the abundance patterns of stoneflies observed in Cement Creek could not be generalized to fish and fishless sections of other streams in the region (Peckarsky et al. 2008), and that predatory stonefly abundance is more likely controlled by physical variables related to bed disturbance (Peckarsky 1991) than by fish predation.

Although the results of both the fish density manipulation and the expansion of the scale of observations were consistent with the hypothesis that consumption by trout did not explain variation in the distribution and abundance of stream insects, neither approach provided a direct test of causation. Therefore, we implemented mechanistic experiments to explain and interpret results from phenomenological studies and fine-tune our new conceptual framework. In one type of experiment we estimated consumption rates by predators to test whether predation could explain abundance of prey or natural loss rates of prey in streams (Kerans et al. 1995, Peckarsky et al. 2008). The second type of experiment measured effects of predators on prey behavior as a test of the hypothesis that trophic cascades in these streams were driven primarily by nonconsumptive mechanisms (Peckarsky and McIntosh 1998). Experiments were carried out in mesocosms (0.8 m²; see figure 9.0, top) using natural stream water. By preventing or allowing predators to consume prey, we could compare the behavioral and consumptive effects of predators on lower trophic levels.

In predation experiments, trout consumption of predatory stoneflies was negligible, consistent with the observed independence of stonefly abundance from the presence of fish (Peckarsky et al. 2008). However, both trout and stoneflies caused significant mayfly mortality in mesocosms, with trout consuming an order of magnitude more mayflies per predator than stoneflies (McPeek and Peckarsky 1998, Peckarsky et al. 2008). Note that estimates of annual prey consumption under natural conditions (three sites in Cement Creek) suggested that mayfly consumption by trout was roughly twice that of stoneflies in natural streams (Allan 1982b). Most interestingly, in predation experiments the effect of both predators foraging together was not additive: trout reduced the effects of stonefly predation in treatments with both predators (Peckarsky et al. 2008). Those results suggested a new hypothesis: Trout do not reduce stonefly abundance but instead modify stonefly foraging behavior, thereby reducing the impact of stoneflies on mayfly prey (a behavioral trophic cascade).

Observations in mesocosms showed that chemical cues from trout suppressed the movement behavior of stoneflies foraging on mayflies (McIntosh and Peckarsky 1999), which also cascaded to decrease the biomass of algae, the basal resource (Peckarsky and McIntosh 1998). Furthermore, stonefly predation rates on mayflies were negligible in the presence of fish (Kerans et al. 1995), but could be high enough to account for natural losses of the mayfly *Baetis bicaudatus* in streams without fish (Peckarsky et al. 2008). Results of both predation and behavioral experiments thus suggested that a traditional trophic cascade in which top predators reduce the density of intermediate predators by consumption, thereby releasing their prey from predation, was not operating.

Persistent inquiry in one place and experimental innovation reinforced a new conceptual model, indicating that we had to focus more on behavioral than on consumptive mechanisms. Furthermore, the combined effects of trout and stoneflies on mayfly mortality could not have been predicted from traditional models of predator-prey interactions, or from the individual effects of each predator. In this case, the top predator reduced predation rates of intermediate predators without affecting their densities. Interactions between the two types of predators reduced the rate of consumption of mayflies by the intermediate predator, even though trout predation on stoneflies was negligible.

Although those ideas were initially met with resistance, our commitment to place prevented us from abandoning them. We conclude that behavioral interactions among multiple predators and shared prey may counteract or override the effects of consumption by predators in open systems where predators and prey are highly mobile. Future studies should compare estimates of predator impacts on prey mortality to predator impacts on prey emigration using a common currency, such as a standardized index of prey loss rate (e.g., Cooper et al. 1990). Such an approach will be challenging, but should allow effect sizes obtained in experiments to be used to explain patterns of abundance of multiple predators and shared prey in natural systems.

Generalization 2: Effects of predation may be obscured or exaggerated by prey immigration or emigration

The unexpected lack of a response of mayfly abundance to the large-scale trout reduction experiment in Cement Creek (Allan 1982a) forecast the surprising patterns obtained when we observed more streams in the upper East River drainage basin (Peckarsky et al. 2001). There we observed that drifting mayflies, the most important prey species for trout (Allan 1978a, 1981) were actually more abundant in trout streams than in fishless streams. Replication (rather than scale) of observations over space and time contributed most strongly to our ability to generalize, and further underscored the necessity for modifying existing models about the role of predators.

Mechanistic experiments carried out in semipermeable field enclosures supported Allan's (1982a) hypothesis that immigration of highly mobile prey swamped the effects of predation. Prey density was reduced in enclosures with stoneflies only when prey migration was not permitted (Peckarsky 1985, Cooper et al. 1990). This pioneering experimental approach demonstrated that prey behavior needed to be taken into account when measuring the effects of predators. Initial resistance to our unorthodox experimental methods has dissipated, because subsequent experiments have so clearly demonstrated that both trout and stoneflies can have profound effects on the behavior of their mayfly prey, particularly by affecting their emigration rates (drift in the water column).

Early studies in Cement Creek documented that large numbers of drifting invertebrates enter the water column during a 24-hour period (Allan 1982a), providing an important food source for trout, which feed on invertebrates from the drift more than from the benthos (stream bottom). Only a fraction of this potential resource is available to trout, however, because only a small percentage of total drift occurs during daylight hours (roughly 10 to 20% based on data in Allan 1978b). Moreover, drift-feeding trout are size-selective, and day-drifting individuals are mainly small taxa or instars (juvenile stages) that are less visible to these visual predators.

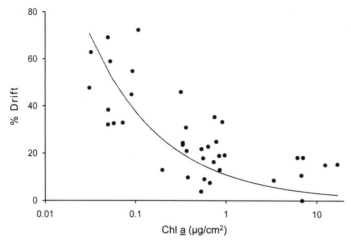

Figure 9.2. The percentage of *Baetis* individuals entering the drift in laboratory streams from stones containing different algal densities, measured as micrograms of chlorophyll a per square cm. Treatments were established during midday, and individuals did not drift until dusk, six to eight hours later. At very low food levels, more than half of all individuals drifted immediately after dusk, whereas few drifted from stones with abundant algae (J. D. Allan, unpublished data).

Unlike the first generalization, in which early observations in one stream were not representative, replicating observations of mayfly drift has yielded consistent results; natural patterns from many fish and fishless streams have shown clearly that mayfly drift is nocturnal in trout streams (Allan 1978b, Allan et al. 1986, Allan 1987, McIntosh et al. 1999), and aperiodic in fishless streams (McIntosh et al. 2002). Those observations might possibly be explained by (1) daytime consumption of drifting mayflies by trout, (2) induction of nocturnal foraging behavior of mayflies by trout (phenotypic plasticity), or (3) evolved responses (Flecker 1992).

Early mechanistic experiments in a laboratory stream were consistent with the third hypothesis, in that a highly mobile grazing mayfly common in the diets of trout (*Baetis*) delayed drifting until nightfall, but showed higher probabilities of emigrating from substrates with low algal biomass (figure 9.2). This response initially suggested hard-wired behaviors including avoidance of daytime foraging, saltatory search behavior during darkness, and more active foraging in response to food shortages. However, both the experimental water and the mayflies were obtained from streams with fish, and thus did not test for induced responses.

Subsequent mechanistic experiments carried out in microcosms (150 cm²) supported the second hypothesis—that *Baetis* activity is sensitive to the presence of trout—because *Baetis* in treatments using water from trout streams showed nocturnal behavior, while individuals in fishless water were more aperiodic (Cowan and Peckarsky 1994). Similarly, experimental introduction of trout odor into stream-side mesocosms (figure 9.0, top) has consistently induced nocturnal drift periodicity and decreased the magnitude of nighttime drift for late-instar *Baetis* obtained from both fish and fishless streams (McIntosh and Peckarsky 1996, 1999; Peckarsky and McIntosh 1998). Furthermore, a phenomenological experiment introducing trout odor into naturally fishless streams caused mayflies to become nocturnal (McIntosh et al. 2004). Interestingly, the addition of high concentrations of trout odor into a natural trout stream (figure 9.0, bottom) reduced nighttime drift of large *Baetis*, but increased nocturnal drift of small *Baetis* larvae (McIntosh et al. 1999). In contrast, there were no effects of added trout odor on daytime drift of *Baetis*, which was low in all treatments in this trout stream.

Taken together, the results of our field and smaller-scale experiments demonstrated that responses of mayflies to trout chemical cues can be induced, and are prey size- and prey species-specific, both of which are potentially related to prey vulnerability. Allan (1978b) has shown that brook trout prefer large over small *Baetis* and consume markedly more *Baetis* than less mobile prey (e.g., *Cinygmula, Epeorus*; Allan 1981). Relative drift densities (scaled by benthic densities) of the mayfly prey species most common in trout diets (*Baetis*) are much higher than those of mayflies less common in trout diets (e.g., *Cinygmula*; Allan 1978a, 1983; Allan et al. 1986; Allan 1987) or stonefly diets (Allan 1982b, Peckarsky, 1985, Peckarsky et al. 1994) in both fish and fishless streams (McIntosh et al. 2002).

Highest drift rates for the most vulnerable mayfly species seemed counterintuitive at first, because we expected more risk-sensitive behavior from more vulnerable prey. However, high mobility of *Baetis* may be critical to its ecological dominance in many systems, providing benefits in both foraging and dispersal but exacting a cost in exposure to predation (Peckarsky 1996). Indeed, drift, which is a saltatory food search behavior, is most nocturnal in the size classes that are most vulnerable to trout predation (Allan 1978b, McIntosh et al. 2002). Likewise, mechanistic experiments in microcosms demonstrated that the magnitude of the response of mayflies to trout chemical cues (nocturnal drift) was greater for species with greater risk of predation (McIntosh et al. 2002). Furthermore, the activity of the most vulnerable prey species was most nocturnal when they were subjected to cues from the most effective predators (native cutthroat, *Oncorhynchus clarki*, versus non-native brook trout) (McIntosh and Peckarsky 2004).

In summary, the effects of predation in streams may not be detected by observing prey abundance because immigration of prey can counteract predation rates. Testing this hypothesis required us to present a cue indicating risk of predation while preventing actual consumption, which we did by introducing "essence of brook trout" into experimental units as small as microcosms and as large as natural streams. This approach provided further support for the conceptual model that in open systems predator effects on prey behavior rather than on prey mortality may better explain the distribution and abundance of highly mobile organisms.

Generalization 3: If top and intermediate predators have opposite effects on emigration of shared prey, then counterintuitive patterns of prey abundance may result when both predators are present

Insights from studies of trout effects on stonefly and mayfly behavior suggested that the effectiveness of predatory stoneflies could also be altered by changes in the behavior of shared prey (mayflies) in the presence of trout. We have observed that while trout cause mayflies to decrease nocturnal drift rates, foraging stoneflies can cause mayflies to increase their nighttime drift in microcosms (Peckarsky 1996, McIntosh et al. 2002) and mesocosms (McIntosh and Peckarsky 1998, McIntosh and Peckarsky 1999) because they drift in response to encounters with stoneflies (Peckarsky 1980). While this behavior could increase mayfly susceptibility to drift-feeding trout, we have observed no evidence that the presence of stoneflies facilitates trout predation (Peckarsky et al. 2008), probably because stonefly-induced *Baetis* drift occurred primarily in the absence of trout and at night (Peckarsky and McIntosh 1998, McIntosh and Peckarsky 1999) when trout could not feed as effectively (McIntosh et al. 2002).

In our first mesocosm experiments we were able to induce the natural periodicity of *Baetis* drift and foraging only when we included stoneflies in both fish and fishless treatments (McIntosh and Peckarsky 1996). Although this interesting bit of natural history was not included in the published version of the study, it indicated that natural diel periodicity of these mayflies is a function of their responses to both predators: high nocturnal drift rates in the absence of fish are stimulated by foraging stoneflies, and low nocturnal drift rates in the presence of fish reflect both threat-sensitive behavior of mayflies and suppressed foraging by stoneflies.

Observations of behavioral responses of mayflies also stimulated our thinking about how nonconsumptive interactions between predators and prey might affect natural patterns of distribution and abundance (figure 9.3). For example, high rates of stonefly-induced mayfly drift (emigration) in fishless streams could reduce their natural abundance compared to that in streams with fish. In contrast, trout reduce mayfly drift (emigration), suppress stonefly consumption of mayflies, and decrease the effects of stoneflies on mayfly drift, all of which could increase mayfly abundance in fish streams.

Net predator effect:

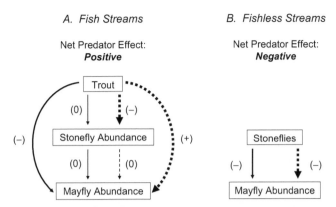

Figure 9.3. Conceptual model of consumptive (solid arrows) and behavioral (dashed arrows) effects of trout on stonefly and mayfly abundance, and of stoneflies on mayfly abundance. Thickness of arrows indicates hypothesized relative effect sizes, and (+), (—), or (0) respectively indicate positive, negative, or negligible effect on prey. (a) In fish streams, trout do not affect stonefly abundance via consumption (thin solid line), but do reduce their foraging activity (thick dashed line). Stoneflies have negligible consumptive (thin solid line) and behavioral effects (thin dashed line) on mayfly abundance. Consumption by trout decreases the abundance of mayflies (solid line), but trout suppress mayfly emigration (drift) and thereby increase their abundance (thick dashed line). Thus, the net effect of predators on mayfly abundance may be positive, depending on the magnitude of consumptive losses relative to positive direct and indirect behavioral effects. (b) In fishless streams, stoneflies are the top predator. Their consumption decreases mayfly abundance (solid line), and their foraging activity increases prey drift or emigration (dashed line). Both consumptive and nonconsumptive effects act in the same direction, resulting in a net negative effect of stoneflies on mayfly abundance.

In streams with both predators, positive behavioral effects may override the negative effects of consumption, providing a plausible mechanism to explain why mayfly abundance is higher in streams with fish than in fishless streams (Peckarsky et al. 2001, Peckarsky et al. 2008). Also, if the direct effect of trout on mayfly drift overrides stonefly-induced drift, trout streams will retain mayflies (Diehl et al. 2000). Consequently, the complex effects of interacting predators may obscure associations between stonefly densities and *Baetis* abundance in natural trout streams (Peckarsky 1991).

Tests of hypotheses involving the interactive consumptive and nonconsumptive effects of predators on prey abundance constitute a work in progress. Extrapolating drift densities observed in the field over 24 hours (from McIntosh et al. 2002) suggests that trout streams may be "sticky" (retaining individuals) and fishless streams may be "leaky" (losing individuals) because the foraging movements of mayflies are not constrained by any risk of fish predation in fishless streams. However, preliminary studies measuring *Baetis* drift into and out of 45-m reaches of 12 different streams (Wilcox et al. 2008) suggest a more complex answer. While benthic density increased directly with the ratio of drift in to drift out (indicating "stickiness") in fish streams, *Baetis* density was independent of this ratio in fishless stream reaches, suggesting that net migration could explain abundance patterns in streams with fish but not in fishless streams. Furthermore, fishless streams were not categorically more "leaky" than fish streams—that is, they did not have lower ratios of drift-in to drift-out.

In summary, although top and intermediate predators have opposite effects on emigration of shared prey, predator-induced prey emigration does not explain the counterintuitive patterns of prey abundance we observed (higher prey densities in fish streams). Even getting to this point was an extraordinary effort involving many intrepid field assistants and more all-nighters than occur in a college dormitory during final exams. And still we have no definitive answer. Nonetheless, we are convinced that we will eventually sort out the alternative hypotheses needed to understand the influence of predator-induced changes in prey movement on the distribution and abundance of mayflies relative to the effects of other factors.

Generalization 4: Predator-induced changes in prey behavior and development can reduce prey fecundity, which may have greater impact on rates of prey population growth than predator-induced mortality

In our earlier work we used observations and mechanistic experiments in microcosms to investigate the nonconsumptive effects of stoneflies on mayfly behavior, morphology, and life histories (Peckarsky et al. 1993, Peckarsky et al. 1994, Peckarsky 1996). We observed that interspecific trade-offs between prey mobility and the extent of morphological defense lead to a variety of prey antipredator strategies arranged along an axis of prey mobility. The most vulnerable mayfly species (*Baetis*) showed reduced feeding in the presence of stoneflies whose mouthparts were glued to prevent feeding (Peckarsky et al. 1993). Behavioral experiments further showed that the mechanism causing reduced *Baetis* feeding in the presence of stoneflies was energetically costly prey dispersal rather than microhabitat shifts (Peckarsky 1996). In longer-term experiments both male and female *Baetis* matured at smaller body sizes, and the fecundity of *Baetis* females was reduced in microcosms with glued stoneflies (Peckarsky et al. 1993). Smaller-bodied *Baetis* females produced fewer eggs; but we have found no disadvantage of small male body size in *Baetis* (Peckarsky et al. 2002a).

Peckarsky might have continued working only on stonefly-mayfly interactions, thereby spinning into a mechanistic black hole, had it not been for inspiration from the perspectives of Allan, McIntosh, and Taylor to add more trophic levels and take a broader view of food webs in the East River system. Building on the classical work of Allan, we have now investigated the influence of salmonid fishes on prey population dynamics—revealing strong effects of brook trout on the behavior of their mayfly prey, especially the movement and foraging of *Baetis* on algal resources, and on the interactions of mayflies with predatory stoneflies (McIntosh and Peckarsky 1996, Peckarsky et al. 1997, McPeek and Peckarsky 1998, Peckarsky and McIntosh 1998, McIntosh and Peckarsky 1999, McIntosh et al. 1999, Dahl and Peckarsky 2002).

First, over multiple years and at multiple streams with and without fish in the upper East River drainage, we measured the size and fecundity of last-instar *Baetis* larvae, as well as other attributes (e.g., densities of predatory stoneflies, potential competitors, algal abundance, and stream physical-chemical characteristics) that could affect their growth or development rates. As a consequence of faster larval development times, the *Baetis* that emerge from trout streams are significantly smaller on average than those that develop in fishless streams (Peckarsky et al. 2001). Thus, *Baetis* reduce their exposure to mortality by emerging sooner from the risky larval environment, at the cost of reduced fecundity. A path analysis using structural equation modeling showed that size variation of *Baetis* that emerge from natural environments was best explained by predators (trout and stoneflies) rather than by variation in resources, resource-mediated competition, or other environmental variations such as temperature, stream size, or water chemistry (Peckarsky et al. 2001).

Field observations did not, however, rule out a consumptive mechanism for size differences in *Baetis* between streams with and without fish (e.g., size-selective predation). Therefore, we conducted mechanistic experiments in stream-side mesocosms to measure the impact of predator avoidance at medium spatial scales on the size and fecundity of *Baetis*. Trout odor and stoneflies (with mouthparts glued) reduced the size of male and female *Baetis* that emerged during the experiment, but only stoneflies delayed *Baetis* emergence, suggesting that stoneflies may have different nonconsumptive effects on *Baetis* life histories than do trout (Peckarsky and McIntosh 1998). We then increased the scale of our experiments, adding fish chemical cues to whole streams that were naturally fishless to test the influence of those cues on prey behavior, life history, and population dynamics as well as other community and ecosystem processes (Peckarsky et al. 2002b, McIntosh et al. 2004). Astoundingly, we were able to induce the development of smaller mayflies from 30-m reaches of fishless streams by introducing fish chemical

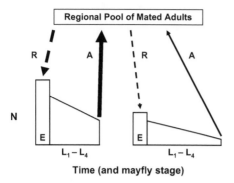

Fish Streams   Fishless Streams

Figure 9.4. Demographic model of possible mechanisms for greater observed densities of *Baetis* in fish streams (left) than in fishless streams (right). Height of polygons indicates relative abundance (N) of eggs (E) and larval stages (L1 to L4). Eggs overwinter in diapause. Development from stages L1 to L4 is accelerated in fish streams; the relative time spent in each stage is unknown. Width of arrows indicates the relative number of eggs laid (dashed) or adults emerging (solid). More mayflies enter the regional pool of mated adults (A) from fish than fishless streams because they develop faster (shorter time from egg through larval developmental stages) and hence are exposed to risk for less time, despite having greater mortality rates. Therefore, demographic changes induced by cues from predatory fish increase the probability of survival in fish streams. Modified from Peckarsky et al. (2008).

cues, demonstrating definitively that observed variations in the life history attributes of *Baetis* in natural streams are indeed responses to predation risk, not to size-selective predation.

As a further test we compared the relative contributions of predator-induced mortality and predator-induced reduction of prey fecundity to the instantaneous rate of *Baetis* population growth. We estimated trout and stonefly predation rates on *Baetis* from mesocosm experiments (Kerans et al. 1995, McIntosh et al. 2002, Peckarsky et al. 2008), and effects of non-feeding predators on *Baetis* fecundity from microcosm and mesocosm experiments (Peckarsky et al. 1993, Peckarsky and McIntosh 1998). Integrating those approaches showed clearly that predation by trout and stoneflies contributed very little to reduction in rates of *Baetis* population growth compared to predator-induced reduction of fecundity (McPeek and Peckarsky 1998).

As "the invisible present" unfolds, we are in the process of modifying our conceptual model to incorporate recruitment as well as post-recruitment processes. Studies in collaboration with Andrea Encalada have demonstrated that regional-scale processes associated with variation in hydrology and geomorphology (independent of predation) can limit recruitment (oviposition) of *Baetis*, thereby having a major influence on their population dynamics (Encalada 2005). Thus, we have proposed a new demographic model to explain the population dynamics of these mayflies (figure 9.4). To evaluate it, future research needs to focus on how recruitment (oviposition) interacts with post-recruitment processes such as trout and stonefly-induced changes in prey behavior, development, and mortality.

Discussion

The value of studying one place

Several key ingredients have contributed to the insights gained by our research group during nearly 40 years of studying one place. First, a long time-series of information on natural prey populations in one drainage basin has enabled us to establish a context for experimental evidence of the influence of predator-induced mortality on prey population dynamics. Second, our desire to understand and explain patterns observed in this system has necessitated the incorporation of processes spanning a wide range of levels of organization, from behavior of individuals to landscape configuration. Third, working in the same place has provided the circumstances for making discoveries. For example, detailed long-term knowledge of one system prepared us to recognize unusual or different patterns, such as the effects of unusually large snow pack and very high flow rates on prey population dynamics (Peckarsky et al. 2000). Fourth, focusing our studies in one place has also fostered the development of innovative methods, such as elaborate plumbing of micro- and mesocosms that use gravity and natural stream water, the presentation of predator cues while preventing prey consumption, whole-stream manipulation experiments (Allan 1982a, Peckarsky et al. 2002, McIntosh et al. 2004) and application of an electrofishing machine to sample invertebrates (Taylor et al. 2001, 2002). Finally, place-based research has fostered more persistence when we were faced with questions or approaches that led us astray or contradicted existing dogma.

Although we are tempted to argue that insights from place-based research can be extrapolated to other places (e.g., Flecker 1992), the validity of such extrapolation is debatable. Our data suggest that extensive prey dispersal in our open systems is what produces stronger nonconsumptive than consumptive predator effects. We have confirmed the logic of this speculation by collaborating with theoreticians to develop general models of the dynamics of prey populations in open systems (e.g., Abrams 2000, McPeek and Peckarsky 1998, Peckarsky et al. 2001). Nonetheless, we need to address whether more reliable generalizations can be made by thoroughly investigating one place than by comparing many systems.

Uncertainty regarding generalization to other systems raises a related question: What is the value of place-based research relative to that of a comparative approach? Place-based research potentially leads to conclusions that may only be valid in one place, and may foster a particular way of thinking about ecology. At worst, long-term study at one site can potentially lead to false generalizations of how natural systems function or are structured on a more global scale. Frequently, more is learned when observations from one location do not agree with predictions derived from another. For example, the weak consumptive effects of trout observed in Rocky Mountain streams do not explain prey distribution and abundance in slower-moving streams with sessile grazers that are eaten by predatory invertebrates more susceptible to predatory fishes (Power 1990, Power et al. 1992). Furthermore, elaborate and innovative studies in another place have emphasized how predators can induce evolved life-history responses in guppies of Trinidad streams (e.g., Reznick et al. 1990) in contrast to our documentation of phenotypic plasticity. We have attempted to avoid potential place-based bias by involving new people in our research program who bring perspectives from other systems (e.g., streams in the Venezuelan piedmont, Sweden, Ecuador, Australia, New Zealand, and Spain), and by testing alternative hypotheses suggested by research on other systems (e.g., Peckarsky et al. 2004).

While it is indisputable that insights are gained from cumulative studies and weight of evidence, we contend that with many studies from many locations it is more tempting to explain away inconsistencies as simple idiosyncrasy. In contrast, place-based research embraces rather than denies the importance of contingency, thereby forcing us to resolve contradictions between observations and theory or across individual experiments and lines of evidence. Over time those resolutions have converged into a powerful body of evidence. In our combined work we have arrived at a number of conclusions that contrasted with initial expectations: that top-down numerical control was weak, which was explained partly by the open nature of the system with donor-controlled prey subsidies, and partly by a suite of risk-sensitive prey behaviors. We argue that this painstaking approach underscores the value of place-based research, and thereby maintain that our time-tested generalizations arising from Rocky Mountain streams are well worth testing in other systems.

Whether our place-based research on the ecology of streams has contributed to ecological theories outside the discipline warrants particular attention because stream ecology has been criticized by one of its most avid practitioners for not contributing to general theory (Fisher 1997). In fact, the lack of general laws in ecology has been a fundamental criticism of many ecologists (e.g., Peters 1991, Lawton 1999; see also Pulliam and Waser, chapter 4). That being said, we suggest that studies on prey exchange rates in streams (Cooper et al. 1990), as well as larval and adult insect movement (Peckarsky et al. 2000, McIntosh et al. 2002) have been critical in shaping our understanding of other open systems (Speirs and Burney 2001, Anderson et al. 2005). Perhaps the most important contribution of our research is the realization that prey exchange rates and antipredator traits modulate consumptive effects of predators in open streams.

Long-term research without continuous monitoring

Another contentious issue related to place-based research is that of distinguishing its value from the well-established advantages of long-term environmental monitoring. Long-term studies are widely recognized as valuable (Strayer et al. 1986, Likens 1989, Cody and Smallwood 1996) and form an important component of the ecology of place because the sheer amount of information needed to understand complex ecological systems can only be accumulated over time. While many place-based research programs are built around continuous long-term monitoring, we argue that place-based research uses time to pursue a question, whereas long-term monitoring measures the effects of time on environmental parameters and interprets observed patterns with respect to historical measurements.

While our research on predator-prey interactions in Rocky Mountain streams has spanned almost four decades, our questions have evolved from the relationships among observations, experiments and ideas of many people rather than the continuous repeated sampling of organisms or processes by one or a few people that is more typical of long-term studies (e.g., Edmondson 1991). In our case, observations and experiments have continued to generate the fodder for future questions. Ironically, our approach has been viewed as an "intellectual weakness" which, according to a summary of a recent proposal funded by an National Science Foundation panel, is "an inductive boiling of hypotheses, with too little effort to place the work in a more general ecological context." On the contrary, we believe that our long-term place-based research, an approach in which questions boil up out of the system, is not only intellectually sound but also has contributed significantly to general ecological understanding.

How we got here

What we have learned may be easier to articulate than how we got here. We think that insights arose partly from the place, and possibly even more from the community of people associated with the place. Clearly, our conceptual growth was motivated by natural history, repeating observations in multiple streams, extracting system components and studying them in replicated arenas where interactions could be observed more directly. Nonetheless, we also attribute our progress to our open and sometimes heated discussions, which were very constructive in the evolution of new ideas and approaches. Ideas that initially seemed ridiculous or impossible (e.g., altering fish odor in whole streams) were tossed around until nonbelievers could be convinced rather than being quickly dismissed. Finally, many of our epiphanies came from being in the field—getting cold fingers pinched by heavy rocks, narrowly avoiding lightening strikes, and tolerating swarms of biting flies—where observations became new discoveries that would not have been possible by reading literature or solving equations. Thus, we got here by letting the place inform us of what was interesting and important, by sharing ideas, and by being open-minded and passionate about the work and the place.

Acknowledgments

We thank the following people for collaborating with us and improving this chapter. Stan Dodson and Jim Kitchell provided conceptual inspiration. Peter Abrams, Scott Cooper, Steve Ellner, Nelson Hairston, Billie Kerans, Dave Lytle, Mark McPeek, and Sandy Walde contributed theory. Maruxa Álvarez, Chester Anderson, Emily Bernhardt, Wendy Brown, Chris Caudill, Cathy Cowan, Jonas Dahl, Andrea Encalada, Brian Feiferick, Alex Flecker, Sean Gross, Matt Harper, Alison Horn, Jane Hughes, A. J. Jenkins, Billie Kerans, Steve Kohler, LeeAnne Martinez, Peter Ode, Todd Palmer, Marge Penton, Paul Reich, Sandy Walde, and Brooke Zanatelle did primary research and contributed new approaches and ideas. Leon Barmuta, Gail Blake, Don Conklin, Marcia Grimshaw, Bryan Horn, Steve Horn, Elise Jones, Menna Jones, Ben Koch, Dave Lytle, Kate Macneale, LeeAnne Mumpy, Eric Odell, Emma Pearce, Tracy Smith, Darcy Tickner, Tom Ticknor, Mark Wallin, Alison Weller, Sheila Wiseman, Lyle Zimmerman, Dave Zweig, and Ann Zweig helped us do our research and contributed new approaches and ideas. Jim Estes, Mary Price, and Ian Billick provided comments that improved earlier drafts of this chapter. Finally we thank Scott Wissinger, billy barr, and the Rabbit for providing inspiration, and our families for their support and indulgence of our fascination with stream ecosystems in the East River Valley.

References

Abrams, P. A. 2000. The impact of habitat selection on the spatial heterogeneity of resources in varying environments. *Ecology* 81:2902–13.

Allan, J. D. The diversity and distributional ecology of benthic insects in Cement Creek, Colorado. *Ecology* 56:1040–53.

———. 1978a. Diet of brook trout (*Salvelinus fontinalis* Mitchill) and brown trout (*Salmo trutta* L.) in an alpine stream. *Verhandlungen der Internationalen Vereinigung für Theoretische und Angewandte Limnologie* 20:2045–50.

———. 1978b. Trout predation and the size composition of stream drift. *Limnology and Oceanography* 23:1231–37.

———. 1981. Determinants of diet of brook trout (*Salvelinus fontinalis*) in a mountain stream. *Canadian Journal of Fisheries and Aquatic Sciences* 38:184–92.

———. 1982a. The effects of reduction in trout density on the invertebrate community of a mountain stream. *Ecology* 63:1444–55.

———. 1982b. Feeding habits and prey consumption of three predaceous stoneflies (Plecoptera) in a mountain stream. *Ecology* 63:26–34.

———. 1983. Food consumption by trout and stoneflies. Pages 371–390 *in* T.D. Fontaine III, and S. M. Bartell, eds. *Dynamics of Lotic Ecosystems.* Ann Arbor Science Publishers, Ann Arbor, Michigan.

———. 1987. Macroinvertebrate drift in a Rocky Mountain stream. *Hydrobiologia* 144:261–68.

Allan, J. D., A. S. Flecker, and N. L. McClintock. 1986. Diel epibenthic activity of mayfly nymphs, and its non-concordance with behavioral drift. *Limnology and Oceanography* 31:1057–65.

Allen, K. R. 1951. The Horokiwi Stream: A study of a trout population. *New Zealand Marine Department Fisheries Bulletin* 10.

Anderson, K. E., R. M. Nisbet, S. Diehl, and S. D. Cooper. 2005. Scaling population responses to spatial environmental variability in advection-dominated systems. *Ecology Letters* 8:933–43.

Brooks, J. L., and S. L. Dodson. 1965. Predation, body size, and composition of plankton. *Science* 150:28–35.

Cody, M. L., and J. A. Smallwood. 1996. *Long-Term Studies of Vertebrate Communities.* Academic Press, New York.

Cooper, S. D., S. J. Walde, and B. L. Peckarsky. 1990. Prey exchange rates and the impact of predators on prey populations in streams. *Ecology* 71:1503–14.

Cowan, C. A, and B. L. Peckarsky. 1994. Diel feeding and positioning periodicity of a grazing mayfly in a trout stream and a fishless stream. *Canadian Journal of Fisheries and Aquatic Sciences* 51:450–59.

Dahl, J., and B. L. Peckarsky. 2002. Induced morphological defenses in the wild: Predator effects on a mayfly, *Drunella coloradensis*. *Ecology* 83:1620–34.

Diehl, S., S. D. Cooper, K. W. Kratz, R. M. Nisbet, S. K. Roll, S. W. Wiseman, and T. M. Jenkins Jr. 2000. Effects of multiple, predator-induced behaviors on short-term producer-grazer dynamics in open systems. *American Naturalist* 156:293–313.

Dodson, S. I., T. F. H. Allen, S. R. Carpenter, A. R. Ives, R. L. Jeanne, J. F. Kitchell, N. E. Langston, and M.G. Turner. 1998. *Ecology.* Oxford University Press, New York.

Dunham, A. E, and S. J. Beaupre. 1998. Ecological experiments: scale, phenomenology, mechanism, and the illusion of generality. Pages 27–49 *in* W. J. Resetarits Jr. and J. Bernardo, eds., *Experimental Ecology: Issues and Perspectives.* Oxford University Press, New York.

Edmondson, W. T., ed. 1991. *The Uses of Ecology: Lake Washington and Beyond.* University of Washington Press, Seattle.

Elton, C. S., and M. Nicholson. 1942. The ten-year cycle in numbers of the lynx in Canada. *Journal of Animal Ecology* 11: 215–44.

Encalada, A. C. 2005. *Patterns and Mechanisms of Selective Oviposition of* Baetis Bicaudatus *(Ephemeroptera; Baetidae) and Its Consequences to Population Dynamics.* PhD dissertation, Cornell University, Ithaca, NY.

Englund, G. 1997. Importance of spatial scale and prey movements in predator caging experiments. *Ecology* 78:2316–25.

Estes, J. A., M. T. Tinker, T. M. Williams, and D. F. Doak. 1998. Killer whale predation on sea otters: Linking oceanic and nearshore processes. *Science* 282:473–76.

Fisher, S. G. 1997. Creativity, idea generation, and the functional morphology of streams. *Journal of the North American Benthological Society* 16:305–18.

Flecker, A. S. 1992. Fish predation and the evolution of invertebrate drift periodicity: Evidence from neotropical streams. *Ecology* 73:438–48.

Kerans, B. L., B. L. Peckarsky, and C. R. Anderson. 1995. Estimates of mayfly mortality: Is stonefly predation a significant source? *Oikos* 74:315–23.

Lawton, J. H. 1999. Are there general laws in ecology? *Oikos* 84:177–92.

Likens, G. E. 1989. *Long-Term Studies in Ecology: Approaches and Alternatives.* Springer, New York.

Magnuson, J. J. 1990. Long-term ecological research and the invisible present. *BioScience* 40:495–501.

McIntosh, A. R., and B. L. Peckarsky. 1996. Differential behavioural responses of mayflies from streams with and without fish to trout odour. *Freshwater Biology* 35:141–48.

———. 1999. Criteria determining behavioural responses to multiple predators by a stream mayfly. *Oikos* 85:554–64.

———. 2004. Are mayfly anti-predator responses to fish odor proportional to risk? *Archiv für Hydrobiologie* 160:145–51.

McIntosh, A. R., B. L. Peckarsky, and B. W. Taylor. 1999. Rapid size-specific changes in mayfly drift caused by alterations in fish odour. *Oecologia* 118:256–64.

———. 2002. The influence of predatory fish on mayfly drift: Extrapolating from experiments to nature. *Freshwater Biology* 47:1497–1513.

———. 2004. Predator-induced resource heterogeneity in a stream food web. *Ecology* 85:2279–90.

McPeek, M. A., and B. L. Peckarsky. 1998. Life histories and the strengths of species interactions: Combining mortality, growth, and fecundity effects. *Ecology* 79:235–47.

Mittelbach, G. G., A. M. Turner, D. J. Hall, J. E. Rettig, and C. W. Osenberg. 1995. Perturbation and resilience: A long-term whole-lake study of predator extinction and reintroduction. *Ecology* 76:2347–60.

Paine, R. T. 1966. Foodweb complexity and species diversity. *American Naturalist* 100:65–75.

Peckarsky, B. L. 1980. Predator-prey interactions between stoneflies and mayflies: Behavioral observations. *Ecology* 61:932–43.

———. 1985. Do predaceous stoneflies and siltation affect the structure of stream insect communities colonizing enclosures? *Canadian Journal of Zoology* 63:1519–30.

———. 1991. Habitat selection by stream-dwelling predatory stoneflies. *Canadian Journal of Fisheries and Aquatic Sciences* 48:1069–76.

———. 1996. Alternative predator avoidance syndromes in stream-dwelling mayflies. *Ecology* 77:1888–1905.

———. 1998. The dual role of experiments in complex and dynamic natural systems. Pages 311–423 *in* W. J. Resetarits Jr. and J. Bernardo, eds., *Experimental Ecology: Issues and Perspectives.* Oxford University Press, New York.

Peckarsky, B. L., C. A. Cowan, M. A. Penton, and C. Anderson. 1993. Sublethal consequences of stream-dwelling predatory stoneflies on mayfly growth and fecundity. *Ecology* 74:1836–46.

Peckarsky, B. L., S. D. Cooper, and A. R. McIntosh. 1997. Extrapolating from individual behavior to populations and communities in streams. *Journal of the North American Benthological Society* 16:375–90.

Peckarsky, B. L., C. A. Cowan, and C. Anderson. 1994. Consequences and plasticity of specialized predatory behavior of stream-dwelling stonefly larvae. *Ecology* 75:166–81.

Peckarsky, B. L., J. M. Hughes, M. Hillyer, and A. C. Encalada. 2004. Are populations of mayflies living in adjacent fish and fishless streams genetically distinct? *Freshwater Biology* 50: 42–51.

Peckarsky, B. L., B. L. Kerans, A. R. McIntosh, and B. W. Taylor. 2008. Predator effects on prey population dynamics in open systems. *Oecologia* 156:431–40.

Peckarsky, B. L., and A. R. McIntosh. 1998. Fitness and community consequences of avoiding multiple predators. *Oecologia* 113:565–76.

Peckarsky, B. L., A. R. McIntosh, C. C. Caudill, and J. Dahl. 2002a. Stabilizing selection on male body size of high altitude populations of *Baetis bicaudatus* (Ephemeroptera: Baetidae). *Behavioral Ecology and Sociobiology* 51:530–37.

Peckarsky, B. L., A. R. McIntosh, B. W. Taylor and J. Dahl. 2002b. Predator chemicals induce changes in mayfly life history traits: A whole-stream manipulation. *Ecology* 83:612–18.

Peckarsky, B. L., B. W. Taylor, and C. C. Caudill. 2000. Hydrologic and behavioral constraints on oviposition of stream insects: Implications for adult dispersal. *Oecologia* 125:186–200.

Peckarsky, B. L., B. W. Taylor, A. R. McIntosh, M. A. McPeek, and D. A. Lytle. 2001. Variation in mayfly size at metamorphosis as a developmental response to risk of predation. *Ecology* 82:740–57.

Peters, R. H. 1991. *A Critique for Ecology.* Cambridge University Press, Cambridge.

Polis, G. A., and S. D. Hurd. 1996. Linking marine and terrestrial food webs: Allochthonous inputs from the ocean supports high secondary productivity on small islands and coastal land communities. *American Naturalist* 147:396–423.

Power, M. E. 1990. Effects of fish in river food webs. *Science* 250:811–14.

———. 1998. Experimentation, observation, and inference in river and watershed investigations. Pages 113–32 *in* W. J. Resetarits, Jr. and J. Bernardo, eds., *Experimental Ecology: Issues and Perspectives.* Oxford University Press, New York.

Power, M. E., J. C. Marks, and M. S. Parker. 1992. Variation in the vulnerability of prey to different predators: Community level consequences. *Ecology* 73:2218–23.

Reiners, W. A., and K. L. Driese. 2001. The propagation of ecological influences through heterogeneous environmental space. *BioScience* 51:939–50.

Reznick, D. N, H. Bryga, and J. A. Endler. 1990. Experimentally induced life-history evolution in a natural population. *Nature* 346:357–59.

Roughgarden, J., S. D. Gaines, and S. Pacala. 1987. Supply side ecology: The role of physical transport processes. Pages 491–518 *in* P. Giller and J. Gee, eds., *Organization of Communities: Past and Present.* Blackwell Publishers, London.

Speirs, D. C., and W. S. C. Gurney. 2001. Population persistence in rivers and estuaries. *Ecology* 82:1219–37.

Stenseth, N. C., W. Falck, O. N. Bjørnstad, and C. J. Krebs. 1997. Population regulation in snowshoe hare and Canadian lynx: Asymmetric food web configurations between hare and lynx. *Proceedings of the National Academy of Sciences (USA)* 94:5147–52.

Strayer, D., J. S. Glitzenstein, C. G. Jones, J. Kolasa, G. E. Likens, M. J. McDonnell, G. G. Parker, and S. T. A. Pickett. 1986. *Longterm Ecological Studies: An Illustrated Account of*

Their Design, Operation, and Importance to Ecology. Occasional Publication of the Institute of Ecosystem Studies, Millbrook, NY, no. 2, 38 pages.

Taylor, B. W, A. R. McIntosh, and B. L. Peckarsky. 2001. Sampling stream invertebrates using electroshocking techniques: Implications for basic and applied research. *Canadian Journal of Fisheries and Aquatic Science* 58:1–9.

———. 2002. Large-scale invertebrate manipulations in stream ecosystems: Invertebrate removal, algal response, and opportunities for innovation. *Limnology and Oceanography* 47:893–99.

Wilcox, A. C, B. L. Peckarsky, B. W. Taylor, and A. C. Encalada. 2008. Hydraulic and geomorphic effects on mayfly drift in high-gradient streams at moderate discharges. *Ecohydrology* 1:176–86.

Note

1. Our approaches often met with resistance. For example, one anonymous reviewer of a proposal submitted by BLP to the National Science Foundation wrote: "At the outset, I confess that I'm not an enthusiastic believer in community organization studies involving manipulations (simplification) of community structure."

10 *The Ecology of Place in Oak Forests*

PROGRESSIVE INTEGRATION OF PAIRWISE INTERACTIONS INTO WEBS

Richard S. Ostfeld and Clive G. Jones

Figure 10.0. Above: a field assistant checking tree seed traps for the presence of acorns and other mast. Data from the seed traps allow us to quantify the acorn "engine" that drives the dynamics of rodent populations, gypsy moths, ticks, spiro-chetes, songbird nesting success, and other forest processes. Below: a field assistant "drag-sampling" for blacklegged ticks to quantify risk of human exposure to Lyme disease. Tick abundance can be predicted from knowledge of acorn production two years earlier—a finding that was facilitated by a place-based approach. Photographs by Kelly Oggenfuss, 2006.

Abstract

Our studies of ecological dynamics in oak forests of southeastern New York State began in 1980 with the monitoring of an introduced insect, the gypsy moth. After observing a complete population cycle and massive defolia-tion event, we developed hypotheses about the role of white-footed mice in influencing moth dynamics via predation during the pupal stage. Long-term monitoring and experimental manipulations of mouse populations, begun in 1991, led to further observations concerning both the causes (vari-able acorn production) and consequences of fluctuating mouse abundance (altered densities of blacklegged ticks, proportion of ticks infected with Lyme disease–causing bacteria, variable nesting success of ground-nesting songbirds, changing prey selection and abundance of hawks and owls, and system feedback effects of moth defoliation on acorn production). We begin the chapter by providing a brief historical background to our research, and then describe the mast-driven, moth-influenced interaction web as we currently understand it. We then describe how we have fostered an increasingly robust and predictive understanding of this system by combining a more reductionist focus on pairwise interactions with a more holistic exploration of additional entities (species and interactions) that provide the ecological contexts in which pairwise interactions take place. We end by summarizing some of the key features of this approach that we found valuable to the understanding of place.

Introduction

For ecologists, the decision about where to undertake field studies, similar to decisions about which taxa or processes on which to focus, can be made in at least two different ways. One can start by identifying the most compelling concept or theory to examine, and then select a field site (or taxon/process) most suitable for addressing the concept. Alternatively, one can choose a habitat type (or taxon/process) that seems inherently interesting or convenient, and then identify the most conceptually compelling questions that can be asked of that system. We suspect that ecologists often pick their field sites on the basis of personal interest and convenience. Convenience certainly played a strong role in our focus on eastern deciduous forests. This habitat is abundant on the protected and accessible property of the Institute of Ecosystem Studies (IES) in southeastern New York State, where we are employed. However, these forests are ecologically important on a global scale (Pacala et al. 2001) and ideal habitats for testing theories in community ecology.

Eastern deciduous forests are highly changeable places. Two of the most strikingly variable phenomena in these forests are periodic, massive defoliation events caused by gypsy moths (*Lymantria dispar*), and episodic production of enormous acorn crops (i.e., masting) by oak trees (*Quercus*). The conspicuousness of these events draws ecological scrutiny, and our studies have revealed that gypsy moth outbreaks and acorn masting are mechanistically linked to each other and to many other taxa in the forest. Here we provide a personal, historical perspective on how our place-based studies developed, we describe the dynamics of key players and processes in oak forests as we currently understand them, and we discuss the approach we have taken to gain an increasingly robust and predictive understanding of this system. Although the history of our studies is unique, for reasons described below, we fully expect that the insights derived from them will be exportable to many other ecological systems.

Historical Background to Our Research

One of us (Jones) arrived in Millbrook, New York, in the summer of 1980 to witness the second year of a major gypsy moth outbreak that defoliated the oak-dominated forests of the Tea House Hill in the center of what is now the Cary Institute property. While he originally planned to conduct studies on how host-plant chemistry might affect this insect, the defoliation event suggested that trying to understand the causes of such outbreaks might be a more fruitful avenue of research. The gypsy moth (*Lymantria dispar*) is the most important defoliator of oak forests of the eastern United States (United States Department of Agriculture Forest Service 1994). This insect was introduced in Medford, Massachusetts, in 1868 or 1869 by Étienne Léopold Trouvelot, a French artist, astronomer, and amateur entomologist who fancifully imagined that he could create a hardy silk moth by crossbreeding the silk moth and the gypsy moth. The latter escaped from captivity and invaded northeastern forests (Liebhold 1989).

The gypsy moth (hereafter "moth") is a generalist folivore, but it prefers oaks. Moth populations tend to undergo population cycles with about 8 to 10 years between peaks (Elkington and Liebhold 1990), and with annual densities varying over five orders of magnitude (figure 10.1). Defoliation is detectable when moth abundance exceeds approximately 500 egg masses per hectare, and is virtually complete above approximately 3,000 masses per hectare (Gottschalk 1993). The consequences of major defoliation events, which can extend for thousands of square kilometers,[1] are profound and include high tree mortality, especially of conifers; heavy economic losses; curtailed growth and seed production of surviving trees; altered nutrient cycling; altered habitat suitability for other species via increased exposure of the forest floor; aesthetic damage to plantings; and even an adverse human health influence via the urticating hairs of larvae (Doane and McManus 1981, Gottschalk 1990, Lovett et al. 2002).

At the high densities associated with defoliation, moth populations are subject to strong density-dependent mortality arising from a combination of food limitation, specialist pathogens (nuclear polyhedrosis virus, *Entomophaga maimaiga* fungus), and parasites (e.g., *Blepharipa pratensi, Parasetigena silvestris, Brachymeria intermedia*). Thus, peak years are normally followed by dramatic population crashes and several years of low moth densities (Elkinton and Liebhold 1990, Dwyer et al. 1998). By the early 1980s much was known about how food limitation and natural enemies contributed to

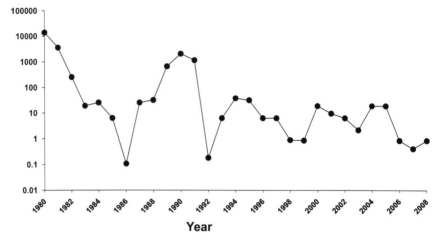

Figure 10.1. Density of gypsy moth egg masses on the long-term monitoring plots at the Cary Institute of Ecosystem Studies in Millbrook, New York. Note the log scale of the *x*-axis. Updated from Goodwin et al. (2005).

moth population crashes. Consequently, we chose to focus on the mystery underlying the release of low-density moth populations (<100 masses per hectare) during the sustained intervals between population peaks. It had long been postulated that small mammals played a key role in this phase (Bess et al.1947), and so our studies began in 1991 with both monitoring and experimental manipulations of a known predator on the pupal stage of the moth, the white-footed mouse, *Peromyscus leucopus* (hereafter "mouse").

To evaluate the connections between mouse and moth population dynamics, we established two large (2.25-ha) live-trapping grids in oak-dominated forests at the Cary Institute in the fall of 1991 (figure 10.2). Almost immediately, our studies began to reveal a strong connection between acorn masting[2] and rodent abundances. Fieldwork that fall was particularly difficult because the copious acorns that littered the forest floor made walking up or down steep hills hazardous. Of course we were experiencing a mast year, and in some areas more than 100 acorns occurred in an average square meter of forest floor. The first indication that masting was influencing mice was that mouse populations had not declined between the fall trapping session in 1991 and resumption of trapping in spring 1992 (typically, mouse populations decline sharply over winter). In addition, juvenile mice began appearing in our traps at least a month earlier in spring of 1992 than in an average year, and by August the mouse population had exploded to more than 100 individuals per hectare.

Although the phenomenon of acorn masting had been described much earlier, the implications of this pulsed resource (Ostfeld and Keesing 2000)

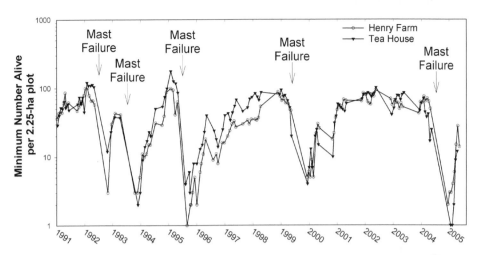

Figure 10.2. Density of white-footed mice on the two long-term monitoring plots at the Cary Institute of Ecosystem Studies in Millbrook, New York. Years of mast failure (<5 acorns m^{-2}) are indicated.

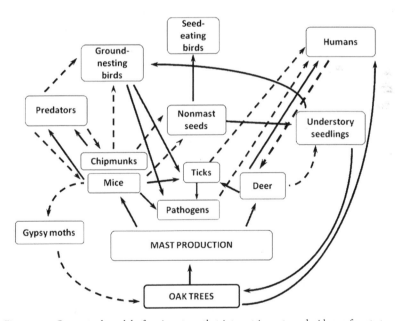

Figure 10.3. Conceptual model of various taxa that interact in eastern deciduous forests to influence forest dynamics, Lyme-disease risk ("pathogens"), nesting success of some ground-nesting songbirds, and gypsy moth dynamics. Arrows indicate the direction of the effect of one taxon on another. Solid lines indicate positive effects (e.g., mast production on mice), and dashed arrows indicate negative effects (e.g., gypsy moths on oak trees).

for the dynamics of terrestrial vertebrate communities were largely unexplored. Beginning with the population responses of mice to variable acorn production, we began to explore community-level responses to masting (figure 10.3).

Current Understanding of the Oak Forest Interaction Web

Our observational and experimental studies and those of others (Wolff 1996, Elkinton et al. 1996) indicate that autumnal acorn production explains a large percentage (as much as 80%) of the variance in summer population density of white-footed mice and eastern chipmunks (*Tamias striatus;* Ostfeld et al. 1996a, 1998, 2001, 2006; Jones et al. 1998). Acorns are important food for rodents because they are nutritious and can be stored; overwinter survival of mice is higher and overwinter breeding is more likely immediately following an autumn of heavy acorn production. Consequently, rodent populations enter the spring breeding season at relatively high densities and reach high mid- to late-summer peaks. Conversely, years of poor acorn production are followed by low spring densities of rodents, which usually result in low summer densities (figure 10.2). The summer densities of mice and chipmunks have profound but species-specific impact on several other key interactions in these forests.

The moth connection

The crashes in mouse population density that typically follow acorn failures (figure 10.2) result in a marked reduction in rates of predation on gypsy moth pupae. Gypsy moth caterpillars are apparently distasteful to most insectivorous vertebrates, but the pupae are large, undefended, and relatively accessible, occurring low on tree trunks or on the forest floor. Our experimental and observational evidence (Ostfeld et al. 1996a, Jones et al. 1998), as well as that of others (e.g., Elkinton et al. 1996), provides strong evidence that relaxed predation on moth pupae can initiate an outbreak—a phase of rapid population growth (ca 1.5 orders of magnitude density increase per year) which, if sustained over a period of two to three years, can result in moth densities that cause defoliation. Relaxed pupal predation by mice is much more likely to initiate a defoliating outbreak if the mouse crash occurs during a year of moderate moth density (10–50 masses per hectare) than if it occurs during a year of low moth density (< 10 masses per hectare). If moth populations are at moderate densities when mouse populations crash following acorn failure, the moths may then reach densities that can escape mouse regulation[3] even if mouse populations then rebound following a masting event. In contrast, if moth populations are at low densities when mice crash following acorn failure, moths may fail to reach densities that can escape regulation by mice before the mouse populations rebound after a masting event. Furthermore, monitoring as well as experimental (Jones et al. 1998) and empirically derived modeling studies (Goodwin et al. 2005, Schauber et al. 2005) all show that moderate to high mouse densities can cause local moth extinction at the scale of less than one hectare if moth densities are already low at the beginning—the significance of which we will return to later. In contrast to the large effects of mice on moths, we have good evidence (E. M. Schauber, C. G. Jones, and R. S. Ostfeld, unpublished data) that chipmunks are not involved in any meaningful trophic interactions with gypsy moths. Finally, and relevant to our later discussions, we know from other studies that the moth can suppress acorn production for time periods of 2 to 10 years, depending on the intensity and frequency of defoliation (Gottschalk 1990).

The Lyme disease connection

Natural enemies of rodents, including predators and parasites, are sometimes capable of regulating rodent densities (Ostfeld and Holt 2004), and the predators and parasites themselves can be regulated by prey or host densities. Our initial monitoring of mouse and chipmunks populations at IES revealed striking levels of parasitism by ticks and infection with the causative agent of Lyme disease, *Borrelia burgdorferi*, prompting us to address the causes and consequences of tick burdens on rodents. The hypothesis that mouse populations were regulated by these ectoparasites was rejected when we found a positive rather than negative correlation between tick burdens and the length of time individual mice were present on our sites (Ostfeld et al. 1996b). But addressing the effects of fluctuating rodent populations on ticks and pathogens was more complicated.

Prior studies had demonstrated that mice and chipmunks were important in determining Lyme disease risk because they were key hosts for immature stages of the tick vector (*Ixodes scapularis* in eastern and central North America) and key reservoirs for *B. burgdorferi* (Mather et al. 1989, Lane et al. 1991). However, no previous research had asked whether temporal fluctuations in rodent abundance influenced the dynamics of tick or *B. burgdorferi* populations. Newly hatched larval ticks feed on a great variety of mammalian and avian hosts, but tend to feed successfully and acquire *B. burgdorferi* most efficiently from these two rodent hosts (LoGiudice et al. 2003). Consequently, we expected that high summer densities of mice and chipmunks would provide ample opportunity for larval ticks to feed on these hosts, which in turn would lead to greater numbers of infected nymphs the following year. Recent analyses of 14 years of monitoring data reveal that interannual variation in the key ecological risk factors for Lyme disease—the proportion of nymphal ticks infected with *B. burgdorferi* and the density of infected nymphal ticks—is caused largely by interannual variation in the abundance of acorns, mice, and chipmunks. Mice and chipmunks have similarly strong, positive effects on the density of nymphs, and acorn abundance is the best predictor of nymphal infection prevalence (Ostfeld et al. 2006).

Our early studies suggested an additional pathway by which acorns might indirectly affect numbers of ticks. White-tailed deer are important consumers of acorns and are attracted to oak-dominated forest stands during times of heavy acorn production, whereas they avoid oak stands in autumns of poor acorn production (McShea and Schwede 1993). Deer are an important host for adult blacklegged ticks, which are active in the fall. Consequently, we expected that heavy acorn production would attract deer to oak stands, causing them to import their burdens of adult ticks into those stands. These oak-dominated areas would then be the primary sites where adult ticks would complete their blood meals, drop off their deer hosts, and oviposit. We therefore predicted that larval ticks should erupt in oak stands during the summer following heavy acorn production. Experimental simulation of masting on three 2.25-ha forest plots resulted in substantial increases in both host-seeking larval ticks and larval ticks attached to mice (Jones et al. 1998), consistent with these expectations. However, long-term monitoring of natural variation in masting, deer abundance (from bow-hunter counts) and activity (from winter browsing intensity), and larval tick populations has revealed that interannual variation in deer abundance and activity has no detectable impact on interannual variation in larval tick abundance (Ostfeld et al. 2006). Perhaps even more important, the abundance of larval ticks in a given year is a poor predictor of nymphal tick abundance the following year (Ostfeld et al. 2006), further weakening any supposed effect of deer distribution in the fall on nymphal tick abundance or distribution two summers later. Apparently the experimental deployment of acorns on comparatively small forest plots attracted deer and their attached adult ticks to these locally enriched areas, resulting in local elevation of larval tick abundance the following summer. But natural masting events, which involve enrichment over enormous geographic areas, do not similarly result in localized increases in abundance of larval ticks because mast-consuming deer are distributed over much wider areas. In moderate or low mast years, when acorn production is limited to scattered individual trees or patches of trees, we expect that deer will find these locales, transport engorged female ticks there, and facilitate localized larval outbreaks the following summer. These local larval hotspots are unlikely to be detectable from large-scale tick monitoring, and they might not have strong epidemiological relevance, particularly if locally

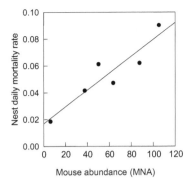

Figure 10.4. Relationship between annual nest daily mortality rate (Veery) and annual mouse abundance (MNA: minimum number alive). Each data point represents one year. Annual MNA was calculated as the mean of six 2.25-ha trapping grids. Modified from Schmidt et al. (2006).

elevated larval densities do not persist into the nymphal stage. Apparently, local abundance of mice and chipmunks is a stronger determinant of subsequent density of nymphs than is abundance of larvae (Ostfeld et al. 2006).

The bird connection

Mice and chipmunks are trophic generalists, feeding primarily on seeds, fruits, fungi, and various invertebrates. Both species, however, will eat unusual prey when they happen upon them—for example, eggs of ground- and shrub-nesting songbirds such as the veery (*Catharus fuscescens*) and wood thrush (*Hylocichla mustelina*)—in a phenomenon we have termed incidental predation (Schmidt et al. 2001). Long-term monitoring of predation rates on natural nests reveals that daily nest mortality rates for veeries are a strongly positive linear function of concurrent rodent density (figure 10.4). In years of high mouse and chipmunk density, nest predation rates are so high that population-level nesting success of veeries remains below the level necessary for replacement, and populations decline. Both these rodent species tend to undergo population crashes in the year following peak density (two summers following heavy acorn production), and this results in relaxed predation rates on songbird nests. In those years, however, hawks and owls tend to exert heavy predation pressure on fledglings, also leading to low reproductive success of songbirds. Apparently, when rodents are scarce, these raptors switch to feeding on songbirds. Some evidence also suggests that raptors are more abundant in the year following high rodent densities, owing to the raptors' high fledging success when food is abundant. This numerical raptor response to rodents, combined with the prey-switching just described, contributes to heavy predation pressure on fledgling and adult songbirds one year following a rodent peak (Schmidt and Ostfeld 2003a, 2003b, 2008). Only in years of moderate rodent densities do the songbirds appear to have positive population growth within the forest. In these years rodent predation on nests is only moderate, and raptor attack rates on fledgling and adult birds are modest (Schmidt and Ostfeld 2003b, 2008).

From the Pairwise to the System-Wide

In many respects, our basic research strategy has been consistent with traditional approaches to community ecology, in which interactions between pairs of species are assessed by observational or experimental studies, and interaction strength is some average measure of the variation accounted for by one of the pair (e.g., acorns on mice, mice on moths, chipmunks on birds). However, we have found this approach to be insufficient for understanding community dynamics in this particular place, and below we articulate some of the reasons why.

Incidental predator-prey relationships

Traditional approaches to predator-prey interactions tend to assume interdependence of the two taxa such that predators have a negative impact on prey population growth and prey have a positive impact on predator population growth. In other words, predators and prey are generally considered to have coupled population dynamics. In our system, however, the key rodent predators (mice and chipmunks) and their prey (gypsy moths and songbirds) interact asymmetrically—that is, they are decoupled. Although rodents can clearly suppress populations of gypsy moths and songbirds, neither of these prey taxa has any detectable direct influence on rodent population dynamics (but see below for indirect influence of moths on rodents). The lack of bottom-up effect of prey on rodent predators is a consequence of two related factors. First, both moth pupae and songbird nests are typically rare, in the sense that a typical mouse or chipmunk home range will likely include only one to several individual prey (a notable exception occurs during gypsy moth outbreaks). Second, because both rodent species are extreme trophic generalists, their dynamics are unlikely to be influenced by any particular prey type (with the notable exception of acorn mast). Neither individual fitness nor population dynamics of predators can be strongly influenced by such rare, incidental prey. Instead, rodent abundance in any given year is influenced strongly by prior acorn production and possibly by mammalian and avian predators. This decoupling of predator and prey forced us to look beyond the pairwise milieu for other factors influencing interspecific dynamics. In this case it required recognizing an independent resource base (acorns) for mice and chipmunks with its own attendant influences and controls (which we do not yet understand) that have profound effects on mice and chipmunks, while the latter, directly at least, have no effect on acorns.[4]

A hub-and-spoke interaction web

Based on the above, we can recharacterize this forest community to encompass the incidental asymmetry and independent resource base as a web of interactions markedly affected by pulsed acorn production (Ostfeld and Keesing 2000). Acorns have their direct effect on rodent consumers; the rodents in turn are at the nexus or "hub" of at least three additional strong interactions or "spokes," such that their summer abundance influences (1) Lyme disease risk; (2) gypsy moth dynamics, and (3) the dynamics of ground-nesting songbirds and possibly raptors. A series of pairwise interactions between the rodents and their prey, parasites, pathogens, and predators allows us to predict response variables of human health (Lyme disease), forest health and economics (gypsy moth defoliation impacts), and conservation importance (songbird abundance) as a function of acorns. This hub-and-spoke analogy is a useful way to integrate the diversity of pairwise interactions (acorns to consumers, consumers to prey and parasites) into a system, but it is still incomplete.

Feedback

Community ecologists recognize that system-level feedbacks arising from indirect connections among players can potentially confound understanding based on linear-pairwise interactions (Wootton 1994). As noted above, gypsy moths can affect acorn production for time periods of 2 to 10 years, depending on defoliation intensity and frequency (Gottschalk 1990). Although we have yet to experience major forest defoliation during the period when we have been simultaneously measuring all the other players and their interactions, from our extant understanding we can nevertheless envisage some ramifications of this connection. Since mouse populations are driven by acorns, moth populations are driven by mice (at least from the perspective of outbreak initiation, suppression, and persistence), and moths affect acorns, these three pairwise connections create a system-level feedback that has fundamental ramifications for the dynamic behavior of the rest of the interaction web (Jones et al. 1998). Since chipmunks, like mice, are driven by acorns, defoliation-driven changes in masting can potentially affect chipmunk abundance. So even though moths are not a significant prey of chipmunks, and even though chipmunks have no detectable influence on moths, moths can nevertheless potentially affect chipmunks. Similarly, over appropriate time scales, the asymmetrical influence of mice on moths can potentially be somewhat offset by a moth influence on mice via acorn production. Finally, by implication, the connections between these rodents, Lyme disease risk, and birds can potentially be affected by moth-induced changes in masting. Thus, in the absence of a moth feedback the community in the oak forest can best be viewed as being donor-controlled by pulsed resources (acorns) with the multiple ramifications being due to the hub rodents and their spoke-like connections to moths, Lyme disease risk, and birds. In contrast, the moth feedback creates a system with joint donor (acorns on mice, mice on moths) and recipient (moths on oaks/acorns) resource control.

Beyond a mean field view

Our initial goals of predicting community dynamics, like many such studies, focused on developing a mean field view of the interconnections. In other words, what, on average, were the relationships in time and space? While our long-term data on players (e.g., masting and mice) were essential for developing predictive relationships, the resulting regression equations represent what is expected on average over time, not what will happen in a particular year or series of years. Similarly, while such relationships generally hold for all the studied plots of approximately two hectares in size within our forest, these plots do not all behave exactly the same. For example, the variation among plots in abundance of mice or chipmunks can exceed a factor of two in any given summer. Both the temporal sequence of events and spatial variation in interactions may have important ramifications (see Grant and Grant, chapter 6), many of which we are only now beginning to explore. In this sense, then, like most community ecologists, we recognize the importance of spatial and temporal heterogeneity, and the need to move beyond a mean field view to specifically encompass spatial and temporal variation within this forest system.

Our first recognition of the need to explicitly address temporal variation came in the design of our acorn-addition and mouse-removal experiments when we added acorns in a mast-failure year and removed mice following a high mast year (Jones et al. 1998). It was clear that adding acorns in a masting year, or removing mice when they were at very low densities following mast failure, would not be conducive to elucidating relationships. Carteblanche application of experimental protocols without regard to temporal context would likely have caused us to falsely reject our hypotheses.

We now know that most of the interactions we have assessed vary strongly from year to year; the nature of the forest changes through time in many important ways. For example, summers following a heavy autumnal mast crop are characterized by abundant rodents, few adult female gypsy moths laying egg masses, and few fledgling birds. The following year tends to be characterized by few rodents, few gypsy moth caterpillars, and little likelihood of defoliation, more fledgling birds, and many infected nymphal ticks. A summer of heavy gypsy moth defoliation is characterized by brighter, hotter conditions on the forest floor, which we suspect may decrease habitat quality for rodents and ticks, but may also increase understory growth and possibly protective cover for ground-nesting songbirds. Finally, the years following defoliation may be characterized by reduced masting frequency and/or intensity. Clearly, oak forests are very different places in some years than in others, and any attempt to predict specific dynamics must necessarily include such considerations. The temporal mean field view, while very valuable, is thus insufficient.

The same basic argument can be made with respect to spatial averaging. Strong spatial heterogeneity exists in this oak forest system. Although masting events tend to extend over thousands of square kilometers (Koenig and Knops 2000), mast production varies considerably among our various ~two-hectare study plots, as do abundances and vital rates of rodents, gypsy moths, ticks, and songbirds. Similarly, we have identified "hot spots" within study plots where rodents concentrate their activity, and in which survival of gypsy moths and bird nestlings is negligible, as well as "cold spots" where rodents are scarce or absent and their prey survive well. These "hot" and "cold" spots persist for varying lengths of time and may or may not reappear in the same locales over time. For both moths and birds, predator-free (or predator-poor) space is heavily exploited by prey either passively (in the case of gypsy moths) or actively (by songbirds, which preferentially nest in these spots; Schmidt et al. 2003a, 2006). The most likely determinants of this spatial variation in rodents' foraging intensity are the rodents' own predation risk (e.g., cover, abundance of predators) and the availability of heterogeneous foods.

Such hot and cold spots may have important ramifications for dynamics at larger spatial scales. For example, mouse predation on moth pupae is so intense that they routinely drive moth populations extinct at spatial scales of less than one hectare. Nevertheless, in 23 years of monitoring, moths have managed to persist at spatial scales of more than one hectare. Our recent studies (Goodwin et al. 2005; Schauber et al. 2007, in review) have shown that persistence of moths in the forest—hence the potential for an acorn feedback from defoliation—may critically depend on the distribution of these hot and cold spots relative to larval dispersal (adult females are flightless). While moths are driven extinct by mice in hot spots, they persist and passively build up in density in mouse cold spots. Although a mouse cold spot can become a hot spot over time (and vice versa), if another cold spot is within the rather limited larval dispersal distance of 30 to 45 m, some moths may reach these safe havens. The end result is overall moth persistence in the forest despite intense local predation. If larval gypsy moths dispersed at greater distances from the hatch site to become more evenly spread out in the forest, their local density would be sufficiently low that they would run a high risk of extinction even in a mouse cool spot. Similarly, if predation risk by mice were spatially more homogeneous, the lack of cold spots would also confer a high risk of extinction (Goodwin et al. 2005). Although we are only just beginning to understand some of the many ramifications of spatial variation for community dynamics, it is nevertheless clear, as was the case with temporal variation, that a spatial mean field view, while valuable, is also insufficient.

Historical context

While community ecologists largely study contemporaneous interactions, a historical perspective is often valuable with respect to generalization and prediction (see Grant and Grant, chapter 6; Estes, chapter 8). In "pristine" systems a tacit historical context is provided by evolution; contemporary interactions are often known or presumed to be the result of evolutionary associations. Thus, in the absence of some major change in such systems, the contemporaneously observed interactions are likely to remain relatively unchanged, even though they may vary due to extrinsic factors (e.g., climatic fluctuations). However, community ecologists are increasingly studying the non-pristine—for example, the effects of invasive species, biodiversity loss, or the impact of altered land use. When contemporary interactions are the result of historical accident, or reflect transient dynamics due to recent change, it is more difficult to assess the likelihood of persistence of the interaction, or to distinguish generality from what might be local idiosyncrasy. Some analysis of the historical context can therefore be of considerable value, and in our work, this is certainly the case.

In many respects, the structure and functioning of present-day oak forests in eastern North America are idiosyncratic consequences of the past. Species extinctions, near extinctions, invasions, and land use changes over the past few centuries have almost certainly altered both the dynamics and consequences of each ecological interaction we have studied. One crucial historical factor was the widespread deforestation of eastern landscapes for subsistence agriculture in the seventeenth through early nineteenth centuries (e.g., see Foster and Motzkin 2003). During precolonial times the landscape of New York and New England was approximately 90% forested, and at the height of subsistence agriculture this value plunged to about 10%. For well over a century, most of the modern-day key players in eastern forests were either scarce or absent—these included oak trees, deer, ticks, *Borrelia burgdorferi*, forest-dwelling songbirds, and of course the introduced gypsy moth. White-footed mice and eastern chipmunks were probably widespread, albeit much less abundant than they are today. Strong evidence suggests that the forests that regrew following the widespread abandonment of agriculture that began in the early to mid-nineteenth century were characterized by tree species composition markedly different from that of precolonial forests (Foster and Motzkin 2003).

The extinction of the passenger pigeon (*Ectopistes migratorius*) and the functional extinction of the American chestnut (*Castanea dentata*), followed by the expansion and increased relative abundance of oaks, almost certainly have caused profound changes to the functioning of eastern forest ecosystems. Passenger pigeons occurred in enormous flocks, sometimes numbering in the millions of individuals (Fuller 1987, Schorger 1955), and they were apparently prodigious consumers of large tree seeds, including those of masting species, before the seeds fell to the forest floor. Their intensive foraging and their ability to move rapidly into areas of heavy seed production almost certainly reduced the availability of abundant mast to forest floor granivores. One would therefore suspect that granivorous mammals such as mice and chipmunks had lower average abundances and reduced interannual variation when passenger pigeons were abundant, and therefore that extinction of these birds has been responsible in part for interannual variation in Lyme disease risk (Blockstein 1998), gypsy moth abundance, and songbird nesting success. The American chestnut, now functionally extinct due to the presence of chestnut blight, *Cryphonectria parasitica*—an exotic fungus accidentally introduced around 1900—produced large, nutritious nuts which, according to human harvest records, apparently varied relatively little in abundance from year to year. Their replacement by oaks, which are notorious for great interannual variation in seed production, also likely contributed strongly to the dramatic interannual variation in rodent abundance and their knock-on effects.

Historical evidence suggests that the abundance of white-tailed deer was extremely low throughout the eastern United States during and after the height of subsistence farming. Only after a multi-decade lag following farm abandonment did deer populations begin to increase markedly (McShea and Healy 2002). Apparently, deer thrive in landscapes that consist of interspersed patches of forests, agricultural land, and suburbia, and they have further benefited from the extirpation of predators such as wolves (*Canis lupus*) and mountain lions (*Felis concolor*). As a consequence, deer abundance in the late twentieth and early twenty-first centuries is unprecedented, with potentially strong impacts on Lyme disease risk. Apparently, the population growth and spread of deer was a necessary (but not sufficient) condition for the subsequent spread of the tick vector of Lyme disease. In addition, because hyperabundant deer can profoundly reduce forest understory complexity and biomass, direct negative impacts on both rodent abundance and ground-nesting songbirds have been detected (McShea 2000). The potential for strong indirect impact on ticks, rodents, and songbirds seems strong, as deer browsing can strongly inhibit recruitment of some tree species (including oaks) and favor others (McShea and Healy 2002).

The intentional importation of the gypsy moth and its accidental escape clearly has had profound consequences for the structure and functioning of eastern deciduous forests. Another, less well-understood introduction may be in the process of altering gypsy moth effects on forest dynamics. The entomopathogenic fungus, *Entomophaga maimaiga*, was originally released near Boston in 1911 to control the gypsy moth. Its disappearance from the forests shortly thereafter suggested that this species would fail as a biocontrol agent. However, *E. maimaiga* "reappeared" in the Northeast in 1989, and it appears capable of reducing gypsy moth numbers at lower density thresholds than many of the other introduced biocontrol agents (perhaps especially in years of moist summer conditions). Population reductions due to this fungus can be sufficient to prevent moth populations from reaching densities that completely defoliate forests. This occurred in 1992 at IES with a curtailed outbreak that resulted in only light, patchy defoliation, and the same phenomenon also occurred elsewhere. However, the degree to which this fungus will reduce the frequency and magnitude of defoliation in the future is largely unknown. Certainly the extensive moth defoliation in 2005 and 2006 in various parts of the Northeast suggests that any expectation of the moth's demise as a major defoliator may be premature.

The forests of New York and New England continue to change. A large-scale decline in the relative abundance of oaks appears to be underway, and some suggest that this decline will be accompanied by increased basal area of red maples (Rodewald 2003). Although the seeds of red maples are consumed by mice and chipmunks (Schnurr et al. 2002, 2004), this species tends to produce fairly consistent seed crops each year and also disperses seeds in spring rather than in autumn. The consequences of the potential replacement of oaks by red maples can only be surmised; one might expect lower-amplitude rodent fluctuations from a predominant seed source that is more constant among years and unavailable over winter (since the rodents cannot cache these seeds). In addition, because of the timing of waves of farm abandonment from the 1800s to the 1930s and the regrowth of oaks and other masting species (e.g., hickories), we may also expect that the changing age structure of these trees will result in changes in forest dynamics and masting. Many oaks and hickories that regrew after the most recent periods of farm abandonment are now approaching the age of peak mast production, whereas in other regions these trees are now or will soon be past the age of peak production. We might therefore expect changes to these mast-driven interaction webs to continue into the future, perhaps with the expectation of reduced mast in areas that were abandoned to forest early, and increased mast where abandonment was late. Last and perhaps most ominously, two exotic pests and pathogens of relevance to our oak forests, recently introduced elsewhere in the United States, are potentially poised to invade in the future: the Asian long-horned beetle (*Anoplophora glabripennis*) that attacks maples, and *Phytophthora ramorum,* the agent that causes "sudden oak death" (Campbell and Schlarbaum 2002).

The above perspective, while necessarily speculative about past events and future trends, shows that the contemporary structure and dynamics of interactions in our oak forest communities is largely a result of history, and is likely to remain so as the result of human actions into the future. The history does not preclude our ability to understand this place at this time, nor does it prevent us from testing predictions regarding how the specific states and dynamic interactions might change in the future. A perspective that combines the historical with the contemporary can inform us about what kinds of real-world changes may be expected to alter the community dynamics, as is illustrated above by some of our speculations about the future. For example, forest change that does not fundamentally alter the pulsed-resource dynamics regulating rodents is far less likely to cause a complete community reconfiguration than is forest change that causes replacement with continuously supplied resources. Similarly, if the moth becomes innocuous and no longer defoliates, then the system will move from donor/recipient control back to donor control.

Our Approach to the Study of Place

We end with some general thoughts on the approaches we have taken that have contributed to our current understanding of place. While our work, like much of community ecology, has focused on questions about pairwise interactions, it has also attempted to connect these pairs into a larger ecological system. Our rationale for so doing was simple. The dynamics of a pairwise interaction that is somehow connected to other parts of a system cannot be fully understood or predicted without recourse to those other connections. The connection of one pair to some other player necessarily reveals a new pair, and so on. What might be originally viewed as a rather mystical indirect effect (i.e., a third species somehow affects one or the other of an interacting pair), now becomes an explicit connection driven by understanding of a particular process (e.g., predation or reservoir competence for disease) that is part of a dynamic system of interactions among multiple

players. Our experience suggests that there are two important further benefits from such an approach. First, because we incrementally expanded the
connections (as opposed to studying pairs presumed to be independent
from other pairs), our understanding of a new pair necessarily provided
context for pairwise interactions that had already examined, thus enhancing
our dynamic understanding. Second, the process of progressive integration
of pairwise interactions started to reveal interesting interactions that initially had not been considered, but nevertheless could now be thought of as
legitimate parts of the interaction web. The progressive establishment of interconnections also helps to delineate the boundaries of interconnectedness:
many things are connected, but not everything is connected to everything.

The above points can be best brought out by example, if we can be permitted to reorder some events for heuristic purposes. We began by asking a
question about the regulation of low-density moth populations (given the
already extensive understanding of dynamics at high moth density). Based
on published precedent, this invoked the consideration of omnivorous small
mammals, particularly mice. Mice are generalists, and data revealed that
despite their profound effect on moths, in low-density moth populations
there was no reasonable expectation that moth pupae could meaningfully
contribute to mouse dynamics. Thus, we needed to find the connection
between mice and the resource that drove their dynamics. Published precedent and the long-term data of a collaborator (Wolff 1996) at another locale indicated acorns, and this was subsequently confirmed by monitoring
and experiments. Invocation of acorns, along with extant recognition that
oak defoliation by moths affected masting, established a potential feedback
that now awaits a major defoliation event for confirmation (although some
studies on forest nutrient cycling effects of defoliation have been carried
out; e.g., Lovett et al. 2002). The dramatic influence of mice on moths revealed by monitoring and experiments raised a new question: How could
moths persist, given intense predation? This revealed the joint importance
of spatial heterogeneity in mouse predation risk and larval dispersal, and it
also generated expectations about the circumstances under which the moth
feedback would be maintained in the system, rather than reverting to acorn
donor control.

The acorn-mouse relationship suggested a connection to another abundant granivore, chipmunks, and also raised the question about the relationship of chimpmunks to moths. This was subsequently shown to be effectively nonexistent, limiting the web connections with respect to moths (i.e.,
acorn-mouse-moth-oak-acorn). Studies on these two mammal species,
together with increasing prevalence of Lyme disease in the area in the late
1980s, the known capacity of the rodents to be reservoir hosts for Lyme,
and the arrival of one of us (Ostfeld) at IES in 1990, logically led to ques-

tions about the role of these species and other mammals in Lyme disease. The concomitant studies of ticks, via their connection to deer, led back to acorns. Finally, the recognition that both mice and chipmunks were generalists capable of incidental use of many food items, including bird eggs, led to studies of ground-nesting songbirds, and the connection of songbirds and small mammals to raptors.

Facilitating the above general approach of progressive pairwise integration were many other important ancillary approaches worth mentioning briefly. We found a focus on natural history to be critical. Despite often being considered unfashionable, it helped us to identify what might be connected to what (e.g., food habits), what the connections might be (e.g., predation), and how they might operate in space and time (e.g., moth and tick life cycles). Long-term monitoring studies were fundamental to our understanding, including the monitoring of multiple spatial locales in the forest over time. They revealed the basic relationships, profoundly informed our understanding of temporal and spatial variation, and told us when and where experiments would be best conducted. We did not jump immediately into experiments. While judicious, timely, appropriately scaled experiments were critical, their design and interpretation were heavily informed by natural history and monitoring.

We found it important to free ourselves from our personal taxonomic preferences and biases, collaborating extensively to consolidate the necessary expertise. No single scientist has the expertise to study the wide array of taxa and processes in our system, along with the combination of empirical and modeling skills. While empirical studies are crucial, modeling has played a central role since many of the phenomena occur at spatial scales or over time frames that are not amenable to direct study. We were acutely aware of the importance of temporal and spatial scale, but we did not attempt to impose a hierarchy of scales. Nor, in fact, did we explicitly study scaling. Rather, we let the interaction processes we were studying define the appropriate time and space scales for study. For example, moth pupation takes two weeks, so the critical interaction between mice and moths only occurs during a brief summer period. Tick life cycles take two years; masting occurs every two to five years; mouse home ranges where foraging occurs are on the order of tens of meters, as is moth larvae dispersal distance; masting and moth outbreaks can occur over areas much larger than our study system. Lastly, we incorporated understanding of the historical context of place so that we could better understand the origins and the possible future of the interaction web.

Our overall approach bears some similarity to that of "progressive contextualization," a term coined in human ecology by Peter Vayda (1983), although we do not subscribe to the notion that such an approach is entirely inductive, theory-free, or spatially unbounded—arguments that were advanced by Vayda and Walters (1999). Perhaps, then, we should call our overall approach "progressive integration." We progressively integrated pairwise interactions into a larger context that encompassed other players and processes, multiple time and space scales, and a historical perspective. We did this incrementally and explicitly, with the answers to one question driving connections to another; we moved both up and down in spatial and temporal scales in addressing questions; we allowed our understanding of history to inform our thinking as each new component was added; and we brought a diversity of other approaches to bear on the problem.

References

Bess H. A., S. H. Spurr, and E. W. Littlefield. 1947. Forest site conditions and the gypsy moth. *Harvard Forest Bulletin* 22:1–26.

Blockstein, D. 1998. Lyme disease and the passenger pigeon? *Science* 279:1831.

Campbell, F. T., and S. E. Schlarbaum. 2002. *Fading Forests II: Trading Away North America's Natural Heritage.* American Lands Alliance, Washington, DC.

Connors, M. J., E. M. Schauber, A. Forbes, C. G. Jones, B. J. Goodwin, and R. S. Ostfeld. 2005. Using track plates to quantify predation risk at small spatial scales. *Journal of Mammalogy* 86:991–96.

Doane, C. D., and M. L. McManus, eds. 1981. *The Gypsy Moth: Research towards Integrated Pest Management.* U.S. Department of Agriculture, Washington, DC.

Dwyer, G., J. S. Elkinton, and A. E. Hajek. 1998. Spatial scale and the spread of a fungal pathogen of gypsy moth. *American Naturalist* 152:485–94.

Elkinton, J. S., W. M. Healy, J. P. Buonaccorsi, G. H. Boettner, A. M. Hazzard, H. R. Smith, and A. M. Liebhold. 1996. Interactions among gypsy moths, white-footed mice, and acorns. *Ecology* 77:2332–42.

Elkinton, J. S., and A. M. Liebhold. 1990. Population dynamics of gypsy moth in North America. *Annual Review of Entomology* 35:571–96.

Foster, D. R., and G. Motzkin. 2003. Interpreting and conserving the openland habitats of coastal New England: Insights from landscape history. *Forest Ecology and Management* 185:127–50.

Goodwin, B. J., C. G. Jones, E. M. Schauber, and R. S. Ostfeld. 2005. Limited dispersal and heterogeneous predation risk synergistically enhance persistence of rare prey. *Ecology* 86:3139–48.

Gottschalk, K. W. 1990. Gypsy moth effects on mast production. Pages 42–50 *in* C. E. McGee, ed., *Proceedings: Southern Appalachian Mast Management.* University of Tennessee, Knoxville.

———. 1993. *Silvicultural guidelines for forest stands threatened by the gypsy moth.* USDA Forest Service Northeast Forest Experiment Station General Technical Report NE-171. 50 pages.

Jones, C. G., R .S. Ostfeld, E. M. Schauber, M. Richard, and J. O. Wolff. 1998. Chain reactions linking acorns to gypsy moth outbreaks and Lyme-disease risk. *Science* 279:1023–26.

Koenig W. D., and J. Knops. 2000. Patterns of annual seed production by northern hemisphere trees: A global perspective. *American Naturalist* 155:59–69.

Lane, R. S., J. Piesman, and W. Burgdorfer. 1991. Lyme borreliosis: Relation of its causative agent to its vectors and hosts in North America and Europe. *Annual Review of Entomology* 36:587–609.

Liebhold, A. M. 1989. Etienne Leopold Trouvelot, perpetrator of our problem. *Gypsy Moth News* 20:8–9.

LoGiudice, K., R. S. Ostfeld, K. A. Schmidt, and F. Keesing. 2003. The ecology of infectious disease: Effects of host diversity and community composition on Lyme disease risk. *Proceedings of the National Academy of Sciences (USA)* 100:567–71.

Lovett, G. M., L. M. Christenson, P. M. Groffman, C. G. Jones, J. E. Hart, and M. J. Mitchell. 2002. Insect defoliation and nitrogen cycling in forests. *BioScience* 52:335–41.

Mather, T. N., M. L. Wilson, S. I. Moore, J. M. C. Ribeiro, and A. Spielman. 1989. Comparing the relative potential of rodents as reservoirs of the Lyme disease spirochete (*Borrelia burgdorferi*). *American Journal of Epidemiology* 130:143–50.

McShea, W. J. 2000. The influence of acorn crops in annual variation in rodent and bird populations. *Ecology* 81:228–38.

McShea, W. J., and W. M. Healy. 2002. *Oak Forest Ecosystems: Ecology and Management for Wildlife*. Johns Hopkins University Press, Baltimore.

McShea, W. J., and G. Schwede. 1993. Variable acorn crops: Responses of white-tailed deer and other mast consumers. *Journal of Mammalogy* 74:999–1006.

Ostfeld, R. S. 1997. The ecology of Lyme-disease risk. *American Scientist* 85:338–46.

Ostfeld, R. S., C. D. Canham, K. Oggenfuss, R. J. Winchcombe, and F. Keesing. 2006. Climate, deer, rodents, and acorns as determinants of variation in Lyme-disease risk. *PLoS Biology* 4:1058–68.

Ostfeld, R. S., C. G. Jones, and J. O. Wolff. 1996a. Of mice and mast: Ecological connections in eastern deciduous forests. *BioScience* 46:323–30.

Ostfeld, R. S., and F. Keesing. 2000. Pulsed resources and community dynamics of consumers in terrestrial ecosystems. *Trends in Ecology and Evolution* 15:232–37.

Ostfeld, R. S., F. Keesing, C. G. Jones, C. D. Canham, and G. M. Lovett. 1998. Integrative ecology and the dynamics of species in oak forests. *Integrative Biology* 1:178–86.

Ostfeld, R. S., M. C. Miller, and K. R. Hazler. 1996b. Causes and consequences of tick (*Ixodes scapularis*) burdens on white-footed mice (*Peromyscus leucopus*). *Journal of Mammalogy* 77:266–73.

Ostfeld, R. S., E. M. Schauber, C. D. Canham, F. Keesing, C. G. Jones, and J. O. Wolff. 2001. Effects of acorn production and mouse abundance on abundance and *Borrelia burgdorferi*-infection prevalence of nymphal *Ixodes scapularis*. *Vector-borne and Zoonotic Diseases* 1:55–64.

Pacala, S. W., G. C. Hurtt, D. Baker, P. Peylin, R. A. Houghton, R. A. Birdsey, L. Heath, E. T. Sundquist, R. F. Stallard, P. Ciais, P. Moorcroft, J. P. Caspersen, and E. Shevliako. 2001. Consistent land- and atmosphere-based US carbon sink estimates. *Science* 292:2316–10.

Rodewald, A. D. 2003. Decline of oak forests and implications for forest wildlife conservation. *Natural Areas Journal* 23:368–71.

Schauber, E. M., B. J. Goodwin, C. G. Jones, and R. S. Ostfeld. 2007. Spatial selection and inheritance: Applying evolutionary concepts to population dynamics in heterogeneous space. *Ecology* 88:1112–18.

Schauber, E. M., R. S. Ostfeld, and A. Evans. 2005. What is the best predictor of annual Lyme disease incidence: Weather, mice, or acorns? *Ecological Applications* 15:575–86.

Schauber, E. M., R. S. Ostfeld, and C. G. Jones. 2004. Type 3 functional response by mice to gypsy moth pupae: Is it stabilizing? *Oikos* 107:592–602.

Schmidt, K. A., J. R. Goheen, R. Naumann, R. S. Ostfeld, E. M. Schauber, and A. R. Berkowitz. 2001. Experimental removal of strong and weak predators: Mice and chipmunks preying on songbird nests. *Ecology* 82:2927–36.

Schmidt, K. A., and R. S. Ostfeld. 2003a. Mice in space: Space use predicts interactions between mice and songbirds. *Ecology* 84:3276–83.

———. 2003b. Songbird populations in fluctuating environments: Nest predator responses to pulsed resources. *Ecology* 84:406–15.

———. 2008. Numerical and behavioral effects within a pulse-driven system: Consequences for direct and indirect interactions among shared prey. *Ecology* 89:636–46.

Schmidt, K. A., R. S. Ostfeld, and K. N. Smyth. 2006. Spatial heterogeneity in predator activity, nest survivorship, and nest-site selection in two forest thrushes. *Oecologia* 148:22–29.

Schnurr, J. L., C. D. Canham, R. S. Ostfeld, and R. S. Inouye. 2004. Neighborhood analyses of small mammal dynamics: Implications for seed predation and seedling establishment. *Ecology* 85:741–55.

Schnurr, J. L., R. S. Ostfeld, and C. D. Canham. 2002. Direct and indirect effects of masting on rodent populations and tree seed survival. *Oikos* 96:402–10.

United States Department of Agriculture Forest Service. 1994. *Northeastern Area Forest Health Report 1992*. Report NA-TP-01-94. United States Department of Agriculture Forest Service, Northeastern Area, Radnor, PA.

Vayda, A. P. 1983. Progressive contextualization: Methods for research in human ecology. *Human Ecology* 11:265–81.

Vayda, A. P., and B. B. Walters. 1999. Against political ecology. *Human Ecology* 27:167–79.

Wolff, J. O. 1996. Population fluctuations of mast-eating rodents are correlated with production of acorns. *Journal of Mammalogy* 77:850–56.

Wootton, J. C. T. 1994. The nature and consequences of indirect effects in ecological communities. *Annual Review of Ecology and Systematics* 25:443–66.

Notes

1. At the peak of the 1981 outbreak, over 12 million acres of forest were defoliated in the eastern United States (USDA Forest Service 1994).

2. Masting is generally defined as the episodic or periodic production of heavy seed or fruit crops by a population of plants.

3. Escape by moths from regulation by mice is a consequence of moth densities reaching levels at which even maximal predation by mice cannot prevent further increases in moth density. Mice have a saturating functional response (Schauber et al. 2004)—i.e., they cannot increase predation rates indefinitely as moth populations increase.

4. Of course, via their predation on defoliating gypsy moths, mice and chipmunks might affect acorn production, but such an effect likely involves very long time lags.

11 *A Sense of Place*

TATOOSH

Robert T. Paine, Timothy Wootton, and Catherine A. Pfister

Figure 11.0. Bob Paine at Tatoosh, with a characteristically massive wave. Photograph by Tim Wootton, 1995.

Abstract

We develop the view that commitment to a single isolated site (Tatoosh Island and its adjacent mainland), experimental manipulation, acquisition of lengthy time-series data on selected species, and what can be called "natural history" have proven to be conceptually rich ecologically. Our loosely bound intellectual consortium of principal investigators and graduate students has focused on central themes of trophic ecology (keystone species, strengths of interaction, indirect effects), disturbance-driven patch dynamics and related successional processes, and the critical connection between individual

variation and population performance. We argue that investigator independence, set within a supportive academic environment, enhances research
creativity and the development of novel ideas and techniques, both of which
are vital if the fog of biological complexity is to be lifted and natural systems
are to be understood.

Introduction

Some ecologists suggest that we should move away from detailed studies at
local sites and instead focus our analyses on data taken across larger geographic scales, because such data promise broad generality (Underwood and
Petraitis 1993, Brown 1995, Lawton 1999). Although large-scale studies are
undoubtedly worth pursuing, primarily because they establish the general
environmental context for more localized studies, we believe it is premature
to dispense with intensive site-based studies, for a number of reasons. First,
detailed knowledge of natural history and historical dynamics allow one to
recognize "surprising" results and identify novel mechanisms that can be
generalized to other ecological systems (see also Doak et al. 2008; Grant and
Grant, chapter 6; Estes, chapter 8; Ostfeld and Jones, chapter 10). It is hard
to know what to be surprised about if the system is studied as a snapshot,
which is generally necessary in broad-scale studies for logistical reasons.
Second, place-based research provides model systems for ecology. Model
systems have proven useful in other areas of biology because they facilitate
the identification and understanding of fundamental mechanisms. Studying
ecological mechanism is of general value in producing a conceptual toolbox
that can be applied elsewhere—for instance, in explicating comparable processes in different ecosystems (Pulliam and Waser, chapter 4). Third, much
ecological theory is based on system dynamics. Linking theory to empirical
data is therefore done most naturally in systems for which dynamics are
documented, and it is best done by detailed long-term studies at defined
sites. In this chapter we illustrate some of these points with examples from
our studies on the rocky intertidal systems of the northwest corner of the
Olympic Peninsula, Washington state, in the United States, and in particular
at Tatoosh Island.

The Major Research Site and Our Collective Approach

Research domains are defined by scale issues: size (in hectares or square kilometers), which is fixed, and duration of study (in years or decades), which
brings the added complications of frequency of observation and specific
interests of the observer(s). Our focal location is Tatoosh Island (48° 24'
N, 124° 44' W), a 17–18-ha cluster of islets, most of which are connected

by land at low tide, about 0.6 km off the northwestern tip of Washington state. A broader definition of "place," however, is needed to include early studies in the Cape Flattery region, all of which took place within about a 15-km radius of Tatoosh. Latitude, climate, and a physical battering by winter waves of 10-m height and more place the area under the influence of the North Pacific Ocean. The occasional winter freezing events, the general lack of herbivorous fishes, the summer fogs that potentially mitigate heat stress, and an intertidal zone enhanced by wave surge have all produced a pattern- and species-rich intertidal assemblage dominated by mussels, barnacles, and kelp. One of us (Paine) began study in June 1963; Wootton first visited Tatoosh in November 1984, and Pfister in July 1988. Frequency of observation throughout more than 40 years has varied, with trips lasting from 1 to 20 days (usually 3–5 days) and occurring at all seasons. Our research focuses on the intertidal zone studied at low tide. Workable winter (October–March) tides occur at dusk; safety concerns generally limit study to two to three tides. The schedule in summer (April–September), when low tides occur during daylight morning hours, is typically two three- to five-day trips per month.

The importance of this single research site, studied in a collegial and cooperative fashion for a long time, derives from several factors. Discovering the idiosyncratic natural histories of some of the species is a collective endeavor. Species are different in behavior and in ecological impact, and observations on the nature of their intra- and interspecific interactions, enemies, quirky behavior, and reproduction accumulate over time. This shared knowledge helps develop a platform around which testable hypotheses can be formulated and impracticable ones discouraged. Of equal significance is that the solid surface of rocky shores has facilitated studies that manipulate species composition and density. Species removals or additions can be implemented manually or with the aid of stainless steel or antifouling paint enclosures or exclosures, tide pools can be added or their equivalents developed, and the rock readily drilled to secure sophisticated electronics, organism mimics, or a wide variety of structures intended to evaluate or encourage larval settlement. In addition, many of the plant and animal species can be individually marked. Such technical knowledge, like natural history detail, accumulates over time; experimental successes have greatly facilitated our understanding of how the community is organized. Our failures are equally important because they identify challenges that, while not necessarily insignificant, cannot be addressed currently.

Our commitment to experiments has yielded, as a very important byproduct, long time-series of observations on both the experimental sites themselves and their controls. Observation in ecology often takes the form of "snapshots"; for instance, a brief, site-specific interlude of intense quantitative sampling, or even a photograph, both being approaches that are insensitive to temporal variation. Experiments, on the other hand, usually have to be maintained, surfaces cleaned or repainted, invaders removed, and materials replaced or replenished. Thus repeated site visits are a natural by-product of experimental maintenance. A number of Paine's Tatoosh time series began in 1968 or the early 1970s; Wootton has tracked species dynamics within mussel beds since 1993 and bird populations since 1985; Pfister's quantitative censuses of tide pool fishes, brown algal growth, and beach cast drift algae date to 1989. Two factors underlie these lengthy and often continuing data sets. First, many intertidal species, including algae, grow slowly and can attain longevities measured in decades or even centuries. Thus, a month or even a year or more of observation is inadequate in duration. Second, even for species with briefer longevity, our long-term commitment allows us to understand temporal variation and the extent to which there is constancy or variability in population dynamics and species interactions. Thus Paine's 25-year *Pisaster* exclusion failed to reveal any gradually developing novel state, and Wootton's seven-year California mussel removals were basic to a test of the predictive ability of Markov models.

Data obviously accumulate with the passage of time, but so do qualitative impressions. Equally, perhaps, intuition sharpens with experience at a site. These aspects of long-term study bear at least three consequences. First, the natural world is intrinsically variable with population numbers waxing and waning, larval recruitment for all species varying in space and time, and the physical setting changing in subtle yet important ways. That is, the ferocity of a winter cold snap, the magnitude of wave-generated disturbances, the timing of a fog that ameliorates intense intertidal heating, and the resultant organism deaths all matter. These local events are superimposed on more global influences such as Pacific decadal oscillations and El Niño events. The net product is one of biological variation. The transmission of such variance demographically and its consequences is enormously significant; the fundamental, species-specific details can only be revealed by long-term monitoring, and hence time series. Second, there is the occasional extreme event. Some events are obvious, such as desiccating, warm winds coinciding with recruitment of algae or animals. Others can be much more subtle, since the human perception of "benign" may not match that of the resident species. Reduced water-column nutrients or heavy rain during an extreme low tide come to mind. Their usual ecological signature once again is increased variation in population numbers, expressed as a steep decline or even a missing year class. However, there can be positive extreme events as well, in which population numbers exceed all known limits. Familiarity with a site enhances recognition of these events. In all cases, a feeling for baseline conditions is required to recognize a legitimate outlier from "normal" variation. Third, as one's perspectives mature along with one's biases, one can extend or challenge older interpretations. Such intellectual ontogeny is to be expected. In an ecological world in which context matters, regional and temporal differences in the outcomes of experimental manipulations should be expected. They signal the importance of explicit, relevant details; they also emphasize how variation in demographic properties can influence ecological interactions.

When observations are prolonged rather than being a snapshot, site experience often helps resolve the ultimate source of extreme mortalities. For instance, in late December 1988 the barge *Nestucca* leaked 168,000 to 231,000 gallons of bunker-grade oil south of Tatoosh. The northward oil trajectory killed thousands of common murres (*Uria aalge*), swept past Tatoosh, and could have caused substantial mussel mortality. In February 1989 extreme cold froze intertidal organisms from Alaska south to Oregon. We were on Tatoosh from 8 to 12 March 1989 and observed numerous new mussel bed patches, one of which was at least 40m in horizontal extent and 1.5 to 3.2m in width. This relatively huge patch had not been present in November 1988. Familiarity with the site and specific clues such as the presence of byssal fibers, tidal height, and the identity of the impacted species (*Mytilus californianus*, rather than *M. trossulus*) indicated extreme cold, not oil-based trauma or winter waves, as the cause of this mortality. Such discrimination bears obvious ecological and economic implications and is made more certain at frequently visited sites.

Finally, a place develops its own ambience and research traditions. Tatoosh is no exception. The island is difficult to access; travel requires helicopters in winter and a 12 km one-way boat trip in spring and summer, with the actual landing made through the surf by inflatable raft. Most supplies and food must be packed from the beach to where we live on the island's top in two old (probably 1930s vintage) abandoned U.S. Weather Bureau buildings, one of which we have refurbished. There are few if any creature comforts; water and firewood must be hauled, and until recently there was no indoor toilet. A solar panel, installed in 1996, generates enough power for evening lights and battery charging in the 3.5 × 5 m secure building. Maximum human occupancy is 10 adults; 6 is more comfortable.

Research support from the U. S. National Science Foundation has been continuous, though hardly lavish, since 1962—initially to Paine and more recently to Wootton and Pfister. Support from the Andrew W. Mellon Foundation since 1995 has proven invaluable. The NSF grants were given to individuals and funded hypothesis-driven, experimental ecology. The product has been 75 to 100 generally sole-authored papers by us and our students in leading journals, and three Mercer Awards (P. K. Dayton, K. P. Sebens, J. T. Wootton) and one Buell Award (S. R. Palumbi) from the Ecological Society of America. These are some of the hallmarks of "small" ecology, in this case equally dictated by circumstance (site isolation, limited group size) and a conviction that research independence rather than a primary group-based endeavor is the mother of innovation.

The above circumstances and attitudes also generate constraints. The positive side of relative geographic isolation is lack of interference with or destruction of our sometimes-conspicuous experimental devices. Further, the island, which now belongs to the Makah Indian Nation, is not subject to the biotic plunder increasingly characteristic of mainland sites. On the negative side are two major issues. First, the island's small size, its isolation, its difficulty of access, and especially its North Pacific maritime environment have limited any attempts at traditional environmental monitoring. Aerosols or salt spray routinely defoliate the island's dominant plant (salmonberry, *Rubus spectablis*), consume our lawn mowers, and have destroyed a U. S. Coast Guard power-generating windmill within two years. One of us (Paine) early on dismissed, perhaps wrongly, any attempt to monitor the island's physical environment, partly because the necessary equipment was expensive and had to be maintained, but primarily because the measurements seemed irrelevant to the microsite variation that dominated individual fitness and survival. Nonetheless, Pfister and Wootton initiated a sampling protocol for water conditions after the 1997 El Niño event. Water temperature, salinity, pH, dissolved oxygen, and water column chlorophyll have been measured essentially continuously from April to September, and these measures were supplemented with monthly water nutrient collections at 10 Tatoosh and 4 offshore sites adjacent to the island (Pfister et al. 2007). As these data have accumulated, they have been a source for unanticipated linkages to global change (Wootton et al. 2008). The second issue on the negative side has been a collective, and traditional, ambivalence about which species to monitor, or even whether to monitor them; the great local species richness (>400 animals, probably >150 benthic macroalgae) only magnifies this dilemma. Clearly, one can at best monitor—let alone identify—a modest fraction of these species in a statistically appropriate fashion. So the task of assembling an island-wide species inventory and a quantitative time series has not been attempted until recently, and even then it is restricted to the well defined but limited conditions of mussel beds on wave-exposed rock benches (Suchanek 1979; Wootton plots, continuously monitored since 1994).

The following paragraphs identify some of our major research themes, and especially their conceptual evolution. One goal is to illustrate the interplay between observation, natural history, hypothesis testing, and experiment. Another theme will be the progressive sophistication of our endeavors; some of our early ideas now seem embarrassingly primitive. We also discuss specific examples of how our site-based research has added to both conceptual and practical ecological endeavors. Some will see this as shame-less advocacy for small science, contra Lawton (1999). A more generous appraisal is that curiosity about the natural world, the esthetic stimulation of working in a truly spectacular setting (Rosenfeld and Paine 2002), and a commitment to manipulation and quantitative assessment have allowed us to challenge and improve our understanding of how nature works.

Keystone Species

The starfish *Pisaster ochraceus* is conspicuous and often abundant on rocky intertidal and shallow subtidal shorelines of western North America. Al-though at times the task can be difficult, and even impossible without dam-aging the starfish, individuals can be pried off the rock surface and their prey determined. The observation is straightforward, and while revealing by itself neither food preference nor rate of consumption, it does identify a diet dom-inated by mussels (two species) and barnacles (five species), supplemented by at least 34 other prey items (Paine 1980). One also observes that on ex-posed and more vertical shores, there is often a conspicuous lower limit to the distribution of the mussel *Mytilus californianus*, and when the barnacle *Balanus glandula* is eaten, a white basal plate persists. These observation en-couraged Paine (1966, 1974) to initiate a five-year removal of *Pisaster* (1963–68) specifically testing the hypothesis that "local species diversity is directly related to the efficiency with which predators prevent the monopolizations of the major environmental requisites by one species" (Paine 1966, p. 65). The immediate ecological consequences could have been anticipated: mus-sels recruited to the removal site, grew, outcompeted other space-requiring species, many of which were not prey species of starfish; and monopolized the spatial resource to the eventual exclusion of all other species capable of attachment to the primary resource, bare rock. Generalization of this result led to the concept of a keystone species (Paine 1969)—basically a statement that certain consumers, by controlling the abundance and distribution of competitively dominant and preferred prey, determine local species com-position. The generality of this conclusion seems robust (Power et al. 1996) under certain conditions. Although the concept is clearly not universal, it has furthered an understanding of the significance of ecological "context," especially the role of consumer density. These studies also motivated the concept of "trophic cascades" (Paine 1980), essentially trophically-linked pairs of strongly-interacting keystone species.

Paine's results assumed that bird influences were inconsequential—that is, they had little impact on the capacity of California mussels to usurp the spatial resources. On the other hand, there was no doubt that glaucous-winged gulls (*Larus glaucescens*), oystercatchers, and crows exploited intertidal resources—especially limpets, predatory gastropods, and even sea urchins (Wootton 1992, 1995, 1997). The critical additional observations were that gulls fed heavily on goose barnacles (*Pollicipes polymerus*), and that these animals often recruited densely into patches formed in mussel beds by winter storms (Paine and Levin 1981). Evidence for gull predation is readily observed: large numbers of gulls tugging on goose barnacles in these patches, patches dominated by "headless" barnacles (these barnacles are stalked), and piles of regurgitated plates.

Paine's view was developed on the assumption that *M. californianus* was ultimately the superior competitor; the goose barnacle observations challenged that assumption. To test this new hypothesis—that on wave-swept shores, goose barnacles are the alpha competitor in the absence of gulls—Wootton had to discover a means of excluding birds. Commercially available vinyl-covered wire letter baskets proved ideal; most survived wave pounding for four years with minimal basket "mortality." Ultimately, the results supported Paine: interestingly, gull predation does not change the end point of successional replacements, but rather accelerates it (Wootton 1993b). Obviously there is a cost to mussels in terms of achieving spatial supremacy by having to compete with goose barnacles, and gulls mitigate it.

We present the above in some detail as an example of evolved and evolving sophistication in both hypothesis generation and the implementation of experimental tests. First, Wootton's studies have made our understanding of *M. californianus*'s competitive superiority increasingly robust, and they complement similar discoveries in Oregon (Menge et al. 1994) and British Columbia (Robles et al. 1995). Second, they were a precursor to and an essential ingredient of Wootton's path-analytic studies of multispecies interactions at Tatoosh (Wootton 1994). Long-term understanding of the resident organisms' biology at a single site encourages such conceptual advances. At the least, it reduces or eliminates a necessary discovery phase, which by itself is a substantial economy.

Interaction Strength

Ecologists should need no introduction to the importance of a population's density. Density can influence behavior, individual growth, survival, and dietary breadth, often in nonlinear ways. Per capita effects of one species on another permeate the dynamic models of Lotka, Volterra, and Gause. MacArthur (1972) captured the sense of these, employing the terms "strong" and "weak"; Paine (1980) extended the interpretation graphically yet qualitatively. Strong interactions could be explored experimentally and recognized by rampant indirect effects. Numerous examples exist. Paine's illustration of how weak effects might be recognized (Paine 1980, figure 1), while correct, was experimentally flawed: how does one distinguish a "no effect = weak interaction" result from a failed manipulation?

Paine (1992) attempted to redress this intellectually stagnant situation. Presence and absence data were clearly inadequate. For instance, *Pisaster* is probably ecologically inconsequential and "just another starfish" at Torch Bay, Alaska (Paine 1980, p. 670), whereas it plays a major ecological role along parts of the Oregon coast (Menge et al. 1994), in Washington (Paine 1966), and in British Columbia (Robles et al. 1995). On the other hand, observations had suggested that macroherbivores (urchins, chitons, and limpets) embodied a better system with which to address per capita effects experimentally. We (Paine and Vadas 1969, Paine 1984) and others (Dayton 1975, Duggins and Dethier 1984) had shown that after sites were scraped and grazers removed, they would initially be dominated by brown algae, *Alaria* spp. The next generation of experiments (Paine 1992, 2002) exploited this feature; scrapings of small areas (<600 square cm), site sterilization with oven cleaner, and barriers of epoxy putty painted with copper-based anti-fouling paint permitted the species composition of the grazer community to be controlled within enclosures. An index of "per capita strength of interaction" was developed. Better indices were subsequently identified.

However, concepts and indices evolve. The claim by one of us (Paine) that interaction strengths were best (and probably only) estimated from experiments was challenged by Wootton (1997), who demonstrated that observation of predator and prey density, calculation of consumption rates, and changes in density of the target species could yield a per capita consumption estimate. A log ratio of interaction strength was identified as the metric of choice, and seems generally accepted today, as the review by Laska and Wootton (1998) suggests. Ruesink's (1998) field microcosm experiments demonstrated that interaction strength could be a strongly nonlinear function of prey abundance. Our conclusions are that (1) empirical estimates of interaction strength remain a critical ingredient of dynamic community models, because they unite consumer density and impact; (2) picking a tractable system is necessary for successful experiments, and depends on familiarity with the manipulated species' natural history; (3) observationally based estimates require a more intimate understanding of the component species' activity periods and density; and (4) important ecological concepts evolve because of interplay, debate, and challenge among researchers.

Indirect Effects

As soon as one acknowledges that natural communities are a highly diverse, multi-trophic-level mix of dynamically interacting species, one must recognize the possible significance of indirect influences. Wootton (1994, p. 151) has defined these as those "that do not involve physical interaction." Thus, in systems in which space limitation is commonplace, when species A competes with or preys on species B, indirect influences on the other species that are associated with species B can be anticipated. Rocky shores are replete with examples. Connell (1961) showed experimentally how predation on one barnacle species reduced the impact of interspecific competition on another, and therefore benefited it. Paine (1966) employed the modifer "indirect" in describing how *Pisaster*, by consistently reducing mussel numbers, permitted many non-prey species to coexist on mutually occupiable space. The subject is hardly ecological "news": it can be traced at least to Darwin (1859). Menge (1995) found that indirect effects accounted for a minimum of 40% of the changes in experimentally manipulated communities. Wootton (1993a) has classified indirect effects into two broad categories: interaction chains and interaction modifications.

What recognition of these often subtle relationships requires is an appreciation for a species' natural history. We provide a number of Cape Flattery and Tatoosh-based examples in order of increasing complexity to illustrate this assertion. The essence of keystone species and trophic cascades resides in indirect influences transmitted across trophic levels. Dayton (1973b) showed how predatory starfish indirectly feed sessile anemones, *Anthopleura xanthogrammica*, by stampeding sea urchins into the mouths of these sit-and-wait consumers. Wootton (1994) expanded a path-analytic approach to test experimentally eleven predictions of the relative importance of indirect effects in a Tatoosh interaction web; all the predictions were supported. Finally, Parrish et al. (2001) could show that common murre decline at Tatoosh was best understood by considering some combination of direct eagle predation supplemented by the indirect consequences of predation on eggs by gulls and crows. Indirect effects must be rampant in nature, and often they are subtle. Ignoring their presence in interpretations of single or multispecies dynamics surely leads to incompleteness and possibly errors. Commitment to old-fashioned observation and sharing of the resulting information will reduce these problems; familiarity with a site facilitates this vital task.

Spatial Ecology

Spatial patterning permeates the organic world, and rocky shores have provided a fertile venue for exploring the underlying processes. Early focus was on characteristic and conspicuous patterns of zonation (Lewis 1964); initial interest in our own case was stimulated by the obvious role that waves play in generating a mosaic of patches in the intertidal of different sizes, ages, and occupant species. Dayton (1971) described 18 patches observed over a two-year interval at three Cape Flattery sites. His report that these patches increased in size (area) by 24% to 4,884% implies a system of ever-expanding patches, which seems improbable. By 1971 Paine had begun following the fate of individually marked patches in these same mussel beds, and by 1972 he had carved experimental gaps and begun quantifying their recovery. Levin and Paine (1974) developed a dynamical model in which the rates of patch formation (birth) and closure (death) were considered functions of initial patch size and age. In essence the approach was demographic; it confronted the fact that two patches the same size need not be of similar age and, conversely, that patches of comparable age could range from a few square centimeters to many square meters in area. The conception was of a nonequilibrial spatial "world" of mussel beds and mussel-less patches, with the mix determined by the intensity and frequency of disturbance. Local species richness would be influenced by patch size, invoking species-area relationships, and patch age, permitting successional patterns to be incorporated. Though such a regional integration was never attempted, Paine and Levin (1981), in a six-year study involving thousands of patches, demonstrated the robustness of the model and its independently estimated parameters.

While the above approach hasn't exactly languished (Levin received the 2005 Kyoto Prize based in part on his research at Tatoosh), other methodologies exist for addressing spatially explicit multispecies relationships. The Levin-Paine approach is analytically demanding and it ignores the effects of species interactions within patches. To probe species interactions, Wooton adopted more traditional stationary Markov models to ask whether transition probabilities similarly identified *M. californianus* as the competitive dominant and therefore plausible endpoint of a successional replacement process. Unsurprisingly, the answer is yes (Wootton 2001a). But the approach has proven conceptually richer than this obvious result. Quantitative shifts in species composition were successfully predicted for vertical rock walls lacking obvious bird predation and characterized by little or very reduced recruitment by dislodged adult mussels. The change in system dynamics documented by Wootton's earlier studies (Wootton 1994), in which bird predation on goose barnacles (*Pollicipes polymerus*) was prevented by letter basket cages, was also successfully predicted. Possibly more significant in a broader ecological context, the experimental tractability of rocky shores permitted Wootton to generate and test novel conditions with the Markov models. When the competitively dominant California mussels were routinely hand-plucked from experimental plots, the multispecies Markov models successfully predicted that the community would be dominated by the coralline alga *Corallina vancouveriensis*. When another mussel (the weakly interacting *M. trossulus*) was removed, *M. californianus* continued to predominate and coralline algae remained a minor community member (Wootton 2004). Such results uniquely show that Markov chain models are robust when challenged to predict the consequences of species removals or absence. Their successful prediction of the equilibrial multispecies mix should complement nicely the Levin-Paine visualization of a simplified and nonequilibrial patchy world in which species presence is determined in part by patch size and age. The direct integration of these very separate approaches has not yet been attempted, but spatially explicit extensions of the Markov chain models to generate cellular automata (Wootton 2001b) provide a compromise approach with encouraging results.

Environmental Variation and Its Contribution to Species Dynamics

An investigator with a long-term presence at a site almost always notices the year-to-year variation in both physical and biotic components. At times the implications of this are obvious, such as the total recruitment failure of sea urchins (*Strongylocentrotus purpuratus*) during the 1982–83 El Niño. At Tatoosh urchins trickle in annually; major recruitment tends to occur during El Niño Southern Oscillation events when near-shore currents move south to north. The 165 mm of rain falling over a 21-hr interval on one of the year's lowest tides in July 1983 substantially reduced both the 1982 and 1983 urchin year classes. But many aspects of environmental variation are more subtle and require established analytical frameworks to understand the implications of a year of poor adult growth, a reduced level of recruitment, or similar events. One of us (Pfister) has focused on the fates of individuals within populations to ask how variation in demographic rates is manifested at the population level. The abundant annual kelp species *Alaria nana* is characterized by highly structured variation among individuals. Individuals have consistencies in growth rates across time intervals such that those that grow well in one interval continue to do so—the phenomenon of growth autocorrelation (Pfister and Stevens 2002, 2003). The implications of this phenomenon are that some individuals get especially large, with increasing reproductive output, and have the potential to contribute disproportionately to future generations. An examination of this pattern over two years shows that growth autocorrelation and size disparities were much less pronounced during an El Niño year (1997), when growth and reproductive output were reduced (Pfister and Wang 2005). Under such conditions, individual fate was much less deterministic, fewer individuals reached a reproductive size, and population growth was much reduced. One can examine the consequences of different individual "pathways" during the El Niño and normal years with matrix projection models (Caswell 2001) and loop analysis, and quantify the much greater contribution to population growth made by good grow-

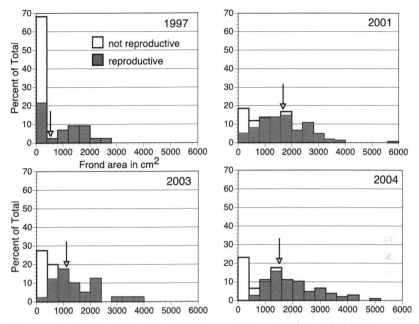

Figure 11.1. The size distribution of the kelp *Alaria nana* in four years of study. The shaded areas denote the presence of individuals within a size grouping that are reproductive; arrows show mean size within a year. Note that mean-sized individuals may not reproduce. The El Niño year (1997) shows a failure of individuals to reach larger size classes. In all years, a relatively small fraction (5%) contributes at least a third of the reproductive output.

ers in good years. These patterns continue to be borne out by the data. 1997 was an aberrant year in kelp demography, not because of mass mortality but because the El Niño conditions allowed relatively few individuals in the population to do persistently well (figure 11.1). As a consequence, only 53% of individuals reproduced in 1997, whereas 81%, 68%, and 77% reproduced in 2001, 2003, and 2004. A look at growth autocorrelation estimates during these four years (estimated from growth rates between May and September), showed that the "good" years of 2001, 2003, and 2004 had estimates of 0.54, 0.68, and 0.57, while the El Niño year was only 0.48. In a related 1997–98 El Niño study, Paine and Trimble (2004) documented that an ENSO-forced brown algal "disaster," coupled to major *M. californianus* recruitment, might lead to the establishment of an intertidal alternative state. Time series dating to 1978 of percentage of cover by the brown alga *Hedophyllum sessile* were used to identify the prevalence of the original state; ongoing *Pisaster* removals to protect the mussels until they become too large to be eaten are now testing this possibility.

Linking Populations to Communities

Communities at local spatial scales are composed of interacting populations. Pfister has used the same analytical techniques described for kelp to ask how recruitment variability among years affects the interaction and coexistence of fishes in a multispecies assemblage. Tidepool sculpins are conspicuous and species-rich in our study area. They are characterized by high site fidelity to a single or group of tidepools, and tidepools rarely contain only one species. Young-of-the-year fishes are numerous during the months from April to June (Pfister 1996, 1997), settling into tidepools with resident adults after a period in the water column that may last from 30 to 60 days. Short-term (two-month long) experiments more than a decade ago showed the competitive dominance of *Oligocottus maculosus* over *Clinocottus globiceps* (Pfister 1995), and the superiority of dynamic regression techniques (state-space models) that relied on multiple census points (also known as time series data) to estimate interaction strengths for replicating this empirical result. Quantifying the competitive interaction set up the prediction that *C. globiceps* must recruit consistently into the system to persist, a prediction that has been borne out from 16 years of annual censuses. The long-term data also did not support an earlier favored hypothesis that variability among species in success of recruitment would provide a mechanism for coexistence via the storage effect. Instead, the relative ranking of species abundances as recruits is remarkably constant, and it shows no evidence that environmental variation, including two El Niño events, has altered this ranking (Pfister 2006).

The above gives us the insight that the links between individuals, populations, and communities in aquatic systems often develop via variability in individual growth. For example, individual fish that recruit relatively early in the spring experience high early growth, and reach a larger size by winter. This head start appears to give them an increased probability of reproducing in their first year (Pfister 1997). Traits of individuals, or small microhabitat advantages, can result in variability in growth that has "memory" through time, resulting in some individuals that become large. In the case of the kelp, *Alaria nana* (figure 11.1), an individual of average size for a population might not reproduce, but attention to variance would reveal that few individuals contribute disproportionately to reproduction. For *A. nana*, the largest 5% of the individuals can contribute as much as a third of the reproductive output of the entire population. In addition to the possible beneficial effects this has for ameliorating population viability, it may be an important aspect of size-based species interactions. In other examples, Paine (1976) has shown how the susceptibility of mussels to *Pisaster* predation depended on their respective sizes; mussels can coexist with their major consumer simply by becoming too large. Wootton (1993b) demonstrated that the original size hierarchies of competitors could influence both the rate of successional change and susceptibility to disturbance. Our experience with this system has shown that variation detected on a scale of weeks has potential implications for generations. In the context of global change, such variation may play a pivotal role. Many of the predictions from global climate models call for changes in mean temperatures, increased variability in environmental variables such as temperature (Intergovernmental Panel on Climate Change 2001), and for an increased frequency of El Niño events (Timmerman 1999).

The Intersection of Independent Studies: Improved Understanding

The study of Dayton (1971) remains an ecological landmark: spatially replicated experiments coupled with superior intuition and observational skills are its hallmarks. Much of the research was conducted in the Cape Flattery region, and only a little at Tatoosh. That paper and its immediate derivatives (Dayton 1973a, 1975) concentrated on spatial patterns and, whenever possible, experimental probes of the underlying causes. One of the conclusions has become ecological dogma. Dayton observed wood splinters embedded in intertidal rock, massive logs commonly deposited in the supratidal, and patches or gaps occurring with regularity in mussel beds. The logical conclusion was that bashing by drift logs had initiated these disturbance events. "Mortality" of two-cm-high nails set in intertidal rock also suggested a role for massive objects like logs. All of Dayton's natural historical observations can be repeated. Our challenge is to understand the generality, not the condition-specific occurrence, of log bashing.

Some evidence is derived from long-term monitoring of the winter extent of mussel bed patch formation, and especially from tracking the fate of devices used in an unrelated algal fertilization experiment described by Wootton et al. (1996). These devices were 15-cm-diameter, 11.5-cm-high clay flower pots cemented to the rock surface at eight sites around the wave-exposed perimeter of Tatoosh. The four experimental periods ranged from less than two to about four months. Pot "survival" was extraordinary: 23 of 24, 28 of 32, 32 of 32, and 31 of 32. Survival of the third series was followed for one year, with the single pot mortality occurring during winter. Figure 11.2 illustrates the 12-year survival pattern of the fourth series. Pot mortality is about two per year (6.3% per year). Our suspicion is that the primary cause is not log bashing in this case, but rather gradual hydration and thus swelling of the cement base leading to deformation and eventual fracture of the pot.

Huge (>8–10 m high; see figure 11.0) waves and drift logs are a fact of life in the northeast Pacific. Here we argue that the former, acting by themselves,

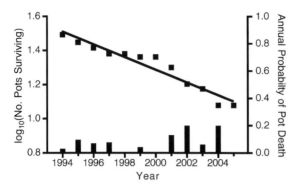

Figure 11.2. Survival of ceramic pots at Tatoosh. The solid line fitted to the actual decline represents a mean mortality of 6.3% per year. Solid bars give the years of pot "death" over 12 years of observation.

are sufficient to disrupt mussel beds. This contention is supported in four ways. First, patches or gaps generally form in the center of mussel beds, or at least rarely include perimeter individuals; Paine and Levin (1981) provide a photograph. Second, Witman and Suchanek (1984) showed that mussel tenacity was higher at the edge than in the middle of mussel beds. Denny (1987), studying mussels at Tatoosh, implicated lift, not shear or drag, as the most important physical force in generating internal patches. Mussel position and tenacity are presumably immaterial factors when and where waterborne logs impact mussels. Third, Dayton's (1971) nails present a surface area estimated at 40 square mm; our more fragile terra-cotta pots at 17,250 square mm in cross-sectional area are not only 430 times larger, but also project five to six times higher into the water column. On both counts they present a much larger target, and hence are presumably more susceptible to waterborne missiles. And fourth, this evolving perspective, coupled with the development of a data set on community dynamics at the same sites spanning seven years (at the time), prompted Wootton (2001b) to develop a cellular automata model of local interactions among individuals, which included the key process of wave-disturbance transmission via attachment of neighbors to each other in the mussel bed. Wooton's model, one of the first empirical parameterizations of this important class of models in any field, was able to successfully predict the large-scale spatial structure of a system generated by disturbance events, thus supporting the wave-impact view.

We present these details because Dayton's observations, measurements, and logic were convincing and remain pertinent to probably all areas in which logs are moved laterally along shorelines. At Tatoosh, longer-term observations on where patches tend to occur, the biomechanical studies of Witman and Suchanek and Denny, and especially the conceptually unrelated pot study, implicate an alternative explanation that a characteristic spatial pattern is generated primarily by hydrodynamic forces.

Generalizing from Single-Site Studies

The quest for ecological generality must progress beyond the demonstration of macroecological patterns. In other words, species identity does matter (e.g., Paine 1980, 1984; Pfister 1995; Wootton 2005); some interactors are "strong" while others are "weak"; some may be keystones, foundation species, engineers, ruderals, and so forth. Such terms, which are jargon to detractors but have implicit meaning to supporters, seem an unavoidable shorthand in the same league as terms like niche, guild, and food web. All disciplines evolve specialized language, as even a cursory examination of mathematics, physics, chemistry, and molecular biology reveals. Ecology's semantic curse is that our vocabulary tends towards the qualitative, driven by the complexity of all natural systems and the intrinsic uncertainty so characteristic of the embedded interactions.

How has our Cape Flattery enterprise fared in the marketplace of ecological ideas? "Keystone species" has survived and seems well established, though it is recognized as not being universally applicable (Power et al. 1996). Tatoosh-based research has contributed in a number of ways to the analysis of spatial patterns; the dynamics of food webs; a deeper understanding, analysis, and nomenclature of indirect effects; and an empirical challenge to the storage effect. The concept of trophic cascades, now recognized in many other ecosystems, was born there. Our individual research programs continue to be based on species studied either alone or when interacting in multispecies complexes.

Intertidal, rocky shore communities pose some challenges to generalization that may be either trivial or highly substantive: sessile plants and animals compete for a basic resource, space. Competition between barnacles and kelp or between sponges and coralline algae, for example, is commonplace. If one considers trophic level or status as a crux of ecological interactions, then the fact of competition across trophic levels can be problematic. We have referred to this as an "amusing fact of intertidal ecology"; others studying terrestrial or pelagic systems may interpret it as an intellectual barrier to generalization. Of equivalent importance are the relative openness of marine systems, the astonishing but spatially limited primary production, and the conspicuousness of trophic interactions. Are intertidal systems outliers because of these differences, and are they thus of little significance to broadly acceptable ecological generalizations? We think the answer is no.

Intense study at a single site also allows ancillary monitoring of a number of variables that would be impossible (and unfundable) over a larger spatial scale. Although some of these variables may be found to be trivial, the prospect of a changing climate and species composition makes ecological "surprises" almost a certainty (Doak et al. 2008). This is well-demonstrated on Tatoosh by the fortuitous placement of a pH probe on an instrument that was intended to measure variables that we hypothesized were of primary importance: chlorophyll a and temperature. The notion that ocean pH might be well-buffered to increases in atmospheric CO_2 has recently been challenged, and our monitoring data from Tatoosh over the last nine years shows an alarming decrease in sea water pH at a more rapid rate than that predicted by physical models (Wootton et al. 2008). Because most pH variation we observe can be explained by associated long-term data on drivers with known mechanistic links to pH, the relevance of our results extends beyond this single site. Furthermore, the concomitant long-term record of species dynamics assembled by Wootton over the same time period suggests effects on dominant organisms with calcium carbonate skeletons, and the extensive history of experimentation and modeling to understand biological interactions at this site provides a necessary perspective to untangle the complex responses of this natural ecosystem that arise from indirect effects. Our unique ability to tie the measured pH decline to in-situ biological effects was the result of a commitment to the study of a place; it is research that would have fared poorly in the funding arena a decade ago when this aspect of climate change was underappreciated. Future understanding of the link between ocean chemistry and biology will require intense study at specific locales. It is unlikely that we will gain full insight from snapshot samples scattered throughout the ocean, or from laboratory experiments divorced from the web of interactions and physical processes that characterize real ecosystems.

Finally, because the Cape Flattery ecosystem is essentially intact trophically, with the possible exception of breeding northern fur seals, we have not ignored the roles of higher trophic levels. Our experiments to date have involved roughly 70 species, a small fraction of the total possible. Our challenge—and more broadly, community ecology's challenge—is how to, or even whether to, confront such species-specific complexity. Resolving this dilemma will involve (1) new analytical techniques; (2) long-term sampling to generate time series data on temporal variation; and, whenever possible, (3) experimental manipulation. Rocky intertidal research has the unique potential to unite these disparate approaches into a cohesive framework. Whether this helps develop common principles through conceptual unification or detracts from this desirable goal remains to be seen.

Summary

The Cape Flattery region has always been a biologically attractive venue. The Makah Indians, presumably attracted by the marine bounty, were the original inhabitants; Rigg and Miller (1949) undertook intermittent surveys from 1911 to 1948; at least 10 students from Oberlin College, Ohio, "in the ecology party of Prof. Lynds Jones" (Tatoosh log book) visited Tatoosh on 24 June 1912. Our research began in 1963 and continues today, with Tatoosh being the primary site.

Observations on natural history and species interactions are easily made, at least at low tide. They have proven to be of fundamental importance to hypothesis formation, and they generate an incomplete but expanding collective wisdom. Spatial patterning is obvious. The solid rock platform has permitted sundry experimental and other devices to be secured. One immediate result has been a proliferation of lengthy time series for many plant and animal species, data often coupled to some specific experimental manipulation. The strength of our spatially myopic efforts—rather than the proverbial devil—lies in the details. Our studies focus on food web dynamics, the interplay between disturbance-generated pattern and successional processes, the significance of variation in recruitment, and how and why growth autocorrelations influence population performance. As each of these themes has matured, they have become increasingly quantitative. Further, by providing details of mechanisms and indirect effects, their integration has made rocky intertidal shores into model study systems. Tatoosh Island is miniscule as ecological places go; our site-based research illustrates well how local knowledge advances conceptual understanding and helps realize the "small" ecology goal of relating explicit process to pattern.

Acknowledgments

Numerous individuals spanning the academic continuum, from undergraduate students to distinguished senior scientists, have contributed both labor and stimulation. Our research has been funded by the National Science and Andrew W. Mellon foundations. We are deeply grateful to the above, but especially to the Makah Nation for permission to conduct research on their spectacular shores.

References

Brown, J. H. 1995. *Macroecology.* University of Chicago Press, Chicago.

Caswell, H. 2001. *Matrix Population Models, Second Edition.* Sinauer Associates, Sunderland, MA.

Connell, J. H. 1961. The influence of interspecific competition and other factors on the distribution of the barnacle *Chthamalus stellatus. Ecology* 42:710–23.

Darwin, C. 1859. *The Origin of Species,* Modern Library edition. Random House, New York.

Dayton, P. K. 1971. Competition, disturbance and community organization: The provision and subsequent utilization of space in an intertidal community. *Ecological Monographs* 41:351–89.

———. 1973a. Dispersion, dispersal, and persistence of the annual intertidal alga, *Postelsia palmaeformis* Ruprecht. *Ecology* 54:433–38.

———. 1973b. Two cases of resource partitioning in an intertidal community: Making the right prediction for the wrong reason. *American Naturalist* 107:662–70.

———. 1975. Experimental evaluation of ecological dominance in a rocky intertidal algal community. *Ecological Monographs* 45:137–59.

Denny, M. W. 1987. Lift as a mechanism of patch initiation in mussel beds. *Journal of Experimental Marine Biology and Ecology* 113:231–45.

Doak, D. F., J. A. Estes, B. S. Halpern, U. Jacob, D. R. Lindberg, J. Lovvorn, D. H. Monson, M. T. Tinker, T. M. Williams, J. T. Wootton, I. Carroll, M. Emmerson, F. Micheli, and M. Novak. 2008. Understanding and predicting ecological dynamics: Are major surprises inevitable? *Ecology* 89:952–61.

Duggins, D. O., and M. N. Dethier. 1985. Experimental studies of herbivory and algal competition in a low intertidal habitat. *Oecologia* 67:183–91.

Intergovernmental Panel on Climate Change. 2001. *Climate Change 2001: The Scientific Basis. Contribution of Working Group 1 to the Third Assessment Report of the Intergovernmental Panel on Climate Change (IPCC).* J. T. Houghton, Y. Ding, D.J. Griggs, M. Noguer, P. J. van der Linden, and D. Xiaosu, eds. Cambridge University Press, Cambridge.

Laska, M. S., and J. T. Wootton. 1998. Theoretical concepts and empirical approaches to measuring interaction strength. *Ecology* 79:461–76.

Lawton, J. H. 1999. Are there general laws in ecology? *Oikos* 84:177–92.

Levin, S. A., and R. T. Paine. 1974. Disturbance, patch formation and community structure. *Proceedings of the National Academy of Sciences (USA)* 71:2744–47.

Lewis, J. R. 1964. *The Ecology of Rocky Shores.* English Universities Press, London.

MacArthur, R. H. 1972. Strong, or weak, interactions? *Transactions of the Connecticut Academy of Arts and Sciences* 44:177–88.

Menge, B. A. 1995. Indirect effects in marine rocky intertidal interaction webs: Patterns and importance. *Ecological Monographs* 65:21–74.

Menge, B.A., E. L. Berlow, C. Blanchette, S. A. Navarrete, and S. B. Yamada. 1994. The keystone species concept: Variation in interaction strength in a rocky intertidal habitat. *Ecological Monographs* 64:249–86.

Paine, R. T. 1966. Food web complexity and species diversity. *American Naturalist* 100:65–75.

———. 1969. A note on trophic complexity and community stability. *American Naturalist* 103:91–93.

———. 1974. Intertidal community structure: Experimental studies on the relationships between a dominant competitor and its principle predator. *Oecologia* 15:93–120.

———. 1976. Size-limited predation: An observational approach with the *Mytilus-Pisaster* interaction. *Ecology* 57:858–873.

———. 1980. Food webs: Linkage, interaction strength, and community infrastructure. *Journal of Animal Ecology* 49:667–685.

———. 1984. Ecological determinism in the competition for space. *Ecology* 65:1339–1348.

———. 1992. Food web analysis through field measurements of per capita interaction strength. *Nature* 355:73–75.

———. 2002. Trophic control of production in a rocky intertidal community. *Science* 296:736–739.

Paine, R. T., and S. A. Levin. 1981. Intertidal landscapes: Disturbance and the dynamics of the pattern. *Ecological Monographs* 51:145–78.

Paine, R. T., and A. C. Trimble. 2004. Abrupt community change on a rocky shore: Biological mechanisms contributing to the potential formation of an alternative state. *Ecology Letters* 7:441–45.

Paine, R. T., and R. L. Vadas. 1969. The effects of grazing by sea urchins, *Strongylocentrotus* spp., on benthic algal populations. *Limnology and Oceanography* 14:710–19.

Parrish, J. K., M. Marvier, and R. T. Paine. 2001. Direct and indirect interactions between bald eagles and common murres. *Ecological Applications* 11:1858–69.

Pfister, C. A. 1995. Estimating competition coefficients from census data: A test with field manipulations of tidepool fishes. *American Naturalist* 146:271–91.

———. 1996. Consequences of recruitment variation in an assemblage of tidepool fishes. *Ecology* 77:1928–41.

———. 1997. Demographic consequences of within-year variation in recruitment. *Marine Ecology Progress Series* 153:229–38.

———. 2006. Concordance between short-term experiments and long-term censuses in tidepool fishes. *Ecology* 87:2905–14.

Pfister, C. A., and F. R. Stevens. 2002. The genesis of size variability in plants and animals. *Ecology* 83:59–72.

———. 2003. Individual variation and environmental stochasticity: Implications for matrix model predictions. *Ecology* 84:496–510.

Pfister, C. A., and M. Wang. 2005. Beyond size: Matrix projection models for populations where size is an incomplete descriptor. *Ecology* 86:2673–83.

Pfister, C. A.., J. T. Wootton, and C. J. Neufeld. 2007. Relative roles of coastal and oceanic processes in determining physical and chemical characteristics of an intensively sampled nearshore system. *Limnology and Oceanography* 52:1767–75.

Power, M. E., D. Tilman, J. A. Estes, B. A. Menge, W. J. Bond, L. S. Mills, G. Daily, J. C. Castilla, J. Lubchenco, and R. T. Paine. 1996. Challenges in the quest for keystones. *BioScience* 46:609–20.

Rigg, G. B., and R. C. Miller. 1949. Intertidal plant and animal zonation in the vicinity of Neah Bay, Washington. *Proceedings of the California Academy of Sciences* 26:323–51.

Robles, C., R. Sherwood-Stephens, and M. Alvarado. 1995. Responses of a key intertidal predator to varying recruitment of its prey. *Ecology* 76:565–79.

Rosenfeld, A. W., and R. T. Paine. 2002. *The Intertidal Wilderness*. University of California Press, Berkeley.

Ruesink, J. R. 1998. Variation in per capita interaction strength: Thresholds due to nonlinear dynamics and nonequilibrium conditions. *Proceedings of the National Academy of Sciences (USA)* 95:6843–47.

Suchanek, T. H. 1979. *The Mytilus Californianus Community: Studies on the Composition, Structure, Organization, and Dynamics of a Mussel Bed*. PhD dissertation, Department of Zoology, University of Washington, Seattle.

Timmerman, A., J. Overhuber, J. Bacher, M. Esch, M. Latif, and E. Roeckner. 1999. Increased El Niño frequency in a climate model forced by future greenhouse warming. *Nature* 398:694–97.

Underwood, A. J., and P. S. Petraitis. 1993. Structure of intertidal assemblages in different locations: How can local processes be compared? Pages 39–51 *in* R. E. Ricklefs and D. Schluter, eds., *Species Diversity in Ecological Communities*. The University of Chicago Press, Chicago.

Witman, J. D., and T. H. Suchanek. 1984. Mussels in flow: Drag and dislodgment by epizoans. *Marine Ecology Progress Series* 16:259–68.

Wootton, J. T. 1992. Indirect effects, prey susceptibility, and habitat selection: Impacts of birds on limpets and algae. *Ecology* 73:981–91.

———. 1993a. Indirect effects and habitat use in an intertidal community: Interaction chains and interaction modifications. *American Naturalist* 141:71–89.

———. 1993b. Size-dependent competition: Effects on the dynamics versus the endpoint of mussel bed succession. *Ecology* 74:195–206.

———. 1994. Predicting direct and indirect effects: An integrated approach using experiments and path analysis. *Ecology* 75:151–65.

———. 1995. Effects of birds on sea urchins and algae: A lower-intertidal trophic cascade. *Écoscience* 2:321–28.

———. 1997. Estimates and tests of per-capita interaction strength: Diet, abundance, and impact of intertidally foraging birds. *Ecological Monographs* 67:45–64.

———. 2001a. Prediction in complex communities: Analysis of empirically-derived Markov models. *Ecology* 82:580–98.

———. 2001b. Local interactions predict large-scale pattern in an empirically-derived cellular automata. *Nature* 413:841–43.

———. 2004. Markov chain models predict the consequences of experimental extinctions. *Ecology Letters* 7:653–60.

———. 2005. Field parameterization and experimental test of the neutral theory of biodiversity. *Nature* 433:309–12.

Wootton, J. T., C. A. Pfister, and J. D. Forester. 2008. Dynamical patterns and ecological impacts of declining ocean pH in a high-resolution multi-year dataset. *Proceedings of the National Academy of Sciences (USA)*105:18848–18853.

Wootton, J. T., M. E. Power, R. T. Paine, and C. A. Pfister. 1996. Effects of productivity, consumers, competitors, and El Niño events on food chain patterns in a rocky intertidal community. *Proceedings of the National Academy of Sciences (USA)* 93:13855–58.

12 *The Ecology of Place and Natural Resource Management*

EXAMPLES FROM MARINE AND TERRESTRIAL ECOSYSTEMS

Charles G. Curtin

Figure 12.0. Collaborative science: Exclosures on McKinney Flats, Gray Ranch, New Mexico. Photo by Charles Curtin, 2006.

Abstract

Given the great environmental challenges that human societies face, there is a pressing need to deploy science more effectively in the solution of applied ecological problems. The weak scientific basis for environmental decision-making is sometimes blamed on inadequacies of general ecological knowledge. However, inadequate use of knowledge is to blame at least as often as inadequate knowledge itself: the application of knowledge is as much a matter of societal goals, investment, and institutional structure as it is a scientific issue. This chapter illustrates, through two case histories—one involving rangelands, one involving fisheries—how a focus on place has facilitated collaboration between local communities of resource users and scientists. Such collaborations create a synergy between "basic" and "applied" science, as basic science helps to redefine approaches to management, and applied problems help to reframe traditional, basic-science questions. The local community brings to the collaboration knowledge of local systems, ecology, and economics. The scientists bring general ecological knowledge and skills in conceptualizing problems and devising research programs to solve them. In both case histories, the goals of the collaboration are to identify the spatial scale on which coupled ecological and economic processes operate, and to develop realistic models on those scales to guide effective natural resource management. In both cases, a desire to conserve the place itself has been instrumental in breaking down cultural barriers among the local community of resource users, resource managers, environmental organizations, and scientists.

Introduction

> ... in a great many places in our society, including academia and most bureaucracies, prestige accrues principally to those who study carefully some aspect of a problem, while discussion of the big picture is relegated to cocktail parties. It is of crucial importance that we learn to supplement those essential specialized studies with what I call a crude look at the whole (Gell-Mann 1995).

Extinction, climate change, pollution, declining crop yields, rangeland degradation, fisheries collapse: environmental problems make daily headlines, and human societies are struggling to solve them. Considerable resources have been invested in the science that is seen as a critical part of the solution. Nonetheless, the scientific basis for environmental decisions is often viewed as inadequate (National Research Council 1986).

Some attribute this to inadequacies in the science itself. They see ecology as a weak science because it has not uncovered general laws that provide predictions precise enough to guide the solution of environmental problems (e.g., see Peters 1991, Shrader-Frechette and McCoy 1993). Others (e.g., National Research Council 1986, Simberloff 2004) counter that it is naive to expect general theory to be able to direct specific management action given the complexity of natural systems; theory must be adapted to fit local situations. They suggest further that ecological knowledge is much broader than general theory; it also includes facts, natural history observations, approaches, models, concepts, and results of prior research. Taken together, they argue, such ecological knowledge provides sufficient guidance to design effective problem-solving strategies.

Scientific knowledge is only part of the equation, however. The knowledge must be applied, and its successful application depends as much on surmounting political, cultural, administrative, or economic hurdles to decision-making as on the adequacy of the scientific knowledge itself (National Research Council 1986, Levin 1999, Dayton 2003).

A focus on place holds considerable promise for overcoming both scientific and social difficulties. As this volume illustrates, place-based research is an effective route to scientific understanding. Place also motivates social and political action by people whose livelihoods are directly tied to the health of ecological systems, and who know a lot about them. It can provide the key to forging common ground between scientists and local communities, thus facilitating a collaborative effort to solve environmental problems. This chapter presents two case histories in which place has played such a role or has the potential to do so.

While one system is wet and the other dry, the two examples are similar in many ways. Both involved the management of natural resources—rangelands in the first case, fisheries in the second—with strategies that relied on controlling the rate of resource exploitation by setting stocking rates or catch quotas. In both cases these management strategies were failing—grasslands were being converted to shrublands and ranches to ranchettes; fish stocks were declining and traditional fishing communities were being supplanted by industrial fishing operations and waterfront developments. In both cases the local community's desire to maintain a traditional way of life provided the incentive for rethinking how the natural resource was being managed. And in both instances the local community's knowledge of natural and cultural history suggested the reason for the failure of management—a mismatch between the spatial scale of management and the spatial scale of ecological and economic processes—and provided the opportunity for the local community to collaborate with scientists and managers to design research programs whose aim was to improve the ecosystem model on which management was based.

Collaborative Science and Conservation in the U.S.–Mexico Borderlands

The borderlands region of southeastern Arizona, southwestern New Mexico, and adjoining Mexico (figure 12.1) is a remarkable place. Located at a crossroads of several biomes, its complex topography, which ranges from grassland valleys at 1400 meters elevation to mountain peaks that reach 2600 meters, supports one of the most diverse biotas in North America (Brown and Kodric-Brown 1995; Curtin and Brown 2001). Its history until the turn of the twentieth century is the stuff of "Wild West" legend, peopled by Apache Indians, mountain men, Mexican and Anglo settlers, cowboys, explorers, miners, entrepreneurs, outlaws, and soldiers (Sheridan 1995). From the late 1800s until the present, cattle have dominated both the economy and the landscapes of the region. Ranching in the borderlands has always

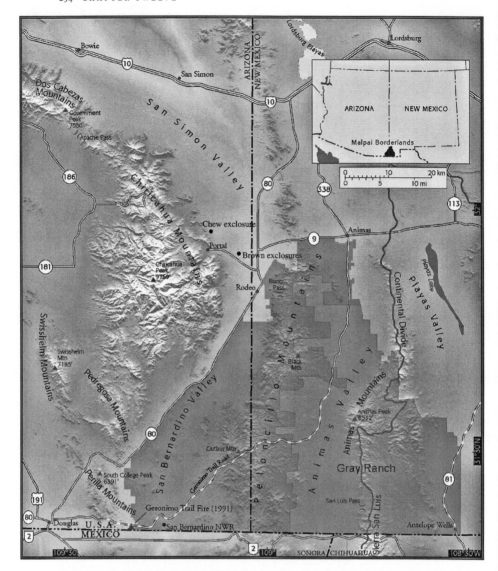

Figure 12.1. The United States–Mexico Borderlands region of southeastern Arizona and southwestern New Mexico. The Malpai Planning Area is the "Malpai Borderlands" shaded area shown in the map inset. Modified from http://www.malpaiborderlandsgroup.org/frontmap.html, courtesy of the Malpai Borderlands Group.

faced variable fortunes, but recent trends in climate, rangeland condition, and public attitudes toward grazing on public lands have raised concerns about the continued viability of this traditional way of life. Those concerns have spawned a pioneering, community-led effort to implement collaborative ecosystem management on the nearly million acres of arid rangeland in the Malpai Planning Area (figure 12.1). The successful collaboration between ranchers, conservationists, scientists, and government agencies grew out of an unusual history of interaction between science and local culture in the borderlands that provided the shared vision and cultural capital necessary for sustained collaborative resource management (e.g., see Berkes and Folke 1998).

In the rest of this section I trace how this collaborative effort came to be. To set the stage, I outline the history of land use, rangeland condition, and science in the region. I then discuss the emergence of the Malpai Border-lands Group (MBG). I trace how the initial model was built from the pooled knowledge of ranchers and scientists, and how it is being tested and refined through ongoing, large-scale experiments whose design again has involved collaboration between scientists and citizens.

Land tenure in the borderlands

At present the Malpai Planning Area (figure 12.1) is a mosaic of 53% private and 47% public lands. The latter are managed by several federal and state public agencies, including the United States Fish and Wildlife Service (USFWS), Forest Service (USFS), and Bureau of Land Management (USBLM); and by Arizona and New Mexico state land agencies. The area is sparsely settled, being inhabited by fewer than 100 ranching families, many of whose ancestors homesteaded the area (McDonald 1995).

This pattern of land ownership is typical of much of the western United States, and is a product of how the area was settled. Prior to the Gadsden Purchase of 1853, the region was part of Mexico, ranched by Mexican settlers. After the purchase, the lands became property of the United States and were explored and surveyed. Some of these public lands were ceded to the states to generate revenue for public education, others to railroad companies to encourage development of transcontinental routes. The Homestead Act of 1862 allowed farmers and ranchers each to obtain title gratis to 160 acres of public land (expanded to 640 acres by the 1877 Desert Land Act), and additional land for $1.25 an acre. Soon nearly every parcel around a water source was acquired; these private parcels served as bases from which ranchers grazed cattle on the unfenced "open range" of surrounding public lands (Sheridan 1995).

In the late 1800s, federal lands policy began to shift toward retaining public lands for the public good, and many states, including Arizona and New Mexico, followed suit (http://www.land.state.az.us/history.htm, http://www.nmstatelands.org/). National forests, parks, and wildlife refuges were withdrawn from homesteading or sale, as were lands rich in oil, gas, or forage. Federal and state agencies were established to manage these public lands for their resource values. By the middle of the twentieth century, the open range had closed; ranchers now had to obtain leases to graze cattle on the public lands surrounding their homesteads.

Land management in the borderlands

During the era of the open range, ranchers in the borderlands were free to decide how many livestock to graze on their privately owned homestead and on surrounding public lands. Today they face a far more complex regulatory environment because public lands are administered by one or another federal or state agency, each of which has different regulations. Furthermore, these regulations change through time as public valuation of the natural resources on public lands evolves and knowledge about those resources accumulates.

In the first half of the twentieth century, agencies focused on managing public lands for their commodities—timber, minerals, game and fish, forage for livestock, and water. With the growth of environmentalism, the goals of management were expanded to include aesthetic and recreational values. Finally, concern over extinction of species added biological diversity to the mix with the Endangered Species Act of 1973.

Ranchers are affected by these changing objectives because agencies achieve them in part by adjusting the intensity, timing, and location of grazing. In the days of "commodity thinking," the metric used to set stocking rates was the productive capacity of the land for cattle; stocking rates were decreased in drought years or in locations experiencing severe soil erosion, under the assumption that reduced grazing pressure would allow the land to recover. However, perceived detrimental effects of cattle on aesthetics, habitat quality, and biodiversity led to increasing public opposition to cattle grazing on public lands. In the 1980s and early 1990s ranchers faced efforts to increase grazing fees and reduce cattle grazing on public lands at a time when they were also facing the stress of deteriorating rangeland condition.

Vegetation change in the borderlands

The 1880s were boom years for the cattle industry in the borderlands. Railroads opened up huge markets for beef, and investors caught wind of a lucrative business opportunity. Stocking rates exploded. The land was stripped of vegetation, leaving it vulnerable to storms that during the 1890s carved gullies in floodplains, washed away fertile topsoil, and channelized riverbeds. In the drought of 1893 as many as 75% of the cattle died, and the cattle industry crashed (Sheridan 1995).The watersheds and vegetation of the region were permanently altered. Perennial streams became intermittent flood channels that drained former marshlands. Water tables declined as infiltration decreased, leaving riparian vegetation high and dry. The perennial grasses did not return; shrublands invaded former grasslands and continued to expand during the twentieth century (Hastings and Turner 1965, Turner et al. 2003).

The causes of this conversion of grassland to shrubland were not immediately apparent. Overgrazing was an obvious factor, but other factors also likely contributed to long-term vegetation change, such as fire suppression, topsoil loss, and changing rainfall patterns (Curtin and Brown 2001). The MBG and their public and private cooperators recognized that understanding the mechanisms of vegetation dynamics was paramount to developing sound strategies for managing public and private lands alike. A long history of science in the borderlands provided an initial context for tackling this issue (Curtin 2005, 2008).

A history of science in the borderlands

Science in the borderlands began with naturalists who accompanied military expeditions or expeditions to survey borders, establish railroad routes, or evaluate the natural resources of newly-acquired lands. For example, frontier photographer C. S. Fly documented parts of the landscape while accompanying General Crook in pursuit of the Apache Chief Geronimo in the 1870s. The remarkable biological diversity recorded by the early naturalists drew others to the area, starting with extensive biological surveys in the late 1800s and early 1900s. In 1955 the Southwest Research Station (SWRS) was established on the eastern side of the Chiricahua Mountains near the hamlet of Portal (figure 12.1). This field station provided a focal point for biological investigation that spread through much of the region.

Some of the research conducted in the borderlands has examined long-term changes in vegetation. Of particular importance to the Malpai borderlands story was the repeat-photography work of R. M. Turner, who matched historical photos with current images to document profound vegetation changes in Arizona in the twentieth century. Photos from desert grasslands and oak woodlands indicated upward migration of vegetation zones and confirmed the shrub invasion that ranchers had been seeing (Hastings and Turner 1965, Turner et al. 2003).

Although Turner's work provided hard data about vegetation change, many ranchers and agency scientists were already convinced of its reality from anecdotal evidence (e.g., Leopold 1924), and projects were initiated to evaluate one factor implicated in vegetation change—cattle grazing. The most common approach has been to observe change following livestock exclusion. The USFS established grazing exclosures and other study plots starting in the 1930s, the New Mexico Department of Game and Fish established a series of exclosures in 1954, and ranchers also "rested" degraded pastures to see whether grasses would recover (Sayre 2005). The first "basic science" study, however, was initiated in 1958 by Robert and Alice Chew. Working out of the SWRS, they established a 9.3 ha cattle exclosure on degraded grassland in the San Simon Valley near Portal that has been monitored ever since (figure 12.1). The Chews' work documented long-term dynamics of plant and animal populations (Chew and Chew 1965; Chew 1977, 1987, 1995), the role of rodents and their interaction with shrubs in organizing desert ecosystems (Chew and Whitford 1992), and the fact that the process of vegetation change following removal of livestock can be extremely slow (Chew 1982, Valone et al. 2003).

Robert and Kathyrn Scholes, in collaboration with ecologists John Ludwig and Will Moir, formed a nonprofit corporation (the BioResearch Ranch) to study natural ecosystem dynamics following removal of livestock in 1972 from a 5,000-acre BLM Research Natural Area (RNA) in the Peloncillo Mountains. The site, located nine miles southeast of Rodeo, New Mexico, at an elevation of approximately 6,000 ft, is composed primarily of semiarid grassland and savanna. Soil and vegetation have been monitored since the early 1970s, and precipitation since 1978 (Moir 1979, Moir et al. 2000). The BioResearch Ranch serves as a critically important reference area and was the first large-scale collaborative experiment in the borderlands. An analysis of BioResearch Ranch data on climate and grazing (e.g., Curtin and Brown 2001) suggested that large grazers are key to sustaining ecosystem function and generated hypotheses being evaluated with larger-scale studies of climate and disturbance begun in the late 1990s.

James H. Brown, Diane Davidson, and O. J. Reichman initiated a third long-term experiment near the Chews' exclosure in 1977 to study interactions among seed-eating small mammals, ants, and birds. Initially scheduled to last four years, the results proved interesting enough so that the study has continued for more than 30 years (Brown et al. 1986, Brown 1998). The study has made numerous important contributions to science in the borderlands, among them an illustration of the power of a well-designed, replicated, multifactorial experiment to distinguish among hypotheses. It demonstrated the importance of long-term study for understanding ecological processes. For example, effects of removing components of the seed-eater community were not apparent until five years into the experiment (e.g., Brown et al. 1986), and effects of climatic variability required 20 years to assess rigorously (e.g., see Brown et al. 1997, Curtin and Brown 2001). Third, and probably of greatest interest to the local ranching community and land managers, it indicated that vegetation responses to climate are different in the presence and absence of small herbivores (Brown et al. 1997, Curtin et al. 2000), and that removal of 50- to 100-gram rodents had far larger impact on the system than more than 20 years of cattle exclusion.

Although the long history of science in the borderlands and the accumulated natural-history knowledge of the local ranchers provided a great deal of information, it was insufficient to answer the questions ranchers and land managers wanted to answer. One reason was that the spatial and temporal scales of field research projects are typically much smaller than the scales on which ranchers and land managers operate. Scale matters; ecologists have increasingly recognized that one cannot extrapolate easily from experimental plots on the scale of meters to landscapes on the scale of kilometers, or from a year-long study to decades (Carpenter 1996, Schindler 1998, Wiens 2002). This is because the hierarchical structure of ecological systems makes phenomena and processes at each spatial or temporal scale dependent on those taking place on the scale above. Attempting to scale up from smaller to larger scales is inherently problematic (Allen and Starr 1982, O'Neill et al. 1986, Allen and Hoekstra 1992).

The appropriate scale for a research project is defined by the questions that are asked, and the questions scientists ask depend on the cultural context and funding environment (Curtin 2005). Left to their own devices, scientists working in the borderlands would not necessarily generate and test hypotheses about how the Malpai ecosystem was functioning on scales relevant to ranchers, conservationists, or public-lands managers (Curtin 2005). It took collaboration with ranchers, who had a long-term stake in the Malpai ecosystem, working together with the scientists and managers, to take the next step and "get the scale right" (Wiens 2002).

The Malpai Borderlands Group and the rise of collaborative science

Beginning in the late 1980s, a small group of ranchers met with environmentalists, Quaker mediator and philosopher Jim Corbett, and scientist Ray Turner at the Malpai Ranch near Douglas, Arizona, to discuss mutual concerns. The group called itself the Malpai Group after the ranch, whose name in turn derived from the volcanic landscape of the area, which is called "malpai," meaning "badland" in Spanish. The ranchers were concerned about their ability to continue their traditional way of life in the face of deteriorating grassland condition, climate variability, and increasing public opposition to grazing on public lands. The environmentalists were concerned about maintaining biodiversity and landscape values in the face of habitat fragmentation and shrub invasion. Ray Turner provided a key link to the scientific perspective on vegetation change and connections to the scientific community, and Jim Corbett provided an overarching vision for human communities living in harmony with the land (Sayre 2005).

A consensus emerged that the existing piecemeal land-management structure was simply not compatible with ecological or economic realities (Curtin et al. 2002). What was needed instead was a community-led private-public partnership to achieve collaborative, landscape-scale management (Curtin 2007). This consensus heralded a shift toward the "radical center," where people with a diversity of perspectives forged common goals for land management and consensus for how to achieve them (McDonald 1995, Wolf 2001, Sayre 2005). The common goals that the group identified are reflected in its mission statement (http://www.malpaiborderlandsgroup.org/): "Our goal is to restore and maintain the natural processes that create and protect a healthy, unfragmented landscape to support a diverse, flourishing community of human, plant and animal life in our borderlands region. Together, we will accomplish this by working to encourage profitable ranching and other traditional livelihoods which will sustain the open space nature of our land for generations to come."

The Malpai Group adopted a concept similar to the land ethic that Aldo Leopold had developed a half a century earlier, in part from his experiences in the Sierra Madre in Mexico just south of the Malpai area (Leopold 1949; Meine 1988; Waller and Flader, chapter 3).

Two events catalyzed further action (McDonald 1995, Sayre 2005). First, in 1991 a small fire (the Geronimo Trail fire; figure 12.1) that was burning in a mosaic of public and private lands was suppressed over the objections of adjacent ranchers. The ranchers argued that fire suppression was partly responsible for shrub invasion, because fire killed young shrubs but not grasses. The federal agency in question acted on a long-standing assumption that fire damaged arid ecosystems (DeBano et al. 1995), and on an obligation to protect people and property. This event caused borderlands ranchers to realize that the only way to reintroduce fire into the landscape was to develop a regional fire management plan in collaboration with private landowners and government agencies.

A second formative event was the purchase of the enormous Gray Ranch (now called the Diamond A Ranch; figure 12.1) by the Nature Conservancy, a national conservation organization, and its subsequent sale to a local ranching family who formed the nonprofit Animas Foundation to manage it as a working ranch. Conservation easements held by the Conservancy ensured that the ranch would not be subdivided and would be managed so as to retain its remarkable biological diversity. Malpai ranchers approved of this experiment in conservation ranching (after all, the Animas Foundation family belonged to the Malpai Group), and asked for the Conservancy's support to help give credibility to the concept of a regional fire plan. In the end, all parties committed to much more than a coordinated fire plan; they committed to the broader concept of adaptive ecosystem management, a novel

TABLE 12.1. Malpai Borderlands Group cooperators (http://www.malpaiborderlands group.org/protection.asp)

Sector	Cooperating organization
Private	Ranchers in Arizona, New Mexico, and Mexico
Public	U.S. Department of Agriculture Forest Service, Coronado National Forest
	Natural Resource Conservation Service of Arizona and New Mexico
	Hidalgo Soil and Water Conservation District
	Whitewater Draw Natural Resource Conservation District
	U.S. Department of the Interior U.S. Bureau of Land Management, Las Cruces and Gila Districts U.S. Fish and Wildlife Service
	U.S. Department of Homeland Security Border Patrol/Customs
	Arizona State Land Department
	University of Arizona Desert Laboratory
	University of New Mexico
	New Mexico State University
	New Mexico Department of Game and Fish
	Arizona Game and Fish Department
Nonprofit	Animas Foundation
	Nature Conservancy
	Fundación San Bernardino

approach that was transforming federal public-lands management policies at the time.

It was logical for the Malpai Group to lead the collaborative effort to forge such a plan. Accordingly it became a 501(c)(3) nonprofit organization, the Malpai Borderlands Group (MBG), and established collaboration with an impressive array of cooperating entities (table 12.1). The MBG and its collaborators recognized that sound management had to be based on scientific understanding of the forces shaping the ecological, economic, and cultural systems of the borderlands, and that scientists trained broadly in landscape and systems ecology, rather than narrowly in natural resource management, could help them to achieve such an integrated understanding. The group accordingly established a science advisory committee and initiated a research program whose first component was the Southwestern Borderlands Ecosystem Management Project, funded by the USFS (Gottfried et al. 2000).

Integrating science and local knowledge to create a shared model of the ecosystem

The first step to a fuller understanding of the borderlands involved distilling the factors of greatest probable importance to vegetation dynamics and biodiversity into a conceptual model of the grassland ecosystem. The model was built with input from scientists and ranchers. The scientists understood broad ecological principles, how to interpret results of previous research, and how to design a research program. The ranchers defined the fundamental questions and contributed knowledge of natural history that helped in identifying candidate processes to include in the model and the scale on which they operated.

The scientists and ranchers adopted a hierarchical approach (e.g., see Allen and Starr 1982), largely because the ranchers understood from personal observation that ecological processes operated across a variety of scales. Fire follows the fuel, crossing vegetation types and jurisdictional boundaries. Cattle impact individual plants but move through the landscape in response to the distribution of water and forage. How removing them affects subsequent vegetation dynamics depends on interacting factors—soils, topography, precipitation, starting vegetation condition, fire and grazing history. The same is true of drought, which affects the entire region, but to different extent in different places depending on a complex interplay among multiple factors including soils, topography, and local rainfall patterns. At a time when conservation biology was focusing on preserving single-species populations, and when both conservation and management were focusing on single-factor solutions (Holling and Meffe 1996), the MBG chose a goal of understanding the effects of multiple interacting processes on scales relevant to management of entire ecosystems (Curtin 2002). They wanted to know whether, and how, the borderlands could be managed for sustainable ranching; they would invest in the best available science and "let the chips fall where they may" (Wolf 2001).

The model incorporated factors of likely importance that could be managed and therefore were amenable to experiment—fire, grazing, species composition of the herbivore community—as well as factors that were not amenable to experimental manipulation on large scales, such as drought or starting condition of the soils and vegetation. These latter factors could be addressed by locating experimental plots on environmental gradients and adopting a research design that allowed for comparison with other studies. Most of these latter studies had been conducted on degraded grasslands on public lands that responded slowly, if at all, to experimental treatments. Many suspected that this slow response occurred because these systems had undergone a "state transition" (Westoby et al. 1989) that was difficult or perhaps impossible to reverse. The MBG therefore decided to focus its research efforts in areas where management might prevent a shift in state, such as areas where shrubs were just beginning to encroach on grasslands. The private lands of the Gray Ranch provided large expanses of such relatively undamaged grassland.

Although much of the science and monitoring in the borderlands is motivated by applied management questions, a number of projects have also been designed to address fundamental questions about the role of general ecological processes such as disturbance or climatic variation in molding biotic communities at a landscape scale. The most ambitious, replicated experimental project has been located on a large pasture, McKinney Flats, on the Gray Ranch (figure 12.1). Four research plots were established, each one km^2 in size and divided into four subplots that received a factorial combination of two treatments: fire and livestock grazing. A pronounced rainfall gradient across McKinney Flats allowed the interaction of climate with fire and herbivory to be examined. Smaller 36 × 36–m exclosures were established within the replicated grazed treatments to exclude livestock, large native grazers such as mule deer (*Odocoileus hemionus*) and antelope (*Antilocapra americana*), folivores such as rabbit (*Sylvilagus auduboni, Lepus californicus*) and javelina (*Pecari tajacu*), and small granivorous rodents. These small exclosures allowed researchers to determine more precisely the relative impact of these organisms on grassland vegetation composition. In addition to the fire and grazing treatments, four black-tailed prairie dog (*Cynomys ludovicianus*) reintroductions were conducted, one in each pasture, to assess their ecological effects. Within each research area the distribution and abundance of lizards, small rodents, and vegetation was measured by field crews that included a combination of professional biologists and local residents. These guilds of organisms were selected because they were suspected of responding in different ways to the treatments, and also because they represented either a fundamental consumer group or, in the case of vegetation, the primary biomass within the system.

After 10 years, the McKinney Flats study is beginning to produce very different results from previous work. For example, whereas small mammal exclusion has had a stronger effect on the composition of vegetation than cattle exclusion in degraded grasslands (e.g., Brown 1998, Curtin and Brown 2001), the opposite appears to be true in less degraded grassland: small-mammal exclusion has little impact, and native grazers have a pronounced impact. (Curtin 2008). In addition, climatic variation appears to exert a stronger effect on vegetation dynamics than does either fire or livestock grazing, and prairie dogs interact positively with cattle to enhance grasses (Curtin 2006).

The McKinney Flats study is being supplemented by monitoring of vegetation and sensitive animal species before and after large prescribed fires, of the condition of vegetation and target animal groups on conservation easements, and of the effects of erosion control structures on watersheds. This coordinated research effort holds great promise for answering questions about which processes are involved in vegetation change, and their relative importance under different environmental conditions in the Malpai Borderlands. Results will inform basic understanding of southwestern rangelands as well as management decisions by ranchers and other land stewards. Especially since most of the federally funded long-term studies in the Southwest either are on degraded sites or do not allow landscape-level application grazing or fire, private-sector studies are essential to understanding the impact of climate and disturbance on functioning native ecosystems, and to providing a crucial baseline for understanding the biotic potential of Southwestern ecosystems (Curtin 2008). Malpai is an exemplar of how a place-based ecosystem model on the appropriate spatial scale can be built and tested by collaboration between a local community and scientists (Curtin 2008).

Community, Cod, and the Commons: Interactions of Fisheries Science and Local Knowledge in the Gulf of Maine

Another case of collaborative science has arisen in the Gulf of Maine (figure 12.2) to improve management of bottom-dwelling fish (groundfish), such as cod (*Gadus morhua*) and haddock (*Melanogrammus aeglefinus*), once a mainstay of fishing communities in the northeastern United States. As with rangelands, the issue here is how to establish institutional structures that "tighten the feedback loops" between the manager and the system to be managed, and that base management on knowledge of the system (Levin 1999). Key to this process, the place itself provides common ground for building linkages between local and science-based knowledge, and also for cooperative management.

Fisheries in the Northwest Atlantic region in general, and in the Gulf of Maine in particular, have been dealing with substantially depleted stocks of groundfish. As in the Malpai Borderlands Area, resource depletion has not only put an economic squeeze on the communities but has also threatened the identities of communities that have long defined themselves in terms of their economic relationship with an ecosystem.

In this section I review the history of groundfishing in the region, highlighting practices that have led to collapse of the fisheries, including the failings of top-down attempts at resource management. I then explore how efforts to integrate local and scientific knowledge have modified the models that can guide management and serve as the foundation for more effective, collaborative management of the system.

History of groundfishing

Groundfish species vary in their life histories, but cod serve as a reasonable model. Male and female cod congregate at spawning grounds to reproduce. There they broadcast sperm or eggs; fertilization occurs in open water. Larvae

Figure 12.2. The Gulf of Maine. Modified from http://research.usm.maine.edu/gulfofmaine
census/wp-content/images/bathymetry/map-gulfofmaine.jpg, courtesy of Gulf of Maine Area–
Census of Marine Life and USGS Digital Bathymetry.

remain in the water column before eventually settling near the ocean bottom where they feed as adults, often far from the spawning areas. Adult cod
display a diversity of movement patterns. While some fish remain in a single
well-defined place that includes feeding and spawning areas, others move,
apparently in response to changing water temperatures, food availability, or
reproductive status.

Groundfish such as cod have been economically and socially important
in the Northwest Atlantic region for more than 500 years (Kurlansky 1997).
By the seventeenth century, fishing stations were developed by European

powers on islands along the Maine coast. Early colonists also exploited inshore fisheries for their own subsistence. Near the end of the seventeenth century, New England fishermen began to fish offshore on the shallow continental shelves, or "banks," in the Gulf of Maine. Coastal waters of Penobscot Bay (south of Bangor, Maine; figure 12.2) and eastern Maine were found to be highly productive, resulting in eastern Maine having large numbers of smaller, inshore groundfish vessels (O'Leary 1996). These fishing grounds sustained local communities for nearly two centuries (Ames 1994). From 1831 to 1865, eastern Maine supplied the majority of U.S. cod landings, peaking with three-quarters of the total U.S. catch in 1861 (O'Leary 1996). By the mid-1800s, nearly every coastal village was engaged with major inshore fisheries (Kurlanksy 1997).

Groundfishing was originally conducted with handlines from sail-powered vessels, which were replaced by tub trawling and handlining from dories by the 1860s. The purse seine came into use in the 1870s, and in 1893 the first otter trawl vessel was introduced by the U.S. Fisheries Commission. By 1918, steel-hulled otter trawlers were being built and a Massachusetts-based trawl fleet began to grow. In the twentieth century, technological advances and commercial demand caught up with ecological constraints as "rocky hopper" gear allowed nets to be pulled into increasingly rough and rocky habitats that had previously served as refugia for fish. By the 1950s the Maine and Massachusetts fleets had depleted eastern Maine's coastal groundfish stocks and eliminated large numbers of coastal spawning grounds. Vessels large enough to handle open water left the coastal shelf to fish the offshore banks, while small and mid-sized boats left fishing altogether or shifted to other species (Ames 2004). By the late 1960s to early 1970s, catches had declined steeply but technological advances continued. Sonar allowed fishermen to locate fish directly, while loran and global positioning systems allowed fishermen to mark the location of catches precisely. As one fisherman recently remarked, "With modern technology we can now catch every last damn fish in the ocean; the challenge lies not in catching fish but in having the restraint to let them live."

Unlike in the Malpai example, a single federal government entity, the National Marine Fisheries Service, manages fisheries in offshore waters. Research and management have effectively been dictated from the top down, although efforts have been made to legislate collaborative management. Regional management was implemented with passage of the 1976 Fisheries Conservation and Management Act, and eight councils were formed to develop regionally appropriate plans to protect fisheries. The Gulf of Maine is included in the New England Fishery Management Council. In an attempt to build collaborations, the councils consist of government managers and scientists as well as representatives from interest groups including environmentalists, commercial fishermen, sport fishermen, and fleet owners. Unfortunately the process has not been particularly collaborative; disproportionate amounts of power lie in government agencies that are influenced by large commercial interests (Kurlanski 2008). Fishermen argue that the time—which can take two days of travel in some places—and expense involved in attending council meetings are prohibitive for family-owned boats outside of large ports in southern Maine and Massachusetts. Consequently few small-boat owners can afford to participate.

The Fisheries Conservation and Management Act, along with other economic and political factors, has contributed to the substantial decline of fisheries. In addition to establishing the regional councils, the act fundamentally rescaled the fishery by extending the exclusive economic fishing zone of the United States from 22 km (12 nautical miles) to 370 km (200 nautical miles). Canada passed similar legislation that limited the access of both distant-water and U. S. fleets to Canadian waters. When the Hague Line (the Canadian/U.S. maritime boundary in the Gulf of Maine and on Georges Bank; figure 12.2) was established in 1984, U.S. fishing in Canadian waters was terminated. Loss of access to Canadian waters intensified pressure on all New England groundfish stocks. Simultaneously, federal loan programs led to overcapitalization of the fisheries and record catches were quickly followed by collapse of the fisheries and federal buybacks of fishing boats (Kurlanski 2008).

In 1986 the Northeast Multispecies Fishery Management Plan (NMFMP) institutionalized large-scale regional management of groundfish stocks, but stocks continued to decline. In response to this evidence of regulatory failure, and also to a series of successful lawsuits by environmentalists, management plans were amended to create more accountability. This has not improved how the fishery is managed, however: fundamental differences remain between fishermen, scientists, and regulators in their understanding of the ecosystem.

Understanding the ecosystem

Many of the problems with fisheries management occur because policies are based on an inadequate model of how the ecosystem works (Smith 1995, Wilson 2006). Biologists, economists, and managers have modeled the fishery as a simple system; the Gulf of Maine is considered a homogeneous mass of water in which fish move about at random (Curtin et al. 2005). This "bathtub model" describes population dynamics with differential equations, assuming that fish species do not interact and that mortality and recruitment are simple functions of the stock and fishing rate. Management is predicated on the concept of maximum sustainable yield, in which one adjusts the fishing effort to achieve a stock (population) size that maximizes the replenishment rate of the population (Kurlanski 2008).

In contrast, fishermen view the system as involving complex interactions among multiple species that are affected by numerous external factors such as pollution, storms, and water temperature. This perspective views a fishery as a dynamic system in which numerous variables interact in a nonlinear and often chaotic fashion (e.g., see Smith 1995, Holland 1998, Wilson 2006). While the approach of the fishermen lacks the experimental rigor that allows statistical inference, it does lead to a deep understanding of how the fisheries work.

In an effort to improve the models of the managers and scientists, the Gulf of Maine Council on the Marine Environment (an independent, binational committee of fishermen, academics, and resource managers) launched an initiative in 1995 to document local knowledge of groundfish spawning habitats in the Gulf of Maine. Ted Ames, a former fisherman from Maine, interviewed approximately 40 retired fishing boat captains to map locations of cod and haddock spawning grounds in the U.S. portion of the gulf. Almost identical work was also done by the Canadian Department of Fisheries and Oceans.

Both studies indicated that the cod fishery consists of a number of discrete populations attached to distinct spawning grounds, most of which have been lost. The results have in turn been supported by genetic and tagging studies that also suggest numerous small and discrete populations (Ruzzante et al. 2000, Robichaud and Rose 2004). This information suggested that cod stocks should be modeled as multiple populations, each with its own largely independent dynamics.

This change in assumptions about population subdivision has important implications for the management of the fishery. If one assumes that there is a single stock for each species—the bathtub model—then the present spatial distribution is attributable to a contraction of a once-abundant stock into the best areas, the "core range." Under this model, reduced catch quotas should reduce fishing mortality throughout the Gulf and allow stocks to increase to the point that populations recolonize less ideal habitat. Alternatively, a model that assumes local, largely independent populations leads to different conclusions. The current management approach sets a goal for how many fish should be taken within the entire Gulf based upon estimates of the sustainable yield in a homogeneous population. To achieve these quotas, permits are issued for a certain number of boat-days of effort. This encourages highly mobile fishing strategies dominated by relatively large and fast boats that move to follow the fish so as to maximize catch within the allocation of boat-days. The result is that small local populations with lower maximum sustainable yields than the average in the gulf as a whole can be fished more intensively than is sustainable (e.g., see Wilson 2006). Craig Pendleton, a longtime fisherman who represented the community-based Northwest Atlantic Marine Alliance (NAMA), summarized the situation thus: "We feel that the scale at which fishing takes place in New England is out of whack . . . we're hoping to find a solution before the rest of us (i.e., small and mid-sized boats) go out of business" (Jansen 2007).

While local knowledge has started to change how managers think about this system, collaborative groups are just emerging to help translate that knowledge into improved management strategies. The Downeast Initiative linked science and policy through a coalition between local fishermen, University of Maine biologists and economists, and conservation organizations. NAMA focuses on developing more effective fisheries policy. Local community organizations such as the Coobscook Bay Resource Center and Penobscot East Resource Center provide a focus on building the capacity of fishing communities for community-based science and resource management, while the Island Institute emphasizes the health of island communities. Finally, the Midcoast Fishermen's Association presents fishermen's interests in a diversified fishery. All of these organizations have come together to form the Area Management Coalition, which advocates taking an explicitly place-based approach to fisheries management based on a combination of science and fisherman-based understanding. It has asked the New England Fishery Management Council to substitute local management zones for the current days-at-sea regulatory system. Despite the new understanding of the system that reflects the fishermen's informal understanding and, increasingly, that in the peer-reviewed literature, the political dynamics of the council process are highly resistant to change. It remains to be seen how effective these organizations will be in getting the top-down management system to institute truly collaborative ecosystem management on a more appropriate local scale. To date the place-based approach is still being rejected in favor of conventional management, in spite of that system's pattern of failure.

Conclusion

The two cases just discussed illustrate how a focus on place holds promise for improved management of natural resources. The place-based focus provides a means of linking the scale of management with the scale of ecosystem processes by integrating local and science-based knowledge. In the case of the Malpai the fortuitous alignment of numerous factors has facilitated development of a common understanding of the ecosystem: a history of science in the region and sustained interest by researchers and conservationists; the vision and leadership of a few people with connections to the ranching, scientific, and conservation communities; the resources available to the ranching family that formed the Animas Foundation; the decision of the Nature Conservancy to trust the Animas Foundation to ranch in an ecologically sustainable fashion; and the federal government's adoption of ecosystem management and public-private partnerships.

The long-term success of the Malpai Borderlands Group will be judged, however, by whether ranching can be maintained as a viable industry within a healthy landscape. To do so, the knowledge that emerges from their science and monitoring programs must be transformed into meaningful management decisions. The process of developing a shared understanding of the landscape has increased the level of trust among diverse stakeholders and led to their increased willingness to cooperatively manage the landscape. While such cooperation is not sufficient by itself to achieve the goals of the MBG, it is necessary for a region in which natural processes like fire easily cross the boundaries of individual land-owners and government entities. Keeping the landscape open so that these natural processes can operate will be key to sustaining ecological and social systems over the long term.

The Gulf of Maine fisheries face a different set of challenges and opportunities for collaborative science. Instead of being managed by multiple agencies, offshore U.S. waters are managed as a single set of resources by a single government agency that has research, policy, and regulatory responsibilities. This eliminates the need to coordinate among agencies, but the monolithic structure can create barriers to communication and collaboration with local communities. Approaches such as the Area Management Coalition hold much promise for improving collaborative science, but political barriers must still be surmounted. Some scientists and the federal fisheries management system still view the knowledge of fishermen as anecdotal and less reliable than data collected systematically. And some fishermen still view the scientists and managers with suspicion. The example of the Malpai Borderlands Group suggests, however, that the best antidote is to let the place work its magic through shared hours spent on the water or in the saddle.

References

Allen, T. F. H., and T. W. Hoekstra. 1992. *Toward a Unified Ecology*. Columbia University Press, New York.

Allen, T. F. H., and T. B. Starr. 1982. *Hierarchy: Perspectives for Ecological Complexity*. University of Chicago Press, Chicago.

Ames, E. P. 2004. Atlantic cod structure in the Gulf of Maine. *Fisheries* 29:10–28.

Berkes, F., and C. Folke. 1998. *Linking Ecological and Social Systems*. Cambridge University Press, Cambridge.

Brown, J. H. 1998. The desert granivory experiments at Portal. Pages 71–95 *in* W. J. Resetarits and J. Bernardo, eds., *Experimental Ecology*. Oxford University Press, Oxford.

Brown, J. H., D. W. Davidson, J. C. Munger, and R. S. Inouye. 1986. Experimental community ecology: The desert granivory system. Pages. 41–62 *in* J. Diamond and T. J. Case, eds., *Community Ecology*. Harper and Row, New York.

Brown, J.. H., and A. Kodric-Brown. 1995. Biodiversity in the borderlands. *Natural History* 105:58–61.

Brown, J. T., T. J. Valone, and C. G. Curtin. 1997. Reorganization of an arid ecosystem in response to recent climate change. *Proceedings of the National Academy of Sciences (USA)* 94:9725–33.

Carpenter, S. R. 1996. Microcosm experiments have limited relevance for community and ecosystem ecology. *Ecology* 77:677–80.

Chew, R. M. 1977. Some ecological characteristics of the ants of a desert-shrub community in southwestern Arizona. *American Midland Naturalist* 98:33–49.

———. 1982. Changes in herbaceous and suffrutescent perennials in grazed and ungrazed desertified grassland in southeastern Arizona, 1958–1978. *American Midland Naturalist* 108:159–69.

———. 1987. Population dynamics of colonies of three ant species of ants in desertified grassland, southeastern Arizona, 1958–1981. *American Midland Naturalist* 118:177–88.

———. 1995. Aspects of the ecology of ants (*Myrmecocystus spp., Aphaenogaster spp.*) in desertified grasslands in southeastern Arizona, 1958–1993. *American Midland Naturalist* 134:75–83.

Chew, R. M., and A. E. Chew. 1965. The primary productivity of a desert shrubland (*Larrea tridentata*) community. *Ecological Monographs* 35:355–75.

Chew, R. M., and W. G. Whitford. 1992. A long-term positive effect of kangaroo rats (*Dipodomys spectabilis*) on creosotebushes (*Larrea tridentata*). *Journal of Arid Environments* 22:375–86.

Curtin, C. G. 2002. Integration of science and community-based conservation in the Mexico-US borderlands. *Conservation Biology* 16:880–86.

———. 2005. Complexity, conservation, and culture in the Mexico/U.S. borderlands. Pages 237–58 *in* M. West Lyman and B. Child, eds., *Natural Resources as Community Assets: Lessons from Two Continents*. Sand County Foundation and Aspen Institute. Monona, WI.

——— 2006. Ecological implication of prairie dogs in Chihuahuan Desert grasslands: Initial results from long-term studies. Preservation of Desert Grasslands Conference Proceedings. *USDA Forest Service Proceedings* RMRS-P-40.

———. 2007. Landscape perspective and experimental science in collaborative ecosystem management. *Conservation Biology* 24:1117–19.

———. 2008. Emergent properties of the interplay of climate, fire, and grazing in a desert grassland. *Desert Plants* 24, no. 2.

Curtin, C. G., and J. H. Brown. 2001. Climate and herbivory in structuring the vegetation of the Malpai borderlands. Pages. 84–94 *in* C. J. Bahre, and G. L. Webster, eds., *Vegetation and Flora of La Frontera: Vegetation Change along the United States-Mexico Boundary*. University of New Mexico Press, Albuquerque.

Curtin, C. G., D. A. Kelt, T. C. Frey, and J. H. Brown. 2000. On the role of small mammals in mediating climatically driven vegetation change. *Ecology Letters* 3:309–17.

Curtin, C. G., N. F. Sayre, and B. D. Lane. 2002. Transformations of the Chihuahuan borderlands: grazing, fragmentation, and biodiversity conservation in desert grasslands. *Science and Policy* 5:55–68.

Dayton, P. K. 2003. The importance of the natural sciences to conservation. *The American Naturalist* 152:1–13.

DeBano, L., P. F. Folliott, A. Rubio-Ortega, G. J. Gottfried, R. H. Hamre, and C. B. Edminster. 1995. *Biodiversity and Management of the Madrean Archipelago: The Sky Islands of Southwestern United States and Northwestern Mexico.* General Technical Report RM-GTR-264. U.S. Department of Agriculture, Rocky Mountain Forest and Range Experimental Station, Fort Collins, CO.

Gell Mann, M. 1995. *The Quark and the Jaguar: Adventures in the Simple and the Complex.* Holt Publishing Company, New York.

Gottfried, G. J., C. B. Edminster, R. J. Bemis, L. S. Allen, and C. G. Curtin. 2000. Research support for land management in the southwestern borderlands. Pages 330–34 *in* M. Dillion, ed., *Land Management and Stewardship in the 21st Century.* U.S.D.A. Forest Service Proceedings RMRS-P-13.

Gunderson, L. H., and C. S. Holling. 2002. *Panarchy.* Island Press, Washington, DC.

Gunderson, L. H., C. S. Holling, and S. S. Light. 1995. *Barriers and Bridges to the Renewal of Ecosystems and Institutions.* Cambridge University Press, Cambridge.

Hastings, J. R., and R. M. Turner. 1965. *The Changing Mile: An Ecological Study of Vegetation with Time in the Lower Mile of an Arid and Semiarid Region.* University of Arizona Press, Tucson.

Holland, J. 1998. *Emergence.* Perseus Books, Cambridge.

Holling, C. S., and G. K. Meffe. 1996. Command and control and the pathology of natural resource management. *Conservation Biology* 10:328–37.

Jansen, B. 2007. *Portland Press Herald,* final edition. March 25, 2007.

Leopold, A. 1924. Grass, brush, timber, and fire in southern Arizona. *Journal of Forestry* 22: 1–10.

Kurlanski, M. 1997. *Cod: A Biography of the Fish that Changed the World.* Penguin Books, New York.

———. 2008. *The Last Fish Tale.* Riverhead Books, New York.

Levin, S. A. 1999. *Fragile Dominion: Complexity and the Commons.* Perseus Publishing, Cambridge, MA.

McDonald, B. 1995. The formation and history of the Malpai Borderlands Group. Pages 483–86 *in* L. DeBano, P. F. Ffolliott, A. Rubio-Ortega, G. J. Gottfried, R. H. Hamre, and C. B. Edminster, eds., *Biodiversity and Management of the Madrean Archipelago: The Sky Islands of Southwestern United States and Northwestern Mexico.* General Technical Report RM-GTR-264. U.S. Department of Agriculture, Rocky Mountain Forest and Range Experimental Station, Fort Collins, CO.

Meine, C. 1988. *Aldo Leopold: His Life and Work.* University of Wisconsin Press, Madison.

Moir, W. H. 1979. Soil-vegetation patterns in the central Peloncillo Mountains, New Mexico. *American Midland Naturalist* 102:317–31.

Moir, W. H., J. A. Ludwig, and R. T. Scholes. 2000. Soil erosion and vegetation in grasslands of the Peloncillo Mountains, New Mexico. *Soil Science Society of America Journal* 64: 1055–67.

National Research Council, Committee on the Applications of Ecological Theory to Environmental Problems (G. H. Orians, chair). 1986. *Ecological Knowledge and Environmental Problem Solving: Concepts and Case Studies.* National Academies Press, Washington, DC.

O'Leary, W. M. 1996. *Maine Sea Fisheries: The Rise and Fall of a Native Industry 1830–1890.* Northeastern University Press, Boston.

O'Neill, A., R. Johnson, and A. W. King. 1986. A hierarchical framework for the analysis of scale. *Landscape Ecology* 3:193–205.

Peters, R. H. 1991. *A Critique of Ecology*. Cambridge University Press, UK.

Robichaud, D., and G. A. Rose. 2004. Migratory behavior and range in Atlantic cod: Inference from a century of tagging. *Fish and Fisheries* 5:185–214.

Ruzzante, D.E., C. T. Taggart, S. Lang, and D. Cook. 2000. Mixed stock analysis of Atlantic cod near the Gulf of St. Lawrence based on microsatellite DNA. *Ecological Applications* 10:109–110.

Sayre, N. F. 2005. *Working Wilderness: The Malpai Borderlands and the Future of the Western Range*. Rio Nuevo Press, Tucson.

Schindler, D.W. 1998. Replication versus realism: The need for ecosystem-scale experiments. *Ecosystems* 1:323–34.

Schrader-Frechette, K. S., and E. D. McCoy. 1993. *Method in Ecology: Strategies for Conservation*. Cambridge University Press, Cambridge.

Simberloff, D. 2004. Community ecology: Is it time to move on? *The American Naturalist* 163:787–99.

Sheridan, T. E. 1995. *Arizona: A History*. University of Arizona Press, Tucson.

Turner, R. M., R. H. Webb, J. E. Bowers, and J. R. Hastings. 2003. *The Changing Mile Revisited*. University of Arizona Press, Tucson.

Valone, T. J., M. Meyer, J. H. Brown, and R. M. Chew. 2002. Timescale of perennial grass recovery in desertified arid grasslands following livestock removal. *Conservation Biology* 16:995–1002.

Westoby, M., B. Walker, and I. Noy-Meir. 1989. Opportunistic management for rangelands not at equilibrium. *Journal of Range Management* 42:266–74.

Wiens, J. A. 2002. Central concepts and issues in landscape ecology. Pages 3–21 *in* K. J. Gutzviller, ed., *Applying Landscape Ecology in Biological Conservation*. Springer, New York.

Wilson, J. 2006. Matching social and ecological systems in complex ocean fisheries. *Ecology and Society* 11(1): 9. [online] http://www.ecologyandsociety.org/vol11/iss1/art9/

Wolf, T. 2001. The Malpai Borderlands Group: Science, community, and collaborative management. Workshop on Collaborative Resource Management in the Interior West. Red Lodge Clearing House, Boulder, CO. http://www.redlodge.org/proceedings.html.

Part 4

THE INTERACTION BETWEEN LOCAL AND GENERAL UNDERSTANDING

Mary V. Price and Ian Billick

Part 4 considers the interplay between knowledge of a particular system and general understanding. Part 3 focused on how an understanding of place is built; we argued that such understanding is best considered as a conceptual model that explains characteristics of the system. Place-based models interact with general understanding, which is embodied in theories

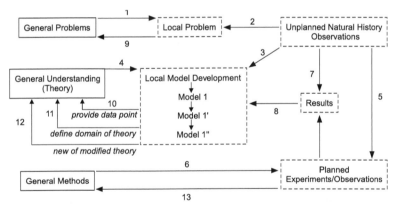

Figure p4.1. The dynamic interplay between local and general understanding. A specific local problem or question can arise from general problems discussed in the literature (path 1), or from natural history observation (path 2). Local understanding consists of a model of how the system works that is built progressively, stimulated by the local problem. Unplanned natural history observation (path 3), along with current general theory (path 4), contributes hypotheses to be tested during model development. Natural history (path 5), along with generally recognized methodology (path 6), contributes to the design of planned experiments or observations. Natural history also contributes to interpretation of results (path 7), which feeds back into model development (path 8). Local understanding contributes to general understanding by defining general problems (path 9), contributing data points for correlative generalizations (path 10), specifying the domain of applicability of general theory (path 11), and suggesting new or modified theory (path 12). Place-based research programs can also contribute to general methods for approaching general problems (path 13).

that account for properties of ecological systems more globally. In these introductory remarks we suggest that there is a dynamic interplay between the specific and general in place-based research that follows multiple, bi-directional pathways. We first consider ways in which general theories contribute to the understanding of specific systems. We then discuss how generality can emerge from particular case studies. We end by discussing the premise, implicit in the three chapters of this part, that progress in ecology depends on the dynamic interplay between local and general understanding that is fostered by sustained place-based research. Pathways of interaction between general and local understanding are summarized in figure p4.1.

How General Understanding Informs Local Understanding

To what extent does general understanding inform local understanding? Some ecologists (e.g., Shrader-Frechette and McCoy 1993, Lawton 1999) would say not much. They would argue that neither predictive nor explanatory generalizations in ecology provide much useful understanding of particular systems. Most predictive generalizations (e.g., latitudinal trends in species diversity, allometric scaling of maximum lifespan) have variances around the empirical pattern that are too large to predict with any precision the properties of particular systems. Furthermore, theories that comprise the second, explanatory, sort of generalization (e.g., that the number of co-existing competitors is explained by diet breadth, diet overlap, and the diversity of food resources) are so lacking in biological detail that they are of little use in accounting for the observed properties of particular systems. Many ecologists would agree that ecological generalizations have these problems, but would disagree with the conclusion that they therefore cannot inform local understanding.

Explanatory theories are conceptual models. They postulate causal relationships among the variables and forms those causal relationships can take on. Because one cannot build a theory that is simultaneously general, precise, and realistic (Levins 1966), general theories necessarily lack realism or precision. They couch the forms of the causal relationships in very general terms, such as whether they are direct or indirect, positive or negative, additive or nonadditive, unimodal or multimodal, accelerating or decelerating. These general forms can have many specific manifestations, which means that while general theories cannot be expected to fit when taken "off the shelf," they can be tailored to fit particular systems simply by substituting a specific function for the general one in the general model (Turchin 2001, Berryman 2003). As they are shaped in this way, the theories become less general—they become specific cases—and they become more realistic and precise with respect to the system they are tailored to fit.

General theories provide useful starting points for solving a local problem, as the game of twenty questions illustrates. In this game one player thinks of an object—a cow, for example—and the other tries to guess what it is by asking a series of questions (up to twenty) that can be answered yes or no. Given the diversity of possibilities, one is certain to lose the game if one asks questions without reference to a conceptual hierarchical grouping of things into non-overlapping sets. The classical first question of a winning strategy is to ask whether the thing is an organism. If the answer is yes, a logical next question separates broad classes of living things. If the answer is no, then the next question separates broad classes of inanimate things, and so on. It's surprising how close to the correct answer one can get with twenty such hierarchical yes/no questions!

General theories represent broad classes of hypotheses about how a local system might operate. By analogy with the winning strategy in the game of twenty questions, it is far more efficient, in the absence of prior information, to start with general hypotheses than with their many possible specific manifestations. For this reason, the initial hypotheses illustrated in figure p3.1a in part 3 are broad ones (e.g., hypothesis 1ABC) that include several specific subhypotheses (1A, 1B, 1C). On average, fewer steps are required to identify the correct specific hypothesis than if each specific subhypothesis is tested individually.

Krebs (chapter 13, figure 13.1) illustrates this hierarchical process with the example of what limits primary production in lakes. Previous work suggested that most lakes are phosphorus-limited, and so a reasonable first experiment to do on any particular lake is to add phosphorus. If that does not increase productivity, then one should try the next most likely factor, and so on. Virtually all of the studies in this book started off by testing broad hypotheses (e.g., top-down vs. bottom-up control, extrinsic versus intrinsic control of tolerance to herbivory, competition versus herbivore limitation of plant distributions) before testing more system-specific versions of those hypotheses.

Another way in which general understanding informs local understanding is by providing a conceptual lens through which local systems are viewed. Often what we perceive as an interesting question to pursue with the local system is defined by the problems that are considered to be of general importance or interest at the time. Louda (chapter 14), for example, first asked how plant secondary compounds affect levels of herbivory in bittercress—a hot topic at the time—and only later expanded the investigation to ask why bittercress is restricted to willow habitats. Paine (chapter 11) began by asking about how competition affects species diversity in the intertidal, competition theory being big at the time. Armitage (chapter 7) provides an extreme example: he chose the question before finding a tractable system for answering it!

Given the efficiency of going from general to specific, it is no surprise that general problems and general concepts often provide the starting point for place-based research, whether the goal of the research is basic understanding (e.g., the chapters in parts 2, 3, and 4) or the solution of an applied problem (e.g., Curtin, chapter 12; National Research Council 1986; Young et al. 2005). These linkages are illustrated by paths 1, 4, and 6 in figure p4.1.

How Local Understanding Contributes to General Understanding

Turning the question around, we can ask how local understanding can contribute to general understanding. One obvious way is through the simple inductive process that underlies the inaugural statement in MacArthur's (1972) seminal book *Geographical Ecology*: "To do science is to search for repeated patterns . . . ." Each place that is studied provides a data point that can be used in cross-study analyses of ecological patterns or outcomes of tests of general theory (figure p4.1, path 10). Here generality consists of statements about averages, trends, or frequencies that can be used to predict the properties of a previously unstudied system. For example, the statement "primary production of most lakes is phosphorus limited" (Krebs, chapter 13) comes from the results of nutrient-addition experiments conducted in many particular lakes. If this were the only route to general understanding, however, funding agencies would hardly consider it worthwhile to invest in career-long studies at single locations of finches, marmots, stream insects, intertidal communities, and the like. They would instead invest in less detailed studies at a series of locations, since these could still generate the data points needed to define a general pattern or provide cases for meta-analytical tests of a general theory's validity. Place-based studies must therefore contribute in more substantive ways to general understanding.

As Krebs (chapter 13) and Kingsland (chapter 2) suggest, one route to generality is to pursue local problems that become recognized as new general problems or that cause old problems to be seen in a new light (figure p4.1, path 9). One has only to think of the influence of seminal empirical studies of single systems to appreciate this dynamic. Cowles's (1899) study of vegetation patterns on the shores of Lake Michigan, for example, allowed plant ecologists to see the dynamic nature of plant communities (Kingsland, chapter 2). Paine's (chapter 11) starfish-exclusion experiment demonstrated that individual "keystone" species can have profound effects on communities. Turesson's (1922) transplant experiments uncovered the phenomenon of local adaptation in plants, a concept further developed by the studies of Clausen, Keck, and Hiesey (1940).[1]

Case studies not only have the potential to define new general problems, but can also serve as exemplars for successful ways of approaching general problems (figure p4.1, path 13). Pulliam and Waser (chapter 4), for example, point out how differences between sites in the distribution of a species along an environmental gradient can be used to identify ecological interactions that prevent species from occupying parts of their fundamental niche or that have modified the fundamental niche. Similarly, Peckarsky et al. (chapter 9) invented experimental ways of assessing the relative strengths of consumptive versus nonconsumptive impacts of predators on prey; and work on Tatoosh Island inspired ways of thinking about, and path-analytical approaches to estimating, direct versus indirect species interactions (Paine et al., chapter 11).

Perhaps the most important route from local to general understanding is via theory: place-based research illuminates the domain of theory and how it is manifested in particular situations (figure p4.1, paths 11 and 12). Each of the case studies in this volume identifies features of ecological context or natural history thought to be important for properties of the system: openness and behavioral flexibility (Peckarsky et al., chapter 9), space limitation and disturbance (Paine et al., chapter 11), pulsed resource input (Ostfeld and Jones, chapter 10), seasonal precipitation patterns (Rundel, chapter 5), history of human exploitation (Estes, chapter 8), temporal and spatial heterogeneity in seed hardness and seed production (Grant and Grant, chapter 6), spatial clumping of hibernation burrows, and short growing season (Armitage, chapter 7). Attention to these contextual driving factors and to basic natural history can suggest why a particular system does or does not fit a general pattern, or why a general theory does or does not apply to a particular system. Furthermore, elaboration of a local model suggests how general theory must be modified to fit systems with those properties. In this way, place-based research provides guidance for the study of new systems with similar or different properties. For example, Peckarsky et al.'s study (chapter 9) might guide further research as follows: "Nonconsumptive predator effects proved important in an open stream-system where prey react to predators; my system is open and my organisms respond behaviorally to predation risk; therefore I will plan to test whether nonlethal as well as consumptive effects should be included in my working model."

In summary, place-based case studies contribute to progress in general ecological understanding in ways that are far more conceptually important than adding points to a statistical analysis: they also provide templates for further research by suggesting new general problems (figure p4.1, path 10) and profitable research approaches (path 13), by illuminating the domain of applicability of theories (path 11), and by modifying theory to fit particular systems or creating new theory (path 12).

Progress in Ecology Depends on Exchange between Local and General Knowledge

In place-based research programs, the paths between local and general do not all lead in one direction or the other; there is instead movement in both directions. The three chapters in this part discuss and illustrate the importance of this exchange for progress in ecology. Several common themes emerge.

First, progress is enhanced when research addresses a problem of general interest. The case studies in this volume illustrate this point. While local problems may have provided the initial motivation for some of the studies (e.g., Ostfeld and Jones, chapter 10; Curtin, chapter 12), in all cases the research was designed, interpreted, and communicated in light of prevailing ideas about general problems, such as consequences of trophic interactions for prey numbers and community composition, determinants of distribution and abundance, causes of periodic insect outbreaks, causes of adaptive radiation, or physiological responses involved in plant-herbivore interactions.

Second, progress is enhanced when diverse strands of local knowledge, ranging from natural history anecdote to outcomes of previous experiments, are deployed in research design. Any general ecological theory that cuts across organisms and environments necessarily lacks site-specific details, such as who is the predator and who the prey, or which resources are limiting. To test such theories, one needs to use local knowledge to express general problems in a locally relevant way (figure p4.1, path 2), to translate general theory into a locally relevant initial model (path 3), and to design rigorous tests of the local model (e.g., Bartholomew 1986; Weiner 1995; Polis et al. 1998; see also figure p4.1, path 5).

Progress also is enhanced when diverse strands of local knowledge are used (figure p4.1, path 7) to interpret results of planned experiments or observations (path 8). This increases the chance that one will recognize results that don't fit the original conceptual framework—results that are "surprising" because they aren't consistent with any known hypothesis—and to see novel hypotheses that may fit observations better than do the others, or novel questions and lines of investigation.

These feedbacks between local and general knowledge are central to the model-building process by which the understanding of place develops (see introductory pemarks to part 3, figure p3.1b). They also contribute to general understanding by providing insight into the domain of application of a general theory, suggesting ways in which a general theory can manifest itself in particular circumstances, or proposing completely new theory.

Krebs (chapter 13) traces the majority of 16 major conceptual advances over the last 50 years to this interplay between local knowledge and what he calls general principles. He discusses the nitty-gritty decisions involved in designing a field study that render impossible the statistician's dream of random sampling, and how this makes purely inductive inference problematic. He argues that a purely hypothesis-testing conceptual framework is less conducive to progress than a process-oriented dialectic between the modeler's push toward simplification and generality on the one hand, and the empiricist's push toward the complexities of the real world on the other. Krebs also points to instances of failed management as underutilized opportunities to foster this dialectic (also see Curtin, chapter 12).

Louda and Higley (chapter 14) illustrate similar themes with five examples of sustained study of interactions between plants and insect herbivores. They emphasize that progress is enhanced when investigators keep general problems always in mind, use natural history information in the design and interpretation of experiments, and follow up on surprising or anomalous results with new questions and changes in the direction of research.

Hubbell (chapter 15) recounts the history of the forest dynamics project on Barro Colorado Island, a long-term effort to understand processes responsible for the astonishing diversity of tree species in tropical forests. This story illustrates particularly well how long-term, multifaceted, place-based studies generate new ways of thinking about classical general problems, and lead to new theory. It also illustrates the value of testing theoretical expectations from multiple hypotheses against diverse types of data. Observations of functional convergence among tree species, the small spatial scale of density dependence, strong dispersal and recruitment limitation, and patterns of temporal changes in species abundances caused Hubbell to question the relevance of classical equilibrium models for species coexistence on large spatial scales, and also to develop new "neutral" theories for community assembly. There is no way these insights could have been gained without the sort of dialectic between theory and place-based empiricism that Krebs advocates.

References

Bartholomew, G. A. 1986. The role of natural history in contemporary biology. *BioScience* 36:324–29.

Clausen, J. J., D. C. Keck, and W. M. Hiesey. 1940. *Experimental Studies on the Nature of Species. I. Effect of Varied Environments on Western North American Plants.* Publication no. 520. Carnegie Institution of Washington, Washington, DC.

Langlet, O. 1971. Two hundred years of genecology. *Taxon* 20:653–721.

Lawton, J. H. 1999. Are there general laws in ecology? *Oikos* 84:177–92.

Levins R. 1966. The strategy of model building in population biology. *American Scientist* 54:421–31.

MacArthur, R. H. 1972. *Geographical Ecology.* Harper and Row, New York.

National Research Council, Committee on the Applications of Ecological Theory to Environmental Problems (G. H. Orians, Chair). 1986. *Ecological Knowledge and Environmental Problem Solving: Concepts and Case Studies.* National Academies Press, Washington, DC.

Polis G. A., D. H. Wise, S. D. Hurd, F. Sanchez-Piñero, J. D. Wagner, C. T. Jackson, and J. D. Barnes. 1998. The interplay between natural history and field experimentation. Pages 254–80 *in* W. J. Resetarits, Jr. and J. Bernardo, eds., *Experimental Ecology: Issues and Perspectives.* Oxford University Press, New York.

Platt, J. R. 1964. Strong Inference. *Science* 146:347–53.

Shrader-Frechette K.S., and E. D. McCoy. 1993. *Method in Ecology.* Cambridge University Press, Cambridge.

Turchin, P. 2001. Does ecology have general laws? *Oikos* 94:17–26.

Turesson, G. 1922. The genotypical response of the plant species to the the the habitat. *Hereditas* 3:211–350.

Young, T. P., D. A. Petersen, and J. J. Clary. 2005. The ecology of restoration: Historical links, emerging issues and unexplored realms. *Ecology Letters* 8:662–73.

Note

1. The history of transplant studies is actually much deeper; see Langlet 1971.

13 *Case Studies and Ecological Understanding*

Charles J. Krebs

Figure 13.0. Rudy Boonstra on snowshoes, trapping snowshoe hares in boreal forest, Kluane, Alaska. Photograph by Alice Kenney, 1988.

Abstract

Does ecology develop as a science mainly inductively, through case studies that lead to theory? Or does it develop deductively by abstract mathematical theory that is then analyzed empirically? Since philosophers of science have long discredited empirical induction, how does ecology really develop? Are case studies just a pleasant outdoor way of "stamp collecting" to validate mathematical theory? I identify 15 major conceptual advances made in ecology during the last 50 years, and attempt to judge what contributions mathematical theory and empirical studies have made to these major advances.

Four of the advances could be classed as having arisen primarily from theoretical work, and I have judged 10 to be primarily empirical in origin. One advance arose from a nearly equal combination of both approaches. Mathematical theory in ecology has described a complex world during the last 40 years, but we have too few empirical evaluations of whether the theoretical world now in place is built on sand or rock. Empirical case studies firmly rooted in place have led to valuable ecological theory whose test is that it is useful for natural resource management. Case studies will continue to enrich ecological theory and practice for the near future.

Introduction

Errors using inadequate data are much less than those using no data at all.
—*Charles Babbage (1792–1871)*

All ecologists, politicians, and business people are in favor of progress, and view time's arrow as pointing in the direction of progress. Anyone who dares to say that we are not making progress in an area, as Peters (1991) did for ecology, is condemned for writing "an essay written by a dreadfully earnest, but ill-informed, poorly read undergraduate" (Lawton 1991). But in every science progress is uneven, reversals occur and are quickly buried and forgotten. The question we need to raise concerns the rate of progress, and whether there are any shortcuts we can follow to speed it up.

The recipe for progress in science is fairly simple: find a problem, designate multiple alternative hypotheses, and test them by searching for evidence that contradicts the predictions of each hypothesis. But as every practicing scientist knows, applying this recipe is complicated by a whole set of decisions and assumptions that are typically unstated in the resulting scientific papers. Among the first of these decisions is the question of place: Where shall I carry out this research? But the location or place of the research carries with it a whole array of assumptions and additional decisions that are rarely considered explicitly. In the first part of this chapter I explore some of these assumptions and decisions with respect to ecological science, and discuss in particular how we might move from site-specific studies to general knowledge. In the second part of this chapter I discuss ecological advances and the role of place-based research in producing progress in ecological understanding.

I will not here discuss evolutionary ecology and its handmaids, physiological ecology and behavioral ecology. These areas have made great advances in recent years because they deal with relatively simple problems with solutions that are known because of evolutionary theory. These areas work in what Kuhn (1970) has called normal science, filling in important gaps in understanding while guided by well-established theory. The rest of ecology, mechanistic ecology, does not have the luxury of an established theory like evolution by natural selection, and so it is much harder to do. This does not mean that mechanistic ecology ignores microevolutionary changes in populations, as there are many examples of how both population and community interactions have changed because of microevolution (Carroll et al. 2007). But if you wish to know why a population stops growing, or why the composition of a community is changing rapidly, the theory of evolution will not tell you a priori which mechanistic processes you should investigate. There is no "optimal foraging theory" for population dynamics or plant succession. It is for this reason that mechanistic ecology is much more difficult than physiological or behavioral ecology.

Assumptions Underpinning Ecological Studies

All good ecology is founded on a detailed knowledge of the natural history of the organisms being studied. The vagaries of species natural history are a challenge to the field ecologist trying to understand natural systems as much as they are a menace to modelers who assume that the world is simple and, if not linear, at least organized in a few simple patterns. I begin with the often unstated background supposition that we have good natural history information on the systems under study. The great progress that ecology has made in the last century rests firmly on this foundation of natural history.

The following is a list of assumptions and decisions that are implicit or explicit in every ecological study. In most published papers you will find little discussion of these assumptions, and in bringing them forward here I am trying to make more explicit the logical skeleton of ecological progress.

A problem has been identified

This is a key step that is rarely discussed. A problem is typically a question, or an issue that needs attention. Problems may be local and specific or general. Local problems may be specific as to place as well as time, and if they are so constrained, they normally are of interest to applied ecologists for practical management matters, but are of little wider interest. General problems are a key to broader scientific progress, and so ecologists should strive to address them to maximize progress. The conceptual basis underpinning a study is an important identifier of a general problem. Applied ecologists can often address what appear to be local problems in ways that contribute to the definition and solution of general problems. A solution to a general problem is what we call a general principle.

General ecological problems can be recognized only if there is sufficient background information from natural history studies to know that an issue is broadly applicable. There is also no easy way to know whether a general problem will be of wide or narrow interest. For example, the general problem of whether biotic communities are controlled from the top down by predation or from the bottom up by nutrients is a central issue of the present time, and of broad interest (see Estes, chapter 8; Peckarsky et al., chapter 9). The answer to this question is critical for legislative controls on polluting nutrients (Schindler 1988) as well as for basic fisheries management (Walters and Martell 2004). The top-down/bottom-up issue will always be a general one for ecologists to analyze because some systems will show top-down controls and others bottom-up controls, so the answer will be case-specific. The level of generality of the answer will not be "all systems are top-down," but only some lower level of generality, such as "Insectivorous bird communities are controlled bottom-up." It is only after the fact that problems are recognized as general, and science is littered with approaches that once appeared to be of great general interest but did not develop. The converse is also true: problems originally thought to be local have at times blossomed into more general issues of wide relevance.

The typical pattern in the evolution of general problems is illustrated in figure 13.1. A problem is recognized, such as: What are the factors that control primary production in lakes? From prior knowledge (e.g., agricultural research) or data from a set of prior studies, a series of hypotheses is set up. A hypothesis that has a reasonable amount of support is what we refer to as a general principle. One can view these hypotheses as "straw men" in the sense that many variables affect any ecological process, and all explanations should be multifactorial. But it is not very useful at this stage to say that many factors are involved and that the issue is complex. Ecologists should introduce complexity only when necessary. Often it is useful to view a hypothesis as answering a practical question: What variable might I change as a manager to make the largest impact on the selected process? Ecologists should sort out the large effects before they worry about the small effects. Large effects may arise from interactions between factors that by themselves are thought to be of small importance. Good natural history is a vital ingredient here because it helps us to make educated guesses about what factors might be capable of producing large effects.

It is nearly universal that once a hypothesis is stated and some data are found that are consistent with the suggested explanation, someone will find a contrary example. For example, although most freshwater lakes are phosphorous-limited, some are micronutrient-limited (e.g., by molybdenum; Goldman 1967; see also Elser et al. 2007). The question then resolves into one of how often the original suggestion is correct and how often it is incorrect, and one or another set of hypotheses should be supported. Although statisticians may be happy with a hypothesis that 87% of temperate lakes are

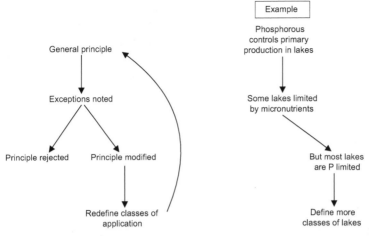

Figure 13.1. A schematic illustration of how generality is treated in ecological research. A simplified example from the controversy over the nutrients responsible for eutrophication in temperate freshwater lakes (Schindler 1977) is used to illustrate the progression from very general principles to more specific principles that are invariant. Statistical principles such as "primary productivity in 72% of freshwater lakes are controlled by phosphorus" are not very useful for management; we try to reach universal principles, although we may never achieve this ideal.

phosphorous-limited, ecologists would prefer to define two (or more) categories of lakes in relation to the factors limiting primary production. We do this in order to produce some form of predictability for the occasion when we are faced with a new lake: are there criteria by which we can judge which factors might be limiting this particular lake? Can we establish criteria that allow near-absolute predictability? Some might argue for a statistical cutoff, such as 80% correct predictability, at which point we should be content with the generalization. But the general approach of rigorous science is to concentrate on those cases in which the prediction fails, so that by explaining contrary instances we can strengthen the generalization. Clearly, though, we cannot investigate all the lakes in the world to achieve complete predictability, so this takes us back to the problem of place.

The statistical population has been delimited

Ecologists often drive statisticians to distraction. We assume that place does not matter, so that, for example, if we wish to study the predator/prey dynamics of aphids and ladybird beetles on cabbage, we can do it anywhere that cabbage is grown. This is a gigantic assumption, but a necessary one in the early stages of an investigation in which we must assume simplicity until there is evidence against it. This assumption about the irrelevance of the place or location where we do our studies is often coupled with the assumption of time irrelevance, so we make the joint assumption that our findings are independent of time and space. Statisticians try to capture these assumptions in the idea of a "statistical population."

Statisticians request that one should define the particular unit of study for which one is trying to make some conclusion the "statistical population." I have not found a single ecological paper that defines the statistical units to which the study is supposed to apply, except in the very general sense that a given study is being done in the rocky intertidal zone, or in the boreal forest, or on a particular island. We do this deliberately because we do not know the extent of application of any conclusions we make in ecology. When in doubt, apply your results to the entire universe of the rocky intertidal zone or the boreal forest. This type of global generalization can be defended as a conjecture that is designed for further testing and subsequent revision. Critics may argue that such broad conclusions are too simplistic, but such a criticism ignores Ockham's razor and the need to embrace simplicity and introduce complexity only when needed. But the issue of defining a statistical population brings us back to asking how a particular site is chosen for a particular study.

Where most of the highly influential ecological field studies have been carried out is almost an accident of history. The presence of field stations, people in particular universities, the location of protected areas, and arrangements of travel all combine to determine where a field study is carried out. A pure statistician would be horrified at such a lack of random sampling, and we are in the anomalous intellectual position of basing our most important ecological contributions on non-random sampling. But of course this is not a problem if you can make the assumption that no matter where you have carried out a particular investigation, you will get the same result. This rescue of generality can be done only if one views the ecological world as invariant in its properties and dynamics over space and time. This is a critical assumption. System dynamics may be invariant over space, but not over time.

There are now good studies that show how the assumption of time invariance is incorrect. Grant and Grant (chapter 6) illustrate this difficulty with two episodes of natural selection on Darwin's finches. Range managers have faced the same problem by not recognizing multiple stable states, so that removing cattle grazing does not necessarily reset the system to its initial conditions (van de Koppel et al. 1997). We need to be aware of the assumption of time invariance, and it may be a mistake to assume that, if a particular study was done from 1970 to 1980, the same results would have been observed from 1995 to 2005.

The assumption of spatial invariance, as Pulliam and Waser discuss (chapter 4), has never been popular in ecology because the abundance of resources, predators, and diseases are well known to vary spatially. Much of modern ecology has focused on trying to explain spatial variation in processes. Plant ecologists discarded the Clementsian monoclimax view of ecological communities and replaced it with the continuum concept of a community (Austin and Smith 1989, Crawley 1997). Animal ecologists recognized keystone species, which showed that a single species could have major community consequences (Paine et al., chapter 11). The exact dynamics of a community may be greatly affected by the species present, their interaction strengths, and their relative abundances. We do not yet know how much variation can occur in community composition before new rules or principles come into play.

The result is that we almost never specify a statistical population in any ecological research program, and we issue a vague statement of the generality of our findings without defining the units to which it should apply. This is not a problem in experimental design if we can repeat our findings in another ecosystem to test their generality. The key to generality is to predict correctly what we will find when we study another ecological system in another place. For the present, ecologists should retain a dose of humility by continually testing the limits of generality of their ideas rather than believing that they have found scientific laws.

Random sampling is applied

In the chosen area of study, we now observe or apply some treatments to obtain the data that will test an array of alternative hypotheses. In the case of observational experiments the sample units are defined by nature, and our job in random sampling is to locate them, number them, and select those for treatment at random. For manipulative experiments we define the sample units and apply a similar random selection of them for each treatment. Most ecological field experiments have a small number of replicates, and Hurlbert (1984) has discussed what can happen if treatments are defined randomly. All our control or experimental plots may end up, for example, on north-facing slopes. Hurlbert recommended maintaining an interspersion of treatments so that both treatments and controls are spread spatially around the study zone.

Consequently a good biologist almost never follows the instructions from the pure statistician for three reasons. First, they may not be practical. The major reason such random assignments may not be practical is that transportation to the sites may limit choices. Not everyone can access field sites by helicopter, and roads typically determine which study units can be used (table 13.1). Second, places for study may need to be in a protected nature reserve or an area in which the private owner welcomes ecologists to use his or her land. Since nature reserves in particular are often put in landscapes that cannot be used economically for agriculture or farming, there is an immediate bias in the location of our experimental units. Third, field stations or other sites where research has been carried out in the past have a legacy of information that draws ecologists to them for very good reasons (Aigner and Koehler, chapter 16; Billick, chapter 17), although this compounds the nonrandomness of choice of field sites.

The consequence of these problems is the practical advice to randomize when possible on a local scale, and to hope that generality can emerge from nonrandom sampling on a regional or global scale.

Transient dynamics are not dominant

The time scale of ecological system responses is assumed to lie within the time frame of our studies. Thus, if we manipulate vole or lemming populations that have several generations per year, we assume that our manipulations will be effective within a year. But what if fast variables like vole numbers interact with slow variables like soil nutrient dynamics or climate change?

The time lags in system response that are inherent in transient dynamics can be found only by longer-term studies (e.g., Grant and Grant, chapter 6), and at present we are guided in these matters only by our intuition, which is based on natural history knowledge and process-based (i.e., mechanistic) models that can explore our assumptions about system dynamics. Process-based models are a vital component of our search for generality because they can become general principles waiting for further testing (e.g., see King and Schaffer 2001; Pulliam and Waser, chapter 4). The important limitation of process-based models is to determine how much structure is essential to understanding the system of study. Too much detail leaves empirical scientists with little ability to discern which factors are more important, and too little detail leaves out biological factors that are critical.

The predictions being tested follow from the hypotheses

Ecological hypotheses typically are less clearly structured logically than might be desirable. In particular the background assumptions that are necessary to support deductions from a particular hypothesis are rarely stated, with the net result that there is a somewhat tenuous connection between hypotheses and predictions. The only remedy for this problem is to demand

TABLE 13.1. Experimental manipulations in the Kluane Boreal Forest Ecosystem Project and the decisions that led to their site placement. Each experimental unit was 1 km². This example illustrates some of the reasons why randomization cannot be achieved in field ecology. Similar forested habitat was the first constraint on area selection, and access in summer and winter was the secondary determinant of location, followed by the need to spread treatments over the 350 km² study area. Details are given in Krebs et al. (2001).

Experimental unit	Treatment	Reasons for location
Fertilizer 1	55 U.S. tons ($\approx$ 50 metric tons) of commercial fertilizer added aerially each year	Along Alaska Highway at north end of study area, 3 km from airstrip used for aircraft loading fertilizer
Fertilizer 2	55 U.S. tons ($\approx$ 50 metric tons) of commercial fertilizer added aerially each year	Along Alaska Highway near north end of study area, 6 km from airstrip, separated from other treatments by at least 1 km
Food addition 1	Commercial rabbit food fed year round	Access by ATV (All Terrain Vehicle) in summer, and minimum 1 km spacing from electric fence treatment
Food addition 2	Commercial rabbit food fed year round	At extreme southern end of study area with ATV access, and 3 km from Control 3
Electric fence	Exclude mammal predators	Along Alaska Highway, access by heavy equipment, relatively flat area
Electric fence and food addition	Exclude mammal predators and add rabbit food	Near Alaska Highway, access by heavy equipment, one side of fence already cleared for old pipeline, relatively flat area
Control 1	None	Along Alaska Highway, spaced 1 km from manipulated areas and 10 km from Control 2
Control 2	None	Along Alaska Highway, spaced 5 km from Control 3 and 7 km from nearest treatment site
Control 3	None	Near southern end of study area, accessed by old gravel road

more rigor in specifying the unstated assumptions that accompany every study.

The relevant variables have been identified

Another difficulty at this stage is that the set of alternative hypotheses proposed as explanations of the identified problem may not include the correct explanation. For example, the three alternative hypotheses for freshwater lakes—that primary production is limited by nitrogen, phosphorus, or carbon—may all be wrong if a lake's production is limited by micronutrients such as molybdenum. There is no simple way out of this problem, except to identify as wide an array of possible explanations as current knowledge will permit, and to recognize always that in the future a new hypothesis may arise that we had not considered.

By diversifying one's observations at a variety of places, we can minimize the probability of failing to see and include a relevant variable. Diversifying means carrying out similar studies in several different places. This is one of the strongest recommendations that one can make about ecological science: we should systematically diversify our observations in different places to help identify relevant variables. If we have missed an important variable, it will be picked up when management actions flow from our ecological findings, because those actions will not achieve their predicted results. Practical management can be used as the touchstone of ecological ideas and as a valuable test for missing variables. This will occur only when management actions have a firm foundation in ecological ideas and—if management actions fail—when time and money are made available to find the source of the failure. Ignoring failures of predictions is a sure way to reduce progress in scientific understanding (Popper 1963).

Ecological Advances

The past century has seen an enormous growth in ecological knowledge, and I will attempt to identify some of the key ideas that have emerged during this interval. The central problems of ecological science are to decipher the reasons for the distribution and abundance of organisms (Krebs 2009), and within this broad umbrella I recognize four key areas of advance. All of these areas have advanced because of multiple studies with many different empirical approaches, whether they be experimental, observational, field-based, or laboratory-based.

The limitation of geographical distributions

This is one of the oldest areas of research in ecology, interfacing with biogeography, which began developing insights in the eighteenth century. Advances in our understanding of the factors limiting geographic ranges have resulted from the recognition that answers are scale-dependent. A forest herb studied at the scale of 1 m^2 may be limited in distribution by competition from other plants, at a scale of 1 ha by the feeding habits of its herbivores, at a scale of 1 km^2 by the vagaries of seed dispersal, and at the scale of continents by its evolutionary origins in relation to continental drift. All these scales bring the concept of *place* into strong focus because we have to specify exactly which place we are studying. This was pointed out by Forman (1964), but the whole issue of scale-dependence in ecological analysis has only slowly developed. Recent advances in understanding range limitations have built on the basic hypothesis that, aside from dispersal limitation and human translocations, climate is the master factor limiting geographical ranges, and climate change is a process that will test this hypothesis. Data showing range changes that have accompanied global warming help to advance the need for action on climate change (Parmesan and Yohe 2003).

Botanists have been at the forefront in analyzing the factors limiting distributions within the geographic range. The classic studies of sand dune succession on the shores of Lake Michigan by Cowles (1899), and the more recent work by Daubenmire (1954, 1956) stand out as major advances. The seminal work of Connell (1961) on barnacle distribution in the Scottish rocky intertidal zone is another classic case of mechanistic hypotheses applied to the problem of geographic distributions.

The recognition of the problem of invasive species has been another achievement of this branch of ecology. The general topic of invasions is not new (Elton 1958), but the details of how particular species have wreaked havoc in ecosystems after deliberate or accidental introductions have led ecologists toward insights on introductions now synthesized in the "enemy release hypothesis" (Keane and Crawley 2002). The spread of the zebra mussel (*Dreissena polymorpha*) in North America (Johnson et al. 2006) and the spread of the cane toad in Australia (Sutherst *et al.* 1995) are good case studies of invasive species expanding their geographic ranges after introduction.

Island biogeographic theory was a bold attempt by MacArthur and Wilson (1967) to tie the loose empirical data on island distribution patterns to a conceptual theory centered on dispersal. As such it was a major conceptual advance, moving from empirical patterns to an elegant theory that had significant ramifications for further research on island populations, metapopulations, and conservation biology (Brown and Lomolino 2000). Its weakness was in fact part of the reason for its success: it dealt only with the numbers of species and did not specify the details of which exact species would occur where—the details that are important for practical conservation decisions.

Macroecology attempts to describe large-scale patterns in the distribution and abundance of species, often adopting as its measure of place a whole country or continent. One of the early insights of the field of macroecology was that the sizes of distributional ranges were correlated positively with abundance, albeit with considerable scatter (Maurer and Taper 2002). This insight has tied together studies of distribution and abundance at large spatial scales. The macroecological search for patterns has been one way of collating many separate studies on different ecosystems in an attempt to generalize over individual place studies. The limitation has been that the patterns recognized are statistical generalizations rather than mechanistic explanations; hence, they are of limited predictive power for management purposes.

Population regulation and limitation

The problem of population regulation was already well discussed by the 1950s but three major advances have marked its maturation. Mathematical models of populations were greatly elaborated in the 1960s and 1970s with the recognition of how complex population dynamics like chaos could result from simple assumptions. Ricker (1954) first pointed this out; the mathematics was elegantly described by Maynard Smith (1968) and then fully elaborated by Robert May (1974). A second major advance has been the array of detailed studies on single-species populations that has made the issue of regulation more complex but more biologically interesting (Sibly et al. 2003). A good example is the importance of seasonal migration in the determination of population size in wildebeest (Fryxell et al. 1988).

The practical application of population dynamics theory to conservation made a third major advance when Caughley (1994) described the declining population paradigm and applied ideas of population limitation to the practical issue of threatened and endangered species. This coincided with an elaboration of the Leslie matrix approach to population projections, and the development of elasticity and sensitivity analysis for population changes (Caswell 2001). These mathematical advances have proven most useful in the analysis of problems in both pure and applied ecology. Life-history theory has adopted matrix methods as an important tool, and conservation options can be explored with these same methods to determine the best management actions to use for species recovery.

Community trophic dynamics

The approach to understanding community dynamics developed originally from analyses of energy flow through communities, beginning with Lindeman (1942) and carried forward by Odum (1968). These ideas were picked up in an influential paper by Hairston, Smith, and Slobodkin (1967), who pointed out that energy and material flows could be regulated—what we now call "top-down" by predators or "bottom-up" by nutrients and food items (Menge and Sutherland 1987, Hunter and Price 1992). These ideas were an extension of concepts used in population dynamics to determine which regulating factors dominated in stopping population increase, and they led to the important idea of indirect effects on community interactions (Holt 1977; Menge 1995; Paine et al., chapter 11).

A second important development in thinking about community dynamics has been the replacement of equilibrium community dynamics with nonequilibrium dynamics (DeAngelis 1987). This arose in part because disturbances were found to prevent communities from ever achieving a state of equilibrium, and by the 1980s ecological concepts of succession (Connell and Slatyer 1977) and nutrient cycling (Bormann and Likens 1967, Vitousek 2004) had expanded their theoretical formulations to include nonequilibrium dynamics. The classic case here is that of coral reefs, formerly considered the paradigm of equilibrial communities until the detailed data of Connell et al. (1997) showed clearly that they were nonequilibrial systems.

Biodiversity has become a focal point for much research effort, as an extension of the older problem of what controls species diversity in communities (Connell and Orias 1964). In a 1955 theoretical exercise Robert MacArthur raised the issue of how species diversity affected community stability, and this triggered a discursive series of studies on the role of biodiversity in plant and animal communities (Loreau 2000, Worm and Duffy 2003). As the capstone concept of conservation biology, biodiversity has brought the taxonomy of earth's organisms to our attention and stimulated much research and analysis of how biodiversity contributes to human welfare through ecosystem services (Costanza et al. 1997).

Ecosystem dynamics

The recognition of ecosystem services has pushed ecology to the forefront in an attempt to prevent the degradation of the earth's biological systems. Paul Ehrlich coined the idea in 1983 (Ehrlich and Mooney 1983) in an attempt to analyze why humans should be concerned about extinctions. The focus on ecosystem processes has gained renewed energy from the spreading concern about climate change and its impact on natural and human-dominated ecosystems. At present this is the touchstone for ecologists to push their concerns about environmental degradation. The difficulty has been in getting some ecological measurements of ecosystem services, and much effort is now going into the economic evaluation of these services, unfortunately while we know little about their ecology and how they are affected by human actions (Daily 1997).

Landscape ecology emerged in the 1980s as part of a recognition that conservation could not be limited to national parks if we were to preserve large, charismatic species (Newmark 1985). As part of a recognition that ecological studies were typically done at very small scales while fundamental ecosystem problems occur at large scales (Kareiva and Andersen 1988; see also Curtin, chapter 12), ecologists began to consider the landscape context of their analyses. Fragmentation of ecological communities was the earliest recognition of landscape issues, and this evolved into theoretical and practical analyses of the consequences of populations and communities being distributed on habitat islands (Gill 1978, Hanski 1998). Flows between ecosystems are also critically important in landscape function (Anderson and Polis 1998; Estes, chapter 8), and fragmentation has consequences at all levels of integration in ecology.

Landscape ecology begins with the assumption that ecological processes are affected by the landscape configuration. While this is a reasonable premise, it will be correct only at certain spatial scales, and for some organisms or processes it may not be correct at all. Because of the difficulty of manipulating large areas, landscape ecology has either used microcosms to investigate questions of configuration (e.g., Huffaker 1958) or used spatial modeling to investigate the possible impact of different habitat configurations (Levey et al. 2005). Since most models are likely to be wildly wrong in their predictions, there is an urgent need to test these scenarios in the real world. And since much microcosm work has been completely misleading when applied to natural systems (Carpenter 1996), another caution about the perceived advances in landscape ecological understanding is in order. While microcosms can be useful for investigations of simple population- or community-level questions, they are less useful for investigating problems in open systems containing many species in typical landscapes. But on the positive side, the importance of landscape is suspected by many experienced ecologists, even though the detailed data are not in—and landscape ecology shouts to ecologists to consider place as being central to understanding ecological processes.

The concept of metapopulation dynamics is closely connected with landscape ecology, although in a population context it is usually thought of as being associated with dispersal between patches. Place as a spatial location has a critical importance in metapopulation dynamics, as in all of landscape ecology, because it is the configuration of each place or habitat in relation to others that can drive the ecological processes.

The Role of Case Studies

Ecology operates as a dialectic between modelers and theoreticians, who push generality in which place has no position, and place-based empiricists, who study particular problems in particular places and try to understand the processes involved. The dialectic is slow to operate. Alfred Lotka and Vito Volterra produced models of competitive interactions and predator/

prey relationships in the 1920s before there was virtually any detailed study of competitive or predatory systems, or any data available. Only later were the limitations of these models recognized, as detailed laboratory and field studies were carried out in various places around the world from 1950 onward. In an ideal world, theoretical and empirical approaches would merge as model → data → revised model → more data in continuous synergistic feedback loops. We may be in an ideal world 200 years from now, if we adopt the development of chemistry and physics as a guide, but at the present time the prospect of grand synthetic models recedes at the speed of light, partly because exceptions to general principles are continually being found (as is illustrated in figure 13.1).

I argue in this chapter that progress in ecology is limited by the great divide between theoretical research and empirical research that focuses on understanding the workings of real ecosystems in the field. The divide occurs not because theoretical and empirical domains are nonoverlapping. After all, theory includes abstract conceptual models as well as mathematical formulations, theoreticians do live in the real world and get ideas from it, and empiricists typically have some general mental or verbal model of their system of study. Instead, the divide occurs because theoretical and empirical approaches are rarely integrated in ecological research. Anyone who reads the journals *Theoretical Population Biology* and *Ecological Monographs* will clearly understand this lack of integration.

Although there are other empirical ecological approaches that we can recognize, I emphasize place-based, process-oriented field approaches because I think the contributions of these alternatives to ecological knowledge that has practical utility are scarcely visible at the present time. Laboratory microcosm studies can certainly be place-based and process-oriented, but they are a poor and often misleading guide to what happens in natural ecosystems. Macroecology is a recent high-level approach to recognizing and studying large-scale empirical patterns, similar to the pattern analysis often used in plant ecology. I consider both of these approaches empirical because they are ultimately reducible to sets of data obtained from studies in particular places—but their utility is compromised by the separation of data from a real ecological context.

If we accept the empirical/theoretical dichotomy, we can ask which of these approaches has been more important for progress in ecological understanding. My vote goes with Shrader-Frechette and McCoy (1993) who emphasized the key role that case studies should play in the development of ecological theory. As table 13.2 illustrates, most of the major advances in ecological science over the last 50 years have come from the place-focused, empirical side of ecology, and while we could construct a similar table for major advances in mathematical ecological theory, we would be unable to

TABLE 13.2. Major advances in ecological understanding during the past 50 years, with the source of each advance classified as theoretical or empirical place-based research. The references are meant to illustrate the point, not to be comprehensive. This classification is only my opinion, and can be considered as a hypothesis rather than a fact. Theoretical advances arise deductively from a set of assumptions connected via a mathematical model, while empirical advances often arise inductively and are rarely rigorously formulated.

Advance	Source	Reference
Factors limiting ranges depend on scale	empirical	Forman (1964)
Many range limits are set by climate	empirical	Parmesan and Yohe (2003)
Invasive species disrupt ecosystems	empirical	Elton (1958)
Distribution and abundance are correlated	empirical	Brown (1984)
Simple population mathematics can lead to fluctuations or chaos	theoretical	May (1974)
Population regulation can be analyzed quantitatively	empirical	Sibly et al. (2003)
Conservation problems can be solved with the declining-population paradigm	empirical	Caughley (1994)
Communities can be controlled top-down or bottom-up	empirical	Menge and Sutherland (1987)
Indirect effects between species can affect community structure	theoretical	Holt (1977)
Communities may show nonequilibrium dynamics	theoretical	Chesson and Case (1986)
Island species pools are a balance of immigration and extinction	theoretical	MacArthur and Wilson (1967)
Species diversity affects community stability	theoretical and empirical	Elton (1958), MacArthur (1965)
Ecosystem services are essential for human life on earth	empirical	Daily (1997)
Landscape processes affect population and community dynamics	empirical	Lidicker (1995)
Dispersal determines metapopulation dynamics	empirical	Harrison (1991)

determine whether any of that theory was applicable to the real world. This dichotomy between mathematical theory and empiricism in ecology was recognized by Fretwell (1972), but the years have not reduced the gap, in spite of heroic efforts. Theory advances at lightning speed because it does not have to deal with place, while empirical ecology moves slowly, always constrained by place and time, enriching our understanding of how the natural world operates but at the same time burying us in its complexity. We should all rejoice that there is so much left to do in empirical ecology. My recommendation is for ecologists to push place-based empirical research to the fore in an effort to test all the accumulated theory that sits idle in the absence of good field data.

Acknowledgments

I thank Alice Kenney, Jim Estes, Mary Price, Ian Billick, and Dennis Chitty for their comments on this chapter. Their challenges helped sharpen my arguments and conclusions, for which only I am to blame.

References

Anderson, W. B., and G. A. Polis. 1998. Marine subsidies of island communities in the Gulf of California: Evidence from stable carbon and nitrogen isotopes. *Oikos* 81:75–80.

Austin, M. P., and T. M. Smith. 1989. A new model for the continuum concept. *Vegetatio* 83:35–47.

Bormann, F. H., and G. E. Likens. 1967. Nutrient cycling. *Science* 155:424–29.

Brown, J. H. 1984. On the relationship between abundance and distribution of species. *American Naturalist* 124:255–79.

Brown, J. H., and M. V. Lomolino. 2000. Concluding remarks: Historical perspective and the future of island biogeography theory. *Global Ecology and Biogeography* 9:87–92.

Carpenter, S. R. 1996. Microcosm experiments have limited relevance for community and ecosystem ecology. *Ecology* 77:677–80.

Carroll, S. P., A. P. Hendry, D. N. Reznick, and C. W. Fox. 2007. Evolution on ecological time-scales. *Functional Ecology* 21:387–93.

Caughley, G. 1994. Directions in conservation biology. *Journal of Animal Ecology* 63:215–44.

Chesson, P. L., and T. J. Case. 1986. Overview: Nonequilibrium community theories: Chance, variability, history, and coexistence. Pages 229–39 *in* J. Diamond and T. J. Case, eds. *Community Ecology*. Harper and Row, New York.

Connell, J. H. 1961. The influence of interspecific competition and other factors on the distribution of the barnacle *Chthamalus stellatus*. *Ecology* 42:710–32.

Connell, J. H., T. P. Hughes, and C. C. Wallace. 1997. A 30-year study of coral abundance, recruitment, and disturbance at several scales in space and time. *Ecological Monographs* 67:461–88.

Connell, J. H., and E. Orias. 1964. The ecological regulation of species diversity. *American Naturalist* 98:399–414.

Connell, J. H., and R. O. Slatyer. 1977. Mechanisms of succession in natural communities and their role in community stability and organization. *American Naturalist* 111:1119–44.

Costanza, R., R. Darge, R. de Groot, S. Farber, M. Grasso, B. Hannon, K. Limburg, S. Naeem, R. V. O'Neill, J. Paruelo, R. G. Raskin, P. Sutton, and M. van den Belt. 1997. The value of the world's ecosystem services and natural capital. *Nature* 387:253–60.

Cowles, H. C. 1899. The ecological relations of the vegetation on the sand dunes of Lake Michigan. *Botanical Gazette* 27:95–117, 167–202, 281–308, and 361–91.

Crawley, M. J., ed. 1997. *Plant Ecology, Second Edition*. Blackwell Science, Oxford.

Daily, G. C., ed. 1997. *Nature's Services: Societal Dependence on Natural Ecosystems*. Island Press, Washington, DC.

Daubenmire, R. F. 1954. Alpine timberlines in the Americas and their interpretation. *Butler University Botanical Studies* 2:119–36.

———. 1956. Climate as a determinant of vegetation distribution in eastern Washington and northern Idaho. *Ecological Monographs* 26:131–54.

DeAngelis, D. L., and J. C. Waterhouse. 1987. Equilibrium and nonequilibrium concepts in ecological models. *Ecological Monographs* 57:1–21.

Ehrlich, P. R., and H. A. Mooney. 1983. Extinction, substitution, and ecosystem services. *BioScience* 33:248–54.

Elser, J. J., M. E. S. Bracken, E. E. Cleland, D. S. Gruner, W. S. Harpole, H. Hillebrand, J. T. Ngai, E. W. Seabloom, J. B. Shurin, and J. E. Smith. 2007. Global analysis of nitrogen and phosphorus limitation of primary producers in freshwater, marine and terrestrial ecosystems. *Ecology Letters* 10:1135–42.

Elton, C. S. 1958. *The Ecology of Invasions by Animals and Plants*. Methuen, London.

Forman, R. T. T. 1964. Growth under controlled conditions to explain the herarchical distributions of a moss, *Tetraphis pellucida*. *Ecological Monographs* 34:1–25.

Fretwell, S. D. 1972. *Populations in a Seasonal Environment*. Princeton University Press, Princeton, NJ.

Fryxell, J. M., J. Greever, and A. R. E. Sinclair. 1988. Why are migratory ungulates so abundant? *American Naturalist* 131:781–98.

Gill, D. E. 1978. The metapopulation ecology of the red-spotted newt, *Notophthalmus viridescens* (Rafinesque). *Ecological Monographs* 48:145–66.

Goldman, C. R. 1967. Molybdenum as an essential micronutrient and useful water mass marker in Castle Lake, California. Pages 229–38 *in* H. L. Golterman and R. S. Clymo, eds. *Chemical Environment in the Aquatic Habitat*. N. V. Noord-Hollandsche Uitgevers Maatschappij, Amsterdam.

Hanski, I. 1998. Metapopulation dynamics. *Nature* 396:41–49.

Hairston, N. G., F. E. Smith, and L. B. Slobodkin. 1960. Community structure, population control, and competition. *American Naturalist* 94:421–25.

Harrison, S. 1991. Local extinction in a metapopulation context: An empirical evaluation. *Biological Journal of the Linnean Society* 42:73–88.

Holt, R. D. 1977. Predation, apparent competition and the structure of prey communities. *Theoretical Population Biology* 28:181–208.

Huffaker, C. B. 1958. Experimental studies on predation: Dispersion factors and predator-prey oscillations. *Hilgardia* 27:343–83.

Hunter, M. D., and P. W. Price. 1992. Playing chutes and ladders: Heterogeneity and the relative roles of bottom-up and top-down forces in natural communities. *Ecology* 73:724–32.

Hurlbert, S. H. 1984. Pseudoreplication and the design of ecological field experiments. *Ecological Monographs* 54:187–211.

Johnson, L. E., J. M. Bossenbroek, and C. E. Kraft. 2006. Patterns and pathways in the post-establishment spread of non-indigenous aquatic species: The slowing invasion of North American inland lakes by the zebra mussel. *Biological Invasions* 8:475–89.

Kareiva, P. M., and M. Andersen. 1988. Spatial aspects of species interactions: The wedding of models and experiments. Pages 35–50 *in* A. Hastings, ed., *Community Ecology*. Springer, Berlin.

Keane, R. M., and M. C. Crawley. 2002. Exotic plant invasions and the enemy release hypothesis. *Trends in Ecology and Evolution* 17:164–70.

King, A. A., and W. M. Schaffer. 2001. The geometry of a population cycle: A mechanistic model of snowshoe hare demography. *Ecology* 82:814–30.

Krebs, C. J. 2009. Ecology: *The Experimental Analysis of Distribution and Abundance, Sixth Edition*. Benjamin Cummings, San Francisco.

Krebs, C. J., S. Boutin, and R. Boonstra, eds. 2001. *Ecosystem Dynamics of the Boreal Forest: The Kluane Project*. Oxford University Press, New York.

Kuhn, T. 1970. *The Structure of Scientific Revolutions*. University of Chicago Press, Chicago.

Lawton, J. 1991. Predictable plots. *Nature* 354:444.

Levey, D. J., B. M. Bolker, J. J. Tewksbury, S. Sargent, and N. M. Haddad. 2005. Effects of landscape corridors on seed dispersal by birds. *Science* 309:146–48.

Lidicker, W. Z. J. 1995. The landscape concept: Something old, something new. Pages 3–19 *in* W. Z. J. Lidicker, ed., *Landscape Approaches in Mammalian Ecology and Conservation*. University of Minnesota Press, Minneapolis.

Lindeman, R. L. 1942. The trophic-dynamic aspect of ecology. *Ecology* 23:399–418.

Loreau, M. 2000. Biodiversity and ecosystem functioning: recent theoretical advances. *Oikos* 91:3–17.

MacArthur, R. H. 1955. Fluctuations of animal populations, and a measure of community stability. *Ecology* 36:533–36.

———. 1965. Patterns of species diversity. *Biological Reviews* 40:510–33.

MacArthur, R. H., and E. O. Wilson. 1967. *The Theory of Island Biogeography*. Princeton University Press, Princeton, NJ.

Maurer, B. A., and M. L. Taper. 2002. Connecting geographic distributions with population processes. *Ecology Letters* 5:223–31.

May, R. M. 1974. Biological populations with nonoverlapping generations: Stable points, stable cycles, and chaos. *Science* 186:645–47.

Maynard Smith, J. 1968. *Mathematical Ideas in Biology*. Cambridge University Press, New York.

Menge, B. A. 1995. Indirect effects in marine rocky intertidal interaction webs: Patterns and importance. *Ecological Monographs* 65:21–74.

Menge, B. A., and J. P. Sutherland. 1987. Community regulation: Variation disturbance, competition, and predation in relation to environmental stress and recruitment. *American Naturalist* 130:730–57.

Newmark, W. D. 1985. Legal and biotic boundaries of Western North American National Parks: A problem of congruence. *Biological Conservation* 33:197–208.

Odum, E. P. 1968. Energy flow in ecosystems: A historical review. *American Zoologist* 8:11–18.

Parmesan, C., and G. Yohe. 2003. A globally coherent fingerprint of climate change impacts across natural systems. *Nature* 421:37–42.

Peters, R. H. 1991. *A Critique for Ecology*. Cambridge University Press, Cambridge.

Popper, K. R. 1963. *Conjectures and Refutations: The Growth of Scientific Knowledge*. Routledge and Kegan Paul, London.

Ricker, W. E. 1954. Effects of compensatory mortality upon population abundance. *Journal of Wildlife Management* 18:45–51.

Schindler, D. W. 1977. Evolution of phosphorus limitation in lakes. *Science* 195:260–62.

———— 1988. Effects of acid rain on freshwater ecosystems. *Science* 239:149–57.

Shrader-Frechette, K. S., and E. D. McCoy. 1993. *Method in Ecology: Strategies for Conservation*. Cambridge University Press, Cambridge.

Sibly, R. M., J. Hone, and T. H. Clutton-Brock. 2003. *Wildlife Population Growth Rates*. Cambridge University Press, Cambridge.

Van de Koppel, J., M. Rietkerk, and F. J. Weissing. 1997. Catastrophic vegetation shifts and soil degradation in terrestrial grazing systems. *Trends in Ecology and Evolution* 12:352–56.

Vitousek, P. M. 2004. *Nutrient Cycling and Limitation: Hawai'i as a Model System*. Princeton University Press, Princeton, NJ.

Walters, C. J., and S. J. D. Martell. 2004. *Fisheries Ecology and Management*. Princeton University Press, Princeton, NJ.

Worm, B., and J. E. Duffy. 2003. Biodiversity, productivity, and stability in real food webs. *Trends in Ecology and Evolution* 18:628–32.

14 *Responsive Science*

THE INTERPLAY OF THEORY, OBSERVATION, AND EXPERIMENT IN LONG-TERM, PLACE-BASED RESEARCH

Svaťa M. Louda and Leon G. Higley

Figure 14.0. Research on thistle-insect interactions in the western part of the 19,000 km² Sand Hills, the largest continental dune grassland in the western hemisphere. Students Natalie West and Deidra Jacobsen are checking rosettes of Platte thistle (*Cirsium canescens* Nutt.). Photograph by Svaťa Louda, June 2007.

Abstract

Major progress in ecological understanding has emerged from tests of theoretical predictions designed using natural history knowledge of a place and its organisms. Indeed, such knowledge is recognized as being essential for designing and interpreting field tests of theory. However, there is another, less acknowledged contribution from sustained place-based ecological research: its creative role in molding the structure of scientific investigations. The familiarity with a system that emerges from such research allows an investigator to recognize the potential in unexpected experimental results, unplanned observations, and serendipitous events, and flexibility in responding to the unexpected can lead to fundamental changes in research direction, allowing the pursuit of productive new questions and previously unanticipated hypotheses. We refer to this acknowledged but generally undervalued creative feedback between theory, experiment, and unplanned evidence as "responsive science." Using five examples from our work, we highlight the unplanned feedback that occurred and its relationship to natural history knowledge emerging from sustained experience with each system. These case histories illustrate at least three main points. First, natural history information allowed hypotheses and research plans based on theory to be tailored to the system, permitting more powerful tests of general concepts. Second, familiarity with the system plus flexibility in response to unexpected data turned the research toward exciting new questions that led to enhanced conceptual insights. Finally, unanticipated practical information emerged along with improved basic understanding. We conclude that more emphasis needs to be placed on doing, teaching, and facilitating such place-based responsive science.

Introduction

The fundamental goal of science is to develop a better understanding of the natural world. For ecological science, the natural phenomena to be explained are patterns in the characteristics of organisms, the abundance and distribution of species, and the structure and dynamics of populations, communities, and ecosystems. Progress in ecology, as in any science, depends on the rate at which significant problems are identified and potential answers are discovered.

Such progress will be expedited if we can identify and use an efficient research strategy: one that maximizes the rate of discovery, such as the one suggested by Platt's (1964) "strong inference" model. Strong inference results from direct tests of the contrasting explanatory hypotheses that emerge from current theory. This model fits what Kuhn (1996) called "normal science": research conducted within an existing conceptual paradigm. However, the strong-inference strategy does not explain how important new problems are identified, old problems are reconceptualized, or new theory is developed. More discussion of how such "creative science" (Kuhn 1996) is stimulated and structured is merited (see Medawar 1979, Watson 1998).

In this chapter we argue that the novel insights of creative science often emerge when chance meets a curious, prepared, and flexible mind in an underappreciated process that we call "responsive science." By responsive science we mean the research that evolves in the context of an initially normative study as an inquisitive, educated response to anomalous results, unexpected observations, and serendipitous observations (see also Stoskopf

2005). Although we are not the first to suggest that the pursuit of unexpected results is a means to new insights, we argue that responsive science based on the knowledge provided by sustained studies in a system is more important and productive than is often acknowledged, and so it merits more recognition, support, and discussion in the training of the next generation of ecologists.

Characteristics of Responsive Science

Responsive science has at least three essential characteristics. The first is a prepared mind, one familiar with the conceptual structure of the field: the ideas, literature, and theory. Regardless of what motivates a particular study, knowing where the work fits into current theory maximizes the probability that the results will enhance general understanding. Further, a prepared mind is ready to confront theory with data from nature—considering alternative hypotheses, deducing their logical consequences, designing informative experiments, and pursuing subsequent hypotheses. Since progress in understanding nature depends upon effective integration of theoretical development with empirical tests (e.g., see Dethier 1976, Kareiva 1996), familiarity with the conceptual structure of the field represents a crucial first element of responsive science.

The second characteristic of responsive science in ecology is a biologically-based approach by the investigator: the use of natural history knowledge, observation, and experience to inform both design and interpretation of the tests of current theory. Using such an approach encourages the recognition and elimination of unlikely hypotheses, allowing a focus on the more promising alternatives; and it facilitates the translation of general hypotheses into biologically realistic, testable forms, informing the design of experiments to distinguish among the predictions of the most likely competing hypotheses. Finally, an empirical, natural history-based research approach contributes to accurate interpretation of observational and experimental outcomes, and provides a strong foundation for a skeptical, inquisitive response to the data. Although the importance of natural history knowledge in ecological investigations is known (e.g., see Polis et al. 1998), we think that the application of such understanding, as a fundamental element of responsive science, plays a still underappreciated role in the development of new knowledge.

The third characteristic of all responsive science is a respect for the data and an openness to what they reveal, especially when the findings are not anticipated. Conflicting results, anomalous data, unanticipated observations, and unexpected serendipitous events often occur in field research. A typical first reaction to such data is to ignore them in the analysis because they are "anomalous," or to strain their interpretation in an attempt to retain the original conceptual model (see also Estes, chapter 8). A more creative response, however, is to ask: What is nature trying to tell me? Could those results be potentially important clues to unexpected complexity, or to an alternative mechanism not yet considered? We have found, for example, that unexpected results sometimes made it clear that critical assumptions of the existing theory were untenable, or that alternative explanations had to be considered, or that the system had properties that prevailing theory could not explain. With a responsive approach, unanticipated results can alter not just the next experiment but the whole direction of subsequent research.

We have found, in fact, that when these three main characteristics of responsive science were present, scientific understanding was accelerated. Application of theoretical and natural history knowledge, along with openness to unexpected results, often redirected the research in highly productive ways. The resulting shift in perspective generally stimulated important new questions, more relevant hypotheses, more realistic experiments and, we think, insights that both improved basic understanding of ecological phenomena and provided practical information for natural resource conservation and management.

We illustrate the process of responsive science, which we appreciated fully only in hindsight, with examples from our research. These case histories highlight the interplay that occurred between experimental tests of theory, natural history observations, and pursuit of chance observations or anomalous results. They also demonstrate the expansion, alteration, or total revision of the direction of research that resulted, and summarize the type of advances that emerged. Our goals in this are to describe what we have learned about the process, to encourage the prospective use of responsive science, and to stimulate more discussion in graduate student training of the value of flexible responses to unexpected results in creating scientific progress. We conclude that a responsive-science approach in the context of sustained place-based research is a demonstrated, underappreciated path for efficient progress in ecological science.

Case Histories

To illustrate responsive science, we briefly summarize the interplay of ideas, observations, and challenges that arose in five studies in which we grappled with the unexpected. In hindsight, the role played by responsive science in the process of discovery became evident. In the first case, a study of two small coastal shrubs initiated to test the effect of floral herbivory in subsequent plant recruitment, an attention to natural history observations altered the experimental design and provided a new theory to explain plant distributions along environmental gradients. In the second case, a study of a montane crucifer that was initiated to test chemical mediation of herbivore effects on plant fitness, an attention to unexpected observations led to a new linkage between plant physiological response to stress, insect herbivory, and plant habitat restriction. In the third case, a study of prairie thistles initiated to test the generality of the hypothesis that floral herbivory by coevolved insects could limit plant populations, an attention to serendipitous events and a willingness to pursue their implications led to new questions and unanticipated information in two applied fields: invasion biology and biological control. In the fourth and fifth cases, studies of prairie forbs initiated to quantify plant physiological response to insect herbivory, an attention to anomalous results and the use of natural history information led to new theory on the roles of plant physiological status and phenological state in predicting the impact of insect herbivores on plant performance. Cumulatively, we think that these cases highlight the characteristics of responsive science and suggest its power in driving growth in scientific understanding.

Case 1: Goldenbushes and floral herbivores in coastal sage scrub and chaparral

The initial aim of this study was to test the theoretical expectation at the time that herbivory by adapted consumers had little quantitative effect on plant fitness or abundance. The theory, called the "green world hypothesis" (Hairston et al. 1960, Slobodkin et al. 1967), stated that since plants and predators were resource-limited, herbivores were limited by their predators. If so, then herbivores could have no quantitative effect on their host plant populations. Although the idea was contested (Murdoch 1966, Ehrlich et al. 1967), prior to this study we found mention of only one actual small-scale test of the effect of herbivory on plant density (Cantlon 1969).

To critically evaluate the fitness and demographic effects of herbivory, we compared seed production and seedling and juvenile recruitment densities for plants on which floral herbivory ("predispersal seed predation") either was or was not experimentally reduced. The plant species studied were two goldenbushes in coastal San Diego County, California, on which we observed floral herbivory: now known as *Hazardia squarrosa* (Hook. & Arn.) and *Isocoma menziesii* (Hook. & Arn.) G. L. Nelson (Hickman 1993; figure 14.1a). For both species, seed production, seedling establishment, seedling density, and juvenile density were all higher for plants whose inflorescences had been insecticide-treated than for control plants; thus, floral herbivores reduced seed and limited subsequent plant recruitment and density (Louda 1982b, 1983). These results, which strongly contradict the "green world"

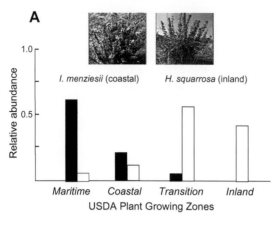

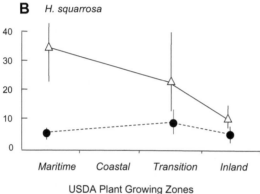

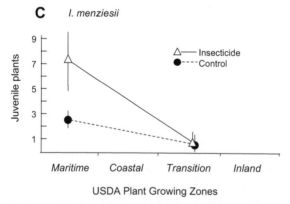

Figure 14.1. A: the distribution of abundance of the coastal goldenbush (*Isocoma menziesii*; filled bars) and the inland goldenbush (*Hazardia squarrosa*; open bars) along the environmental gradient of USDA plant growing zones, from the maritime Pacific coast of California in the west into the inland coastal mountain range to the east; B and C: effects of insect herbivory on the number of seedlings recruited as juveniles (one-year-olds) for the inland goldenbush (B) and the coastal goldenbush (C). Photographs by Svaťa Louda, 1981.

hypothesis, clearly illustrate the power of the direct test of theory, the classical approach to scientific advance.

Yet there is more to the story. Natural history observations showed that densities of these two very similar shrubs varied dramatically along the 80-km gradient from coast to mountains (figure 14.1a)—*I. menziesii* was more common near the ocean, and was replaced by *H. squarrosa* inland (Louda 1989). Prevailing theory predicted that such species replacements along an environmental gradient represented competitive displacements, thus demonstrating the shifting competitive advantage of one species over the other as the environment changed (e.g., Barbour et al. 1987). Interestingly, though, we also observed more floral herbivory on individual *H. squarrosa* plants near the coast, where their density was low, than inland where their density was high. This observation suggested another hypothesis: that greater floral herbivory at the coast shifted the realized distribution of the *H. squarrosa* population inland. To evaluate this hypothesis, we redesigned the initial insect exclusion test for both species by establishing multiple experimental sites along the gradient.

Comparison of the results among sites provided strong evidence that herbivory was important in explaining the species replacement. For the inland *H. squarrosa*, floral herbivory limited viable seed, seedling recruitment, and juvenile density at all sites, but dramatically more so at the coast than inland (figure 14.1b; Louda 1982a). For the coastal *I. menziesii*, floral herbivory also limited seed and seedlings at all sites along the gradient, but an added exclusion experiment showed that vertebrate herbivores limited the plant's juvenile survival more in the inland area than at the coast (figure 14.1c; Louda 1983). So variation in herbivory along the gradient—insect herbivory on *H. squarrosa* inflorescences, plus vertebrate herbivory on *I. menziesii* seedlings—could explain the species replacement (Louda 1989, Louda and Renaud 1989), challenging competitive displacement as the sole explanation for such patterns.

This case not only illustrates the power of the direct experimental test of a theoretical prediction, but also highlights the potential of responsive science stimulated by unexpected natural history observations to expand the research, improve the experimental design, identify shortcomings in prevailing theory, and suggest new theory.

Case 2: A crucifer and adapted foliage-feeding insect herbivores in montane meadows

The initial aim of this research was to test the theory that "secondary" compounds act as plant defenses against herbivory, since little field evidence existed. We studied bittercress, *Cardamine cordifolia* A. Gray (figure 14.2a,

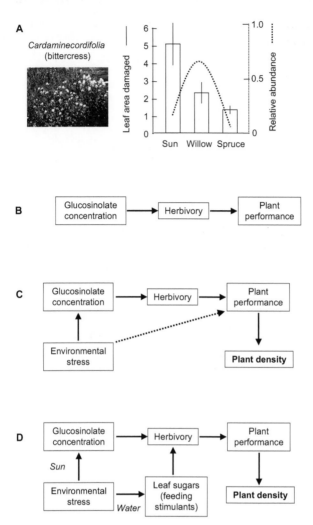

Figure 14.2. A: Bittercress (*Cardamine cordifolia*) in streamside habitat near Rocky Mountain Biological Laboratory, Gothic, Colorado. Photo by Svaťa Louda, 1986. Insect herbivory is highest in sunny meadow habitat, whereas relative abundance of plants peaks in the intermediate shade of willow habitat. B–D: steps in the sequential development of the conceptual model of the role of insect herbivory in bittercress plant performance. B: the initial conceptual model, relating herbivory to glucosinolate concentration in affecting plant performance. C: the modified conceptual model relating plant performance and habitat to both environmental stress and herbivory, including a direct effect (dashed line, not supported) and indirect effect (supported) of environmental stress. D: the emergent final conceptual model suggested by the data, in which increased herbivory results from plant physiological response to sun (increased glucosinolate concentration) and water stress (increased sugar content), and influences plant performance and distribution among habitats.

left), a widespread, native perennial crucifer of moist montane habitats in the Rocky Mountains that produces mustard oils (glucosinolate precursors) and hosts a suite of adapted insect herbivores (e.g., see Chew 1977, Chew and Rodman 1979). If the defense hypothesis held, then herbivory should be inversely related, and plant performance positively related, to glucosinolate concentrations. In fact, we did find that herbivory decreased as glucosinolates increased, as predicted, and that chronic levels of foliage herbivory by insects reduced plant growth and seed production (Louda and Rodman 1983a, 1983b; Louda 1984). Thus, the results provided support for the theory that bittercress secondary compounds functioned defensively (figure 14.2b).

Again, however, observations emerging from sustained research on the system expanded the focus. We observed that bittercress was more abundant in the willows than in either adjacent open meadows or spruce stands (figure 14.2a, right). Conventional theory explained such patterns as plant adaptation to moderate shade (e.g., see Barbour et al. 1987). However, we also recorded much higher levels of herbivory on plants in the sunny open meadow than in the shade (figure 14.2a right; Louda and Rodman 1983a, 1983b). This suggested an alternative hypothesis: that greater foliage herbivory in the sunny meadow restricted bittercress in the sun and shifted the population's distribution into the adjacent, lightly shaded willow habitat. This hypothesis had a precedent. When the exotic chrysomelid beetle, *Chrysolina quadrigemmina*, was introduced as a biocontrol agent against weedy Saint-John's-wort (*Hypericum perforatum*), it reduced plant densities in the sun and restricted the population to shaded areas (Huffaker and Kennett 1957, Harper 1969). However, no evidence existed that native herbivores could produce a similar pattern for a coevolved host plant under natural conditions.

Our experiments, which incorporated shade removal, transplants, and insecticide application, demonstrated that bittercress grew as well in the open sun as in the willow shade only if it was protected against adapted insect herbivores (Louda and Rodman 1996). Greater herbivory reduced plant performance and density in the sunny open meadow, shifting this native plant's realized distribution into the willow shade habitat (Louda and Rodman 1996). This unique result not only contradicted expectations under the "green world" hypothesis but also challenged the role of physiological adaptation as the sole explanation for plant habitat distribution (figure 14.2c, dotted line).

Two unexpected results further stimulated expansion of the research. First, we were surprised that herbivory on bittercress was predicted not by total leaf nitrogen concentrations, since nitrogen generally is limiting for insect herbivores (White 1978), but instead by total nonstructural carbohydrates (leaf sugars; Collinge and Louda 1989). Second, we unexpectedly found that herbivory in the sun varied with soil moisture, increasing on drier soils (Louda 1986). After puzzling over these observations, we hypothesized that environmental stress altered plant performance indirectly through its effect on herbivory (figure 14.2d) rather than directly (figure 14.2c, dotted line), as usually expected. Using diurnal leaf water deficit to quantify plant stress, experiments showed that herbivory increased, even in the shade, when bittercress was stressed by cutting its peripheral roots. However, neither glucosinolates nor total nitrogen increased with water stress, whereas leaf sugars did (Louda and Collinge 1992). These findings contradicted the prevalent theory that insects attacked the most vigorous plants or plant parts (Price 1991). Instead, physiological adjustments by moderately stressed plants, which stimulated increased feeding by adapted herbivores, led indirectly to lower plant performance (figure 14.2d). Thus, including plants' response to environmental growing conditions increased the complexity of the conceptual model, but also enhanced understanding of the interactions in the field.

This case again demonstrates not only the power of a direct experimental test of theory, but also the increase in knowledge that results from the pursuit of new questions in response to unexpected results. The study showed for the first time that ambient levels of herbivory by specialized herbivores could determine the habitat distribution of their host plant under indigenous conditions, and it also provided evidence for alternate mechanistic hypotheses and suggested unanticipated conceptual insights on insect-plant interactions. Among these are the insights that coevolution does not necessarily eliminate strong demographic effects of adapted consumers, that environmental heterogeneity can lead to differential selection pressure, and that increased leaf sugar concentrations associated with plant physiological response to stress can mediate levels of herbivory.

Case 3: Thistles and their native and exotic floral herbivores in prairie grasslands

The initial aim of studying native thistles and their adapted herbivores in prairie grasslands was to test the generality of floral-herbivore impact on plant population dynamics (case 1). Building on previous work with *Cirsium canescens* Nutt. (Platte thistle; Lamp and McCarty 1982; figure 14.3a) and using insecticide exclusion tests, we substantiated that floral herbivory significantly decreased seed production, subsequent seedling density, and eventual adult Platte thistle density (Louda and Potvin 1995). These results also demonstrated that the population was seed-limited and that floral herbivory limited lifetime fitness—a further challenge to the "green world" hypothesis (Hairston et al. 1960).

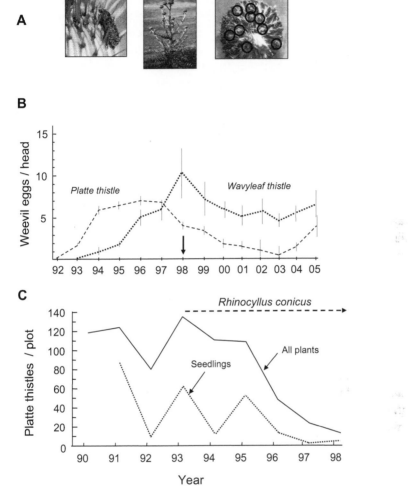

Figure 14.3. A. Left: the invasive biocontrol weevil *Rhinocyllus conicus*. Center: Platte thistle (*Cirsium canescens*). Right: weevil egg cases on Platt thistle. Photographs by Svaťa Louda, 1992 and 1997. B. Increase in the number of biocontrol weevil eggs per flowering head of wavyleaf thistle (*C. undulatum*; dotted line) and Platte thistle (solid line) in sand prairie, Arthur County, Nebraska. Arrow marks the year in which oviposition on wavyleaf thistle became greater than on Platte thistle. C. Change in Platte thistle seedling (dotted line) and overall plant (solid line) densities in 144 m² demography plots, before and after the 1993 invasion of *R. conicus*. Platte thistle abundance was greatly reduced by 1998; the reduction has persisted to date (2008).

Again, however, natural history observations both influenced research design and contributed to the development of unanticipated new knowledge. Reflecting our experience with the system, the experimental design nested blocks within two microhabitats—continuous prairie grassland where Platte thistle was sparse, and wind disturbances ("blowouts") where the thistle was more common. We did this just to be thorough; in looking at the data, however, we discovered that habitat strongly influenced the magnitude of the treatment effect. Specifically, competition in continuous prairie increased the impact of insect-induced seed losses on seedling recruitment (Louda and Potvin 1995). These results made it clear that not only was the floral herbivory critical in plant regeneration, but its importance increased in highly competitive contexts. This was an unexpected result with broad implications.

Serendipity redirected the research dramatically after 1993, when we discovered a new floral herbivore feeding on the native thistles in our Western sand-prairie study sites (figure 14.2a). The insect was *Rhinocyllus conicus* Fröhlich, a Eurasian weevil introduced into North America in 1968 and into eastern Nebraska in 1969 as a biological control agent against exotic thistles, especially musk thistle, *Carduus nutans* (Nutt.) Spreng. (Gassmann and Louda 2001). Because we were in the midst of a long-term demography study (1990–2008), we were quantifying plant survival, growth, herbivory, seed reproduction, recruitment, and population growth (e.g., Louda 1998, Rose et al. 2005). These pre-invasion data provided a unique opportunity to quantify the ecological consequences of invasion of an exotic insect herbivore into a native community (e.g., Louda et al. 1997, 2005a, 2005b).

Our key findings to date can be summarized as follows. Population growth of *R. conicus* was exponential as it invaded a system without its targeted weeds (Louda et al. 1997, Louda 1998; figure 14.2b). The weevil significantly reduced seed production of the native thistle (Louda 2000), which rapidly led to lower densities of Platte thistle, a monocarpic species that dies after flowering once (figure 14.3c; Louda and Arnett 2000, Louda et al. 2003a). The degree of phenological synchrony predicts magnitude of weevil impact on seed production (Russell and Louda 2004). Further, *R. conicus* unexpectedly imposes a greater impact on Platte thistle in the center of its geographic distribution than in peripheral populations closer to populations of the targeted weed (Rand and Louda 2006); this challenges the conventional "center of distribution" protection criterion for conservation planning. Also, a population projection model showed that *R. conicus* has the potential to drive the native Platte thistle toward extinction (Rose et al. 2005). The weevil's demographic effect on the co-occurring native wavyleaf thistle (*C. undulatum*), an iterocarpic species that flowers multiple times before dying, so far is lower (Louda 1998, Louda and Arnett 2000). However, weevil impact on wavyleaf thistle increases when Platte thistle flower heads become scarce (Russell and Louda 2005; figure 14.3b). Further, the invasive weevil negatively affects native insects in the thistle floral-feeding guild (Louda and Arnett 2000, Louda et al. 2008), unexpectedly demonstrating that floral guilds are not necessarily saturated and that indirect effects of invasive species can be large. Finally, spatial scale and environmental context influence weevil numbers on the native thistles (Rand and Louda 2004, Rand et al. 2004, Russell et al. 2007). Thus, sustained, long-term, place-based research on thistle-insect interactions, including new research in response to a serendipitous occurrence, has allowed us to address new questions about insect-plant interactions, the invasion process, and the consequences of invasive species for native communities, providing unique new information that has both basic and applied implications.

A second serendipitous observation led to the discovery that the effects of *R. conicus* are not unique. In searching for *R. conicus* in Colorado, we discovered another Eurasian weevil, *Larinus planus* (F.), feeding on flower heads of a native thistle, *Cirsium undulatum* (Nutt.) Spreng. var. *tracyi* (Rydb.) Welsh (Tracy's thistle). This weevil was deliberately released in 1993 near Gunnison, Colorado, against the exotic Canada thistle, *Cirsium arvense* (L.) Scop. Contemporary host-specificity tests showed that large-headed native thistles, while acceptable, were less preferred than small-headed Canada thistle. This led to the expectation that *L. planus* would not present a risk to native thistles (McClay 1990). Unfortunately, this expectation proved wrong, and *L. planus* now occurs in higher abundance on the native thistle than on the targeted exotic Canada thistle near Gunnison (Louda and O'Brien 2002), and it is having a quantitative negative effect (Dodge 2005). These unexpected results demonstrate that the preference and performance criteria used in standard host-specificity testing provide necessary but not sufficient information to quantify the ecological risk for potential biological control agents (Louda et al. 2003a, 2005a).

Overall, this case provides multiple examples of how the elements of sustained, place-based responsive science—with attention to theoretical expectation and natural history knowledge, plus a willingness to alter research direction in response to unexpected results—can contribute to rapid increases in understanding of basic ecological processes, as well as important information for resource management and conservation.

Case 4: Annual sunflower and checkerspot butterfly larvae along a disturbed stream bank

The initial aim of this work was to characterize the potential influence of different forms of insect herbivory on the primary metabolism of plant hosts, especially photosynthesis (e.g., see Higley 1992, Higley and Peterson 1996).

The research was motivated by debate over the fitness costs of herbivory (e.g., see Jermy 1984, 1988; Bernays and Graham 1988; Crawley 1989) and in part by frustration over the absence of general models to account for variation in agricultural yield losses from insect herbivory (Pedigo et al. 1986). Although Boote (1981) presented a theoretical framework for characterizing possible impact of insect herbivory on plant physiology, relatively few data were available to evaluate it. Most studies reported a reduction in photosynthesis by the tissues remaining after herbivory (Welter 1988), yet some studies reported increases, and our previous research generally showed no differences (Higley 1992, Peterson et al. 1992, Peterson and Higley 1996, Burkness et al. 1999, Hunt et al. 2003). Even when we repeated the studies where rate reductions had been reported, we could not replicate them (Peterson et al. 1996).

Given that gas exchange by plants is highly influenced by environmental factors and can be extremely variable even under controlled conditions, we asked first whether our equipment had the resolution needed to document changes in photosynthesis with leaf-chewing injury. It did; we had demonstrated significant reductions in photosynthesis after feeding by other types of insect herbivores, including leaf skeletonizers (Peterson et al. 1998) and phloem-feeding aphids (Macedo 2003a, 2003b). So next we reasoned that a more likely explanation was that different phenomena had been measured. Specifically, we distinguished between intrinsic vs. extrinsic plant responses to herbivory, whereas others had not (Peterson and Higley 1993). Intrinsic responses are physiological changes under controlled conditions, while extrinsic responses are changes driven by variation in external growing conditions resulting from herbivory, such as increased light penetration through the canopy or improved water relations after herbivore reduction of leaf mass. We found that a plant that showed no differences in gas exchange pre- and post-herbivory under constant light could still show differences under ambient light conditions because more light reached the leaves remaining after herbivory.

Yet the situation was even more complicated. For example, one realization that drove our subsequent research was that most of our experiments were on cultivated legumes, which were limited in genetic variability, artificially selected, grown beyond their areas of evolutionary origin, and interacting with herbivores lacking evolutionary contact with them. Consequently, to generalize we looked for other, natural systems in which we could manipulate levels of infestation by adapted insect herbivores. We found *Helianthus annuus* L. (wild annual sunflower) growing on a disturbed stream bank, as well as large populations of an adapted checkerspot butterfly larva, *Chlosyne gorgone* (Hübner). Because *C. gorgone* aggregate (figure 14.4, top), we had ample sunflowers without any leaf injury that we could infest experimentally.

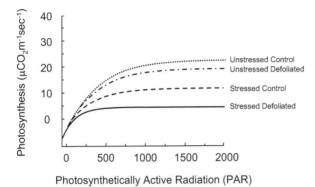

Photosynthetically Active Radiation (PAR)

Figure 14.4. Top: larvae of the checkerspot butterfly, *Chlosyne gorgone*, on a leaf of wild annual sunflower, *Helianthus annuus*. Bottom: photosynthetic response of sunflower to insect herbivory under well-watered (unstressed) and severely water-deprived (stressed) growing conditions. Water stress lowers photosynthetic rate significantly, but herbivory (defoliation) only affects water-stressed plants. Photographs by Leon Higley.

Interestingly, striking reductions in photosynthesis after injury occurred in our initial experiment, but not again over three subsequent years. We did six experiments, and only the first showed photosynthetic reduction in response to herbivory. We looked for instrument errors and analytical errors. We examined the hypothesis that degree of injury varied among experiments. None of these provided an explanation. Although we were tempted to just report the results and ascribe the anomalous test to an indeterminate cause, we did not. We continued to ask: What did the initial, anomalous result suggest? One hypothesis—that we had recorded an interaction between the environment and plant response to herbivory—was strengthened by examining site characteristics. The stream bank was highly eroded and, particularly in the first year, experimental plants were growing near the top, on the steepest portion. It was plausible that water might have been more limiting in the first experiment.

We followed up these observations with more field experiments to evaluate the effect of water stress on photosynthetic response to herbivory. The new experiments revealed that only under severe water stress, near the permanent wilting point, did herbivory cause a major reduction in photo-synthesis (figure 14.4 bottom). Even then, the reduction was transient; as injured tissue healed, reductions were no longer observed (L. Higley et al., unpublished data). Additionally, because sunflowers maintained relatively high photosynthetic rates even under such stress, there was no obvious sig-nal (like dramatically reduced stomatal conductances and gas exchange) to suggest interaction. The initial result, therefore, made sense if the data were from plants under severe water stress, a relatively uncommon circumstance. By following up on an unusual result, we discovered that the initial outcome actually was anomalous, but in a very illuminating way.

This case illustrates four important points. First, it highlights the value of sustained research, with multiple experiments under a range of environ-mental conditions. Second, it illustrates the importance of understanding your study species within its environment. Third, it shows the importance of not giving up on finding an explanation for anomalous results. Fourth, it argues for being open to reevaluation of assumptions underlying current paradigms. Pursuing an anomalous result in relation to the environment led to a major new insight: transient water losses caused by herbivory re-duced photosynthesis, but only under severe stress (Haile 1999). So, while moderate water stress altered herbivory for bittercress (case 2), here we see the other side of the coin—herbivory altered the impact of water stress on plant photosynthesis.

Case 5: Milkweed and native insect herbivores in tallgrass prairie

In another set of studies, we examined the common milkweed, *Asclepias syriaca* L., which has characteristic, well-studied chemical defenses against herbivory (cardenolides). Given theory and the level of herbivory observed (figure 14.5), we hypothesized that the metabolic cost of cardenolide pro-duction in response to herbivory would cause photosynthetic rate reduc-tions. If so, then the result would be important because it would quantify the metabolic cost associated with a specific plant defense. We speculated that the mechanism for photosynthetic limitation could involve reductions in photoelectron transport (from insufficient ADP pools) or limitations in ribulose bisphosphate oxygenase (rubisco) synthesis. We tested these ideas by artificially infesting milkweed plants in tallgrass prairie with larvae of both milkweed tussock moth, *Euchaetes egle* L. (figure 14.5) and monarch butterfly, *Danaus plexippus* L.

Our initial experiments in 1995 showed reductions in photosynthesis in response to herbivory by both insects. However, in subsequent experi-ments designed to identify the mechanisms, we unexpectedly observed a wide variety of photosynthetic responses. In some experiments, profound

Figure 14.5. (a) Insect herbivory on common milkweed, *Asclepias syriaca*. (b) One of the adapted insect herbivores, milkweed tussock moth, *Euchaetes egle* L., on *A. syriaca*. Photographs by Leon Higley.

photosynthetic reductions occurred after herbivory, with little recovery after injury. In others, smaller reductions occurred, with partial to complete recovery; and in yet others, no reduction occurred after herbivory. Additionally, although cardenolide induction with injury had been reported (Malcolm and Zalucki 1996), we did not find any evidence of induction, nor could we associate changes in photosynthesis with concentrations of defensive compounds. We recognized that injury to midribs dramatically reduced photosynthetic rates (Delaney and Higley 2006), while other types of leaf injury caused variable responses; however, the nature of the insect feeding alone did not account for the variation observed.

Further, because cardenolide content varied within and among *Asclepias* species, we hypothesized that the variation in photosynthetic responses might reflect underlying variation in plant defensive chemistry—even if we couldn't find an association between cardenolide concentrations and photosynthesis! Fortunately, we also recognized that our experiments had differed subtly in their timing with respect to flowering phenology. To date, we have conducted 27 experiments to test the role of phenology in plant photosynthetic response to herbivory. These tests span the phenological stages of plant reproduction, from the pre-flowering through seed-forming stages, and they evaluate a variety of herbivores. The responses measured include photosynthetic response by leaves of different ages, photosynthetic recovery after injury, and photosynthesis interactions with water stress. A new picture of plant response to herbivory is emerging from this sustained study. Flowering milkweed plants (those with some seed pod development) show impairment in photosynthetic rates with herbivore injury. In contrast, plants at mid-vegetative or very early reproductive (pre-flowering) stages show strikingly reduced photosynthetic rates in response to herbivory (Delaney 2003). Additionally, we are finding that changes in photosynthetic rate have varied with the amount of herbivory, and these changes differ among *A. syriaca* plants at different phenological stages.

This case illustrates two important features of responsive science. First, if we had remained focused only on our original question (photosynthetic response to herbivory in a heavily defended plant) we might have abandoned work with milkweed in pursuit of a less variable system. Instead, by accepting that this variation might reflect an important underlying biological reality, we were faced with a different question: Does plant response to herbivory vary with plant phenology? Our data demonstrate that phenology can have a huge impact on how plants respond photosynthetically to herbivore injury, and they suggest that our current understanding does not yet accurately incorporate the importance of seasonality and plant growth. Second, this case again illustrates the value of sustained, responsive research within a system. The multiple experiments in one place over a decade have led to a broader array of questions, and to data on variability that is not possible to obtain in short-term studies.

Discussion

The process of discovery illustrated in these cases clearly affirms the power of experiments designed and analyzed with knowledge of place to directly illuminate mechanisms underlying ecological phenomena. These examples provide strong support for the suggestion that clear field tests of explicitly stated, theoretically based hypotheses require knowledge of the study system and its natural history as a foundation (e.g., Polis et al. 1998). However, these examples also suggest another important, less appreciated aspect of sustained, long-term, place-based research. This aspect is the extent to which response to anomalous results, unexpected findings, and serendipitous observations can dramatically alter the direction of research toward unanticipated questions and new areas of research, rather than just lead to refinement of initial ideas and existing theory. Unexpected findings forced us to reconceptualize, often dramatically altering the subsequent research. The new direction was usually more relevant to the interactions being studied; we think it often addressed more conceptually important issues and produced significant, unanticipated advances. Such shifts in direction reflect outcomes that emerge during the process of research, leading us to refer to this subtle, more indirect linkage between observation and experiment as "responsive science." Such science depends upon investigative flexibility and an ability to see new directions in the interaction between observations, theory, and experiment. In ecology we find that responsive science is facilitated by sustained experience with a system—a place, its occupants, and their interactions.

Review of these five case histories suggests at least three general points. First, we conclude that place-based research leads to a relatively fast, accurate "reality check" on theoretical models, literature-based hypotheses, and "dry-labbed" experimental plans. With goldenbushes, for example, the expanded experiment not only challenged the "green world" theory that herbivores were limited by their predators and so could not limit their host plants, but also provided an alternative mechanism to explain species replacement patterns along environmental gradients.

Second, we find that unexpected observations and outcomes often led us to new, broader, and more relevant questions, better mechanistic hypotheses, conceptual insights, and a more comprehensive understanding of biological interactions. With montane bittercress, for example, the unanticipated systematic variation in foliage herbivory among habitats led to questions beyond the role of secondary plant chemistry in explaining level of herbivory; it also led to questions regarding the role of herbivory in plant habitat distribution, and the role of plant physiological response to stress in mediating the levels and effects of herbivory.

Third, we find that unexpected practical insights for resource management often emerge, along with the improved basic understanding, when the research is responsive to knowledge of place. For example, with native thistles the invasion of *R. conicus* into native prairie and its consequences illustrated the value of long-term data in quantifying how an approved, extensively tested biological control agent could still become a threat to a large group of characteristic North American plants, in spite of the host-specificity testing being currently used to assess ecological risk.

Do the specific examples here suggest any overarching general principles? We think so. The power of observation provides a tool to perceive the world, but its weakness is that underlying mechanisms remain ambiguous. The power of experimentation provides a tool to reveal cause and effect, but its weakness is that treatment effects may reflect specific experimental conditions. From our perspective, responsive science is an approach that directly addresses the limitations of both observation and experiment. An openness to contingencies and a willingness to accept challenges to preconceptions permits and encourages a focus on the observations, interactions, and outcome in context, and it also leads to faster scientific advance.

One crucial distinction of science, compared to other ways of understanding, is that data trump theory. Responsive science relies on data—particularly theoretically unexpected data that emerge as a consequence of observation and experiment. Among the advantages of working with species or communities over time is that information on both the basic patterns and their variations accumulates and becomes part of personal knowledge, allowing one to recognize an unusual occurrence in that system. Thus, sustained, long-term, place-based research, by contributing both data and such natural-history knowledge as a basis for responsive science, increases the probability of recognizing data that challenge theoretical expectations. We are not the first to argue that experience provides a foundation for innovation. For example, Pasteur said, "Chance favors the prepared mind." We contend, however, that responsive science is an efficient yet underappreciated research strategy for the development of new theory and paradigms. We think that increased emphasis on responsive science in both research planning and graduate training for the next generation of ecologists is merited.

Responsive science is an indirect route to new knowledge, since it does not follow from the initial question asked. Instead, the creative response to unexpected results and unanticipated field observations leads to new questions, innovative hypotheses, improved experiments, and potentially creative theoretical developments. Adjusting research goals and subsequent work ends up challenging preconceived ideas and previous hypotheses and thus expanding theoretical understanding, often in surprising ways. Again, research that benefits from observations is not a new idea. For example, it is consistent with Louis Agassiz's motto: "Study nature, not books." However, the process of responding to observation and data is seldom explored or discussed for its occurrence, structure, dynamics, or potential for breakthrough insights, even in ecology. Yet knowledge of place—and observer expectations based on sustained experience with place—are the foundation for using serendipitous events to discover new theories and expand understanding of nature. With hindsight we think that our experience, highlighted in the five cases discussed above, illustrates the effectiveness and excitement emerging from the use of responsive science to not only verify existing theory but also lead to new theory and increased basic understanding, as well as provide ecological input for resource management and environmental problem-solving. Thus we encourage expectation of and openness to the use of responsive science in exploring anomalous results or unanticipated observations. We also argue for more discussion in graduate student training of the value of flexible responses to unexpected data as an efficient means toward creative science.

Acknowledgments

We are grateful for the effective help and constructive discussions provided by our valuable and enjoyable collaborators, graduate students, and numerous undergraduate research assistants over the years. We could not have done the work without them. Partial support was provided by NSF DEB-0414777 and USDA 2005-35320-15379 to F. L. Russell and S. M. Louda.

References

Barbour, M. G., J. H. Burk, and W. D. Pitts. 1987. *Terrestrial Plant Ecology*. Benjamin/Cummings, Menlo Park.

Bernays, E., and M. Graham. 1988. On the evolution of phytophagous arthropods. *Ecology* 69:886–92.

Boote, K. J. 1981. Concepts for modeling crop response to pest damage. *ASAE Paper* 81–4007, ASAE, St. Joseph, Missouri.

Burkness, E. C., W. D. Hutchison, and L. G. Higley. 1999. Photosynthetic response of "Carolina" cucumber to simulated and actual striped cucumber beetle (Coleoptera: Chrysomelidae) defoliation. *Entomologia Sinica* 6:29–38.

Cantlon, J. E. 1969. The stability of natural populations and their sensitivity to technology. *Brookhaven Symposium in Biology* 22:197–205.

Chew, F. S. 1977. Coevolution of pierid butterflies and their cruciferous food plants. II. The distribution of eggs on potential food plants. *Evolution* 31:568–79.

Chew, F. S., and J. E. Rodman. 1979. Plant resources for chemical defense. Pages 271–303 *in* Rosenthal, G. A. and D. H. Janzen, eds., *Herbivores: Their Interaction with Plant Secondary Metabolites*. Academic Press, New York.

Collinge, S. K., and S. M. Louda. 1989. Influence of plant phenology on the insect herbivore/bittercress interaction. *Oecologia* 79:111–16.

Crawley, M. J. 1989. Insect herbivores and plant population dynamics. *Annual Review of Entomology* 34:531–64.

Delaney, K. J. 2003. *Milkweed Leaf Photosynthesis Responses to Insect Herbivory: Factors that Influence Photosynthetic Rate Impairment of Injured Leaves*. PhD dissertation, University of Nebraska, Lincoln.

Delaney, K. J., and L. G. Higley. 2006. An insect countermeasure impacts plant physiology: Midrib vein cutting, defoliation and photosynthesis. *Plant, Cell & Environment* 29:1245–58.

Dethier, V. G. 1976. *The Hungry Fly: A Physiological Study of the Behavior Associated with Feeding*. Harvard University Press, Cambridge, MA.

Dodge, G. 2005. *Ecological Effects of the Biocontrol Insects,* Larinus planus *and* Rhinocyllus conicus, *on Native Thistles*. PhD dissertation. University of Maryland, College Park.

Ehrlich, P. R., and L. C. Birch. 1967. The "balance of nature" and "population control." *American Naturalist* 101:97–107.

Gassmann, A., and S. M. Louda. 2001. *Rhinocyllus conicus*: Initial evaluation and subsequent ecological impacts in North America. Pages 147–83 *in* E. Wajnberg, J. K. Scott, and P. C. Quimby, eds., *Evaluating Indirect Ecological Effects of Biological Control*. CABI Publishing, Wallingford, Oxon, UK.

Haile, F. 1999. *Physiology of Plant Tolerance to Arthropod Injury.* PhD dissertation. University of Nebraska, Lincoln.

Hairston, N. G., F. E. Smith, and L. B. Slobodkin. 1960. Community structure, population control, and competition. *American Naturalist* 94:421–25.

Harper, J. L. 1969. The role of predation in vegetational diversity. *Brookhaven Symposium in Biology* 22:48–62.

Hickman, J. C., ed. 1993. *The Jepson Manual: Higher Plants of California.* University of California Press, Berkeley.

Higley, L. G. 1992. New understandings of soybean defoliation and their implications for pest management. Pages 56–65 *in* L. G. Copping, M. B. Green, and R. T. Rees, eds., *Pest Management of Soybean.* Elsevier, London.

Huffaker, C. B., and C. E. Kennett.1957. Ten-year study of vegetational changes associated with biological control of Klamath weed. *Journal of Range Management* 12:69–82.

Hunt, T. E., L. G. Higley, and F. J. Haile. 2003. Imported longhorned weevil (*Coleoptera: Curculionidae*) injury to soybean: Physiological response and injury guild-level economic injury levels. *Journal of Economic Entomology* 96:1168–73.

Jermy, T. 1984. Evolution of insect/host plant relationships. *American Naturalist* 124:609–30.

———. 1988. Can predation lead to narrow food specialization in phytophagous insects? *Ecology* 69:902–4.

Kareiva, P. M. 1996. Developing a predictive ecology for non-indigenous species and ecological invasions. *Ecology* 77:1651–52.

Kuhn, T. S. 1996. *The Structure of Scientific Revolutions.* Third Edition. University of Chicago Press, Chicago.

Lamp, W. O., and M. K. McCarty. 1982. Predispersal seed predation of a native thistle, *Cirsium canescens. Environmental Entomology* 11:847–51.

Louda, S. M. 1982a. Distribution ecology: Variation in plant recruitment in relation to insect seed predation. *Ecological Monographs* 52:25–41.

———. 1982b. Limitation of the recruitment of the shrub *Haplopappus squarrosus* (Asteraceae) by flower- and seed-feeding insects. *Journal of Ecology* 70:43–53.

———. 1983. Seed predation and seedling mortality in the recruitment of a shrub, *Haplopappus venetus* (Asteraceae), along a climatic gradient. *Ecology* 64:511–21.

———. 1984. Herbivore effect on stature, fruiting and leaf dynamics of a native crucifer. *Ecology* 65:1379–86.

———. 1986. Insect herbivory in response to root-cutting and flooding stress on a native crucifer under field conditions. *Acta Oecologia, Oecologia Generalis* 7:37–53.

———. 1989. Predation in the dynamics of seed regeneration. Pages 25–51 *in* M. A. Leck, V. T. Parker, and R. L. Simpson, eds., *Ecology of Soil Seed Banks.* Academic Press, New York.

———. 1998. Population growth of *Rhinocyllus conicus* (Coleoptera: Curculionidae) on two species of native thistles in prairie. *Environmental Entomology* 27:834–41.

———. 2000. Negative ecological effects of the musk thistle biocontrol agent, *Rhinocyllus conicus* Fröl. Pages 215–43 *in* P. A. Follet and J. J. Duan, eds., *Nontarget Effects of Biological Control.* Kluwer Academic Publishers, Boston.

Louda, S. M., and A. E. Arnett. 2000. Predicting non-target ecological effects of biological control agents: Evidence from *Rhinocyllus conicus.* Pages 551–67 *in* N. R. Spencer, ed., *Proceedings of the 10th International Symposium on Biological Control of Weeds.* Montana State University, Bozeman.

Louda, S. M., A. E. Arnett, T. A. Rand, and F. L. Russell. 2003. Invasiveness of some biological control insects challenges adequacy of ecological risk assessment and regulation. *Conservation Biology* 17:1–11.

Louda, S. M., and S. K. Collinge 1992. Plant resistance to insect herbivores: A field test of the environmental stress hypothesis. *Ecology* 73:153–69.

Louda, S. M., D. Kendall, J. Connor, and D. Simberloff. 1997. Ecological effects of an insect introduced for the biological control of weeds. *Science* 277:1088–90.

Louda, S. M., and C. W. O'Brien. 2002. Unexpected ecological effects of distributing the exotic weevil, *Larinus planus* (F.), for the biological control of Canada thistle. *Conservation Biology* 16:717–27.

Louda, S. M., R. W. Pemberton, M. T. Johnson, and P. A. Follett. 2003. Nontarget effects—the Achilles' Heel of biological control? Retrospective analyses to reduce risk associated with biocontrol introductions. *Annual Review of Entomology* 48: 365–96.

Louda, S. M., and M. A. Potvin. 1995. Effect of inflorescence-feeding insects in the demography and lifetime fitness of a native plant. *Ecology* 76: 229–45.

Louda, S. M., T. A. Rand, A. Kula, and A. E. Arnett. In review. Indirect effects of an invasive weevil on native insect floral herbivores in temperate prairie. *Ecological Monographs*.

Louda, S. M., T. A. Rand, F. L. Russell, and A. E. Arnett. 2005. Assessment of ecological risks in biocontrol: Input from retrospective ecological analyses. *Biological Control* 35:253–64.

Louda, S. M., T. A. Rand, A. E. Arnett, A. S. McClay, K. Shea, and A. K. McEachern. 2005. Evaluation of ecological risk to populations of a threatened plant from an invasive biocontrol insect. *Ecological Applications* 15:234–49.

Louda, S. M., and P. Renaud. 1989. Differential predation and pressure: A general mechanism for structuring plant communities along complex environmental gradients. *Trends in Ecology and Evolution* 4:158–59.

Louda, S. M., and J. E. Rodman. 1983a. Ecological patterns in glucosinolate content of a native mustard, *Cardamine cordifolia*, in the Rocky Mountains. *Journal of Chemical Ecology* 9:397–422.

———. 1983b. Concentration of glucosinolates in relation to habitat and insect herbivory for the native crucifer *Cardamine cordifolia*. *Biochemical Systematics and Ecology* 11:199–208.

———. 1996. Insect herbivory as a major factor in the shade distribution of a native crucifer (*Cardamine cordifolia* A. Gray, bittercress). *Journal of Ecology* 84: 229–38.

Macedo, T. B., C. S. Bastos, L.G. Higley, K. R. Ostlie, and S. Madhavan. 2003. Photosynthetic responses of soybean to soybean aphid (Homoptera: Aphididae) injury. *Journal of Economic Entomology* 96:188–93.

Macedo, T. B., L. G. Higley, X. Ni, and S. S. Quisenberry. 2003. Light activation of Russian wheat aphid-elicited physiological responses in susceptible wheat. *Journal of Economic Entomology* 96:194–201.

Malcolm, S. B, and M. P. Zalucki. 1996. Milkweed latex and cardenolide induction may resolve the lethal plant defense paradox. *Entomologia Experimentalis et Applicata* 80:193–96.

McClay, A. S. 1990. The potential of *Larinus planus* (Coleoptera: Curculionidae), an accidentally-introduced insect in North America, for biological control of *Cirsium arvense*. Pages 173–79 *in* E. S. Delfosse, ed., *Proceedings, 7th International Symposium on the Biological Control of Weeds, 6–11 March 1988.* Instituto Sperimentale per la Patologia Vegetale, Rome.

Medawar, P. B. 1979. *Advice to a Young Scientist.* Harper and Row, New York.

Murdoch, W. W. 1966. Community structure, population control, and competition: A critique. *American Naturalist* 100:219–26.

Pedigo, L. P., S. H. Hutchins, and L. G. Higley. 1986. Economic injury levels in theory and practice. *Annual Review of Entomology* 31:341–68.

Peterson, R. K. D., S. D. Danielson, and L. G. Higley. 1992. Photosynthetic responses of alfalfa to actual and simulated alfalfa weevil (Coleoptera: Curculionidae) injury. *Environmental Entomology* 21:501–7.

Peterson, R. K. D., and L. G. Higley. 1993. Arthropod injury and plant gas exchange: Current understandings and approaches for synthesis. *Trends in Agricultural Science and Entomology* 1:93–100.

———. 1996. Temporal changes in soybean gas exchange following simulated insect defoliation. *Agronomy Journal* 88:550–54.

Peterson, R. K. D., L. G. Higley, F. J. Haile, and J. A. Barrigossi. 1998. Mexican bean beetle injury affects photosynthesis of *Glycine max* (L.) Merril and *Phaseolus vulgaris* L. *Environmental Entomology* 27:373–81.

Peterson, R. K. D., L. G. Higley, and S. M. Spomer. 1996. Cecropia moth, *Hyalophora cecropia* (L.) (Lepidoptera: Saturniidae), injury and photosynthetic responses of apple and crabapple. *Environmental Entomology* 25:416–22.

Platt, J. R. 1964. Strong inference. *Science* 146:347–53.

Polis, G. A., D. H. Wise, S. D. Hurd, F. Sanchez-Piñero, J. D. Wagner, C. T. Jackson, and J. D. Barnes. 1998. The interplay between natural history and field experiments. Pages 254–80 *in* W. J. Resetarits Jr. and J. Bernardo, eds., *Experimental Ecology: Issues and Perspectives.* Oxford University Press, New York.

Price, P. W. 1991. The plant vigor hypothesis and herbivore attack. *Oikos* 62:244–51.

Rand, T. A., and S. M. Louda. 2004. Exotic weed invasion increases the susceptibility of native plants to attack by a biocontrol herbivore. *Ecology* 85:1548–54.

———. 2006. Invasive insect abundance varies across the biogeographic distribution of a native host plant. *Ecological Applications* 16:877–90.

Rand, T. A., F. L. Russell, and S. M. Louda. 2004. Local vs. landscape scale indirect effects of an invasive weed on native plants. *Weed Technology* 82:1250–57.

Rose, K. E., S. M. Louda, and M. Rees. 2005. Demographic and evolutionary impacts of native and invasive insect herbivores: A case study with Platte thistle, *Cirsium canescens. Ecology* 86:453–65.

Russell, F. L., and S. M. Louda. 2004. Phenological synchrony affects interaction strength of an exotic weevil with Platte thistle, a native host plant. *Oecologia* 139:525–34.

———. 2005. Indirect effects mediate floral herbivory on a native thistle by an invasive exotic insect. *Oecologia* 146:373–84.

Russell, F. L., S. M. Louda, T. A. Rand, and K. Cassman. 2007. Variation in herbivore-mediated indirect effects of an invasive plant on a native plant. *Ecology* 88:413–23.

Slobodkin, L. B., F. E. Smith, and N. G. Hairston. 1967. Regulation in terrestrial ecosystems, and the implied balance of nature. *American Naturalist* 101:109–24.

Stoskopf, M. K. 2005. Observation and cogitation: How serendipity provides the building blocks of scientific discovery. *Natural Research Council, Institute of Laboratory Animal Resources Journal* 46:332–37.

Watson, J. 1998. *The Double Helix.* Scribner, New York.

Welter, S. C. 1989. Arthropod impact on plant gas exchange. Pages 135–50 *in* E. A. Bernays, ed., *Insect-Plant Interactions. Volume I.* CRC Press, Boca Raton, FL.

White, T. C. R. 1978. The importance of a relative shortage of food in animal ecology. *Oecologia* 33:71–86.

15 *To Know a Tropical Forest*

WHAT MECHANISMS MAINTAIN HIGH TREE DIVERSITY ON BARRO COLORADO ISLAND, PANAMA?

Stephen P. Hubbell

Figure 15.0. Stephen P. Hubbell in the Forest Dynamics Plot, Barro Colorado Island, Panama, with one of 200 seed traps in the background. Photograph by Christian Ziegler, 2004.

Abstract

We are now in the 29th year of a place-based study of the dynamics of a tropical forest on Barro Colorado Island (BCI), Panama. The objective of this long-term project is to understand the fundamental mechanisms that permit more than 300 tree and shrub species to coexist in the BCI forest and, by extension, in tropical forests in general. In this paper I summarize the current evidence regarding ten hypotheses to explain tree species diversity, of which seven are "niche-assembly" and three are "dispersal-assembly" theories. Is the BCI tree community primarily "dispersal-assembled" in the taxonomic nonequilibrium sense of island biogeography theory, or is it primarily "niche-assembled" and more nearly in compositional equilibrium, composed of habitat- or regeneration-niche specialists? By analyzing the patterns of distribution, abundance, and spatial population dynamics of species in the plot, our conclusion is that, despite evidence for some niche differentiation and stabilizing forces, primarily intraspecific density dependence, these mechanisms have not been strong enough to prevent considerable compositional change in relative tree species abundances in the BCI forest over the past quarter-century. The changes in abundances of individual species are inconsistent with stochastic fluctuations around fixed carrying capacities, and are more consistent with a combination of drift and directional change, possibly due to climate change. These conclusions, which are based on relatively short-term forest dynamics, must be regarded as preliminary. We should be prepared for many ecological surprises as the BCI Forest Dynamics Project moves into its second quarter-century, especially in a time of global change.

Introduction and Rationale

Price and Billick (chapter 1) note that all ecological studies in the field are fundamentally "place-based," but this is a book about special places where ecological research has been done in depth, for a long time, often by many people and from many perspectives. These are places where syntheses are more likely to develop from deep understanding of the natural history, structure, and dynamics of particular communities and ecosystems. Barro Colorado Island (BCI) is one such special place (figure 15.1). In a very direct sense, the neutral theory of biodiversity (Hubbell 2001) emerged from the BCI Forest Dynamics Project, and this paper briefly reviews the empirical findings that inspired its development.

In 1979 I approached Robin Foster, then at the University of Chicago, about the possibility of setting up a long-term study of tropical forest dynamics on Barro Colorado Island (BCI), Panama. I was fresh from mapping a plot of dry forest on a private ranch in Guanacaste, Costa Rica (Hubbell 1979), an effort that had had a sad outcome; my plot was cut and burned a few years after I began my study. I did not want this to happen ever again. BCI was an attractive place for such a project because, being a protected site administered by the Smithsonian Institution under treaty with Panama, it guaranteed the long-term security of study plots. It also had two additional special attractions: (1.) a comprehensive flora of BCI had just been published (Croat 1978) and the tree flora was well known by Foster, who had done his PhD work there; (2.) BCI attracted many tropical-forest scientists from around the world, so that the project could potentially find collaborators and benefit from the wealth of research and natural-history information.

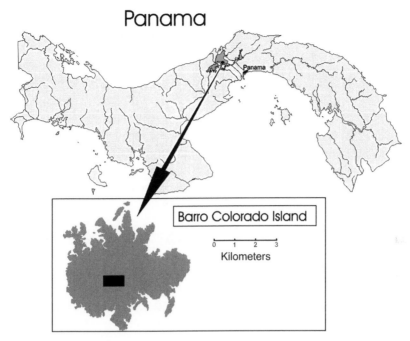

Figure 15.1. Map of Panama showing Barro Colorado Island (BCI), and the 50-ha Forest Dynamics Plot (black rectangle).

Fortunately, Foster agreed. The objective of our study would be to use BCI to test leading theories for the maintenance of high local (alpha) tree diversity in lowland tropical forests. Five considerations guided our project design: (1.) Because most of the theories of diversity involve community-level processes, we would study all the tree species in the community simultaneously. (2.) Because tree species have long lifespans, the dynamical signature of diversity-generating processes would play out slowly, and the study would be long-term, perhaps outliving us. (3.) Because most tropical trees occur at low density, we would need a very large plot to obtain adequate samples for species-level analyses of demography. (4.) Because many of the theories about diversity in sessile plant communities involve explicitly spatial mechanisms, we had to map individual trees, a pioneering effort in spatial ecology. (5.) Finally, because we expected most of the relevant diversifying mechanisms to play out in the trees' early life history, we had to identify and map not just adult trees but also saplings, and possibly trees at even younger stages. We decided to include all free-standing woody stems of more than one cm diameter at breast height (DBH). This size cutoff was a compromise between economics, logistical feasibility, and the desire to include the youngest possible life-history stages. Later, we added a network of seed traps and a grid of seedling plots to include the earliest life history stages. We also added measurements of environmental parameters like soil nutrients so that we could explore the species' niche relationships. For the "main census" of plants of more than one cm DBH, we decided to recensus the plot completely every five years, recording individual tree growth and mortality, and mapping new stems. We also decided to census seed traps every two weeks and seedling plots once a year, because seeds and seedlings turn over much faster than do saplings and trees. Nobody had ever before attempted such a large-scale community-level study of a tropical forest.

Foster and I hiked all over BCI to decide where to put the plot and how big it should be. To do a large plot and meet our other design objectives seemed daunting. So I asked Foster, "Do you think we can do twenty-five hectares?" Without missing a beat or cracking a smile he said, "No, we can do fifty!" So much for having a theoretically driven scientific rationale for choosing the plot size. We had no idea then, really, whether 50 ha would provide adequate species-level samples. In retrospect, they were adequate for only about two-thirds of the species. Indeed, we are still amazed by the large number of really rare BCI tree species; half of them (160 species) collectively constitute less than 1% of all individuals.

Although we did not remotely envision the possibility at the time, the BCI Forest Dynamics Project began an extraordinary international research collaboration on tropical forest dynamics that is still growing. The collaboration, known as the Center for Tropical Forest Science (CTFS), now spans 15 countries in the New and Old Word tropics, involving 22 large plots, most of them 50 ha in size (Condit 1998). Because we now have the luxury of comparative data in the pantropical network of plots, we have begun to test the generality of the BCI results. CTFS is now expanding to include forest plots outside the tropics, and because of its more comprehensive geographic coverage, we are now contemplating calling the network the Global Forest Observatory. Together the plots now contain populations of about one-seventh of all described tree species in the world.

The Theoretical Landscape: Hypotheses about Tropical Tree Diversity

One can identify at least 10 major hypotheses that ecologists have put forward to explain the high alpha-tree diversity in lowland tropical forests (table 15.1) (Wright 2002). All of them, even if inspired by tropical forest diversity, represent general mechanisms for the coexistence of competing species. Hence they are surely not unique to tropical forests, although the diversity of mechanisms and/or their relative strength may be different than in other ecological communities. Most of the hypotheses are not mutually exclusive, and some are variants of one another. The first seven are niche-assembly hypotheses that assume niche differences among species are essential to explaining their coexistence. The last three are dispersal-assembly hypotheses in the spirit of the theory of island biogeography (Hubbell 2001, 2008b) because they hypothesize that species composition of tropical tree communities is determined by processes of colonization and loss of species from points in space.

When the project began in 1980 there were several extant hypotheses, of which three were more prominent than the rest. (1.) The "enemies hypothesis," also known as the Janzen-Connell hypothesis (Janzen 1970, Connell 1971), was the most famous of these. It proposed that seed and seedling mortality from attack by host-specific predators and pathogens interacted with localized seed dispersal to cause higher per-capita mortality near than far from seed parents. If this dependence on density or distance is sufficiently strong, it can limit the maximum local density of a given tree population and allow more species to coexist locally. (2.) The "intermediate disturbance hypothesis" (Connell 1978) proposes that moderate rates of disturbance from, for example, tree-falls that open light gaps will maximize species diversity. In today's terminology, the intermediate-disturbance effect results from a life-history or competition/colonization tradeoff, and we might now include it under the life-history tradeoff hypothesis (see below). In the case of closed-canopy forest, light-demanding, colonizing species are excluded at low rates of disturbance because light gaps are relatively infrequent. At high rates of disturbance, colonizing species predominate because the good competitors—the shade-tolerant, mature-forest species that are poor dispersers—have not yet arrived. Therefore, the richness of tree species should be maximal at intermediate rates of disturbance when the full spectrum of competition/colonization life histories is represented. (3.) The "diversity-stability hypothesis" has had a longer history in ecology than either of the previous two. For decades the idea was that more diverse systems had greater stability (MacArthur 1955), and that diversity and stability reinforced each other in a positive feedback. Although this hypothesis has been challenged theoretically (e.g., see May 1973), more recent experimental evidence in some plant communities supports its plausibility (Tilman and Downing 1994, Loreau et al. 2002).

Since the 1980s, several new hypotheses have been added to the mix. Three of these emphasize various tradeoffs among species in life-history traits, or in traits that determine efficiency of resource use. (4.) The "life-history tradeoff hypothesis," mentioned above in the context of the intermediate disturbance hypothesis, suggests that unavoidable physiological tradeoffs in the ability of species to exploit alternative resources, or in life history traits such as competitive ability versus dispersal ability, prevent the evolution of a single dominant super-species (Tilman 1982, 1988, 1994). In the case of tropical tree species, early manifestations of this hypothesis concentrated on tradeoffs in ability to exploit light gaps (e.g., Ricklefs 1977; Denslow 1980, 1987; Orians 1982). (5.) The "nutrient ratio or R* hypothesis" proposes that tradeoffs in species' minimal requirements (R*) for different nutrients allow coexistence in environments that are heterogeneous in resource renewal rates (Tilman 1982). (6.) The "drought tolerance hypothesis" is another trade-off argument, proposing that plant species vary in their sensitivity to drought and the temporal distribution of water availability; sites that are hydrologically more complex will have more species (Silvertown et al. 1999).

Two of the new hypotheses consider effects of temporal environmental variability. (7.) The "storage hypothesis" proposes that asynchronous recruitment fluctuations in long-lived organisms result in a rare-species recruitment advantage, thus maintaining species in the community (Chesson and Warner 1981). The regeneration niche hypothesis of Grubb (1977) is basically the same idea: species specificity in environmental requirements for seed germination and seedling survival, coupled with temporal and spatial variation in environmental conditions, causes species to compete more strongly with conspecific than heterospecific individuals and to experience reduced competition overall when rare. The predicted phenomenology is more stable species abundances in more diverse communities, and evidence of stabilizing frequency dependence—that is, a rare-species advantage (Wills and Green 1995, Wills 1996).

The last three hypotheses place a much greater emphasis on dispersal assembly. (8.) The "source-sink hypothesis" proposes that populations of some species are not self-sustaining. These "sink" populations are maintained only by continual immigration from "source" populations elsewhere (Pulliam 1988). (9.) The "dispersal/recruitment limitation hypothesis" is that chronic failure of species to reach and establish in all microsites favorable for their growth and survival increases nonequilibrium diversity because many sites are won by default in the absence of better competitors (Tilman 1994, Hurtt and Pacala 1995, Hubbell 2006). (10.) The "symmetric neutrality hypothesis" assumes that species are identical demographically on a per capita basis—at least to a first approximation. This hypothesis is equivalent to asking: What are the properties of ecological communities if all species obey the stochastic average (Hubbell and Foster 1986a, Hubbell 2006)? Like island biogeography theory, dispersal-assembly hypotheses assume that communities are nonequilibrium assemblages of species that coexist only transiently through accidents of colonization and extinction; community stability arises solely from the inertia of large numbers, not the stabilizing effects of niche differentiation (Hubbell 2008, 2009).

How We Are Evaluating These Hypotheses

The first step in evaluating these hypotheses is to ask whether the mechanism is operating—that is, present at all. Accordingly, we have amassed information on spatial distributions, the fate of individual trees as a function of neighborhood species composition and environmental conditions, seed dispersal, recruitment limitation, and temporal changes in species abundance. These empirical data enable us to assess, for example, whether the density dependence or distance dependence in seedling success postulated by the Janzen-Connell hypothesis occurs, and how strong it is. However, because multiple diversity-maintaining processes can operate simultaneously and may vary in their contributions to diversity, we have gone beyond simply asking whether a process is happening at a given point in time; we also look for the signature of each mechanism in long-term population and community dynamics. This has required us to develop new theory. Symmetric neutral theory was the starting point, because it made the fewest initial assumptions. This theory has generated a host of intriguing, nonobvious, and often remarkably accurate predictions about the statistical and dynamic patterns of relative tree species abundance on BCI. Now, however, we are in a new cycle of theory development, adding in species differences and interactions that matter to community dynamics (Volkov et al. 2009).

The Current Evidence from BCI on These Hypotheses

The BCI Forest Dynamics Project is still far from complete, but the evidence accumulated thus far allows us to reach some provisional conclusions about the relative importance of alternative coexistence mechanisms. Table 15.1 summarizes the main conclusions and cites selected papers where they are discussed in greater detail.

The enemies hypothesis

After a quarter-century of studying Janzen-Connell density-dependent effects in the BCI plot, the empirical and theoretical conclusion is that these effects in fact do enhance diversity by increasing the intermingling of tree species growing close together (Chave et al. 2002, Hubbell and Lake 2003, Adler and Muller-Landau 2005). However, density dependence, although very strong and species-specific (e.g., Augspurger 1984, Gilbert et al. 1995, 2001), especially in seed germination (Harms et al. 2000), weakens to background indetectability within about 20 m of any given tree (Hubbell et al. 2001, Ahumada et al. 2004, Uriarte et al. 2004). Theoretical studies of limited-range (spatially local) density dependence show that it is incapable of

TABLE 15.1. Ten contemporary hypotheses about the maintenance of high alpha tree diversity in lowland tropical forests and the current status of the evidence from the BCI Forest Dynamics Project. Hypotheses are listed in the order in which they are discussed in the text.

Hypothesis	BCI results	Key BCI references
Niche assembly hypotheses		
Enemies (Janzen-Connell)	Higher per capita seed and seedling mortality, and reduced sampling growth and survival, within 10 to 20 m of large conspecific adults, but not detectable at larger distances; Janzen-Connell density dependence not strong enough to regulate BCI tree populations at the whole-plot scale.	Augspurger 1984; Harms et al. 2000; Hubbell et al. 2001; Ahumada et al .2004; Comita & Hubbell 2008
Intermediate disturbance	Negative result: tree diversity lower in disturbed areas than in less disturbed areas; large gaps have fewer species than small gaps on a per capita basis; probably due to strong dispersal limitation even in pioneer species, which are absent from many suitable gaps; findings still controversial.	Hubbell et al. 1999; Schnitzer & Carson 2000, 2001
Diversity-stability	No correlation on any special scale in the BCI plot between starting species richness or Fisher's *a* in 1982 and the stability in species composition over the interval 1982–2005, despite large changes in tree species abundances over the 23-year period.	Hubbell 2006
Life-history tradeoffs; habitat filtering	Well-established growth/survival tradeoff in high vs. low light; most BCI species shade tolerant. Evidence of habitat filtering, but no evidence of limiting similarity. No correlation between species richness and diversity of demographic niches.	Harms et al. 2001 Hubbell & Foster 1986b, 1992; Condit et al. 2006
Nutrient Ratio (R*) competition; resource niches	Almost all BCI tree species are nutrient niche generalists. Although not predicted by R* theory, a principal components analysis shows that a combination of nutrients can explain some of the variance in tree species richness in the BCI plot.	John et al. 2007; Hubbell 2008; Hubbell et al. ms.
Hydrology/drought tolerance niches	Some species differentiated in drought tolerance, partially predicting where they grow in relation to topography; however, no evidence of any limiting similarity or competitive exclusion.	Englebrecht et al. 2007
Storage hypothesis; fluctuating recruitment	Strong year-to-year fluctuations in seedling recruitment, but whether these fluctuations affect the abundance of adult trees is unclear due to long juvenile time spans and blurring of age cohorts.	Wright et al. 2005a Wills et al. 2007 Comita & Hubbell ms.
Dispersal-assembly hypotheses		
Source-sink; Allee effects	No hard data. Many rare tree species in the BCI plot may be sink populations.	Hubbell & Foster 1986b; this chapter

TABLE 15.1. (*continued*)

Hypothesis	BCI results	Key BCI references
Dispersal-assembly hypotheses		
Dispersal/ recruitment limitation	Strong and pervasive dispersal and recruitment limitation in all BCI tree species, even among colonizing pioneer tree species.	Hubbell et al. 1999; Dalling et al. 2002; Svening & Wright 2004; Jones et al. 2005
Symmetric neutrality	Neutral models yield a good approximation to the static and dynamic data on BCI relative tree species abundances. However, there is some evidence of non-neutral climate change effects on species composition.	Volkov et al. 2003, 2005; 2007, in press; Hubbell 2008, 2009

regulating populations on scales much larger than its range of effect (Hubbell 2009). This conclusion is supported by the observation that Janzen-Connell effects have not prevented large changes in abundance of tree species, both common and rare, in the BCI plot (table 15.1).

The intermediate disturbance hypothesis

We can test this hypothesis by examining whether natural variation in the disturbance regime causes the predicted changes in species richness. Disturbances to the BCI forest are small in scale and local, consisting of openings in the canopy created by the death of one to several canopy trees; BCI has no history of hurricanes. Since 1983, we have mapped canopy height and light gaps annually on a 5 × 5-m sample grid of 20,000 points (Hubbell and Foster 1986c). These light gaps drive forest regeneration and are sites where one finds the shade-intolerant pioneer species. The intermediate disturbance hypothesis predicts that species richness should be correlated with disturbance frequency. Surprisingly, however, we found no correlation between species richness and the frequency or magnitude of disturbance on a variety of spatial scales from 20 × 20-m to 10-ha quadrats in the BCI plot (Hubbell et al. 1999). Larger gaps contain more variation in the light environment, and if species partition light availability (Denslow 1980, Ricklefs 1977), then large gaps should have more species. However, larger gaps actually have fewer species than small gaps for a given number of stems, contrary to the intermediate disturbance hypothesis (Hubbell 1999). These results are more consistent with dispersal limitation (see below), because large gaps are collectively much rarer and constitute a smaller fraction of total area than small gaps, and so have a smaller seed-source area or "seedshed." Our conclusions

TABLE 15.2. Changes in the abundance of 285 BCI tree species in the 50-ha plot from 1982 to 2005. Percentage change was calculated as the absolute value of the difference in abundance of a species between 2005 and 1982 multiplied by 100 and then divided by the abundance of the species in 1982. The mean percentage of change ± 1 standard deviation over all species was 47.6% ± 67.0%.

% absolute change in abundance, 1982–2005	Number of species	Mean 1982 abundance	Standard deviation in 1982 abundance
0–25%	128	1379.6	4380.7
25–50%	66	568.0	1038.7
50–75%	36	306.4	622.8
75–100%	24	317.5	654.1
>100%	31	83.1	154.9

are controversial (Chazdon et al. 1999; Schnitzer and Carson 2000, 2001), in part because it is obvious that light-demanding species can only be present if there are light gaps. However, if most pioneer species—and tree species in general—are strongly dispersal-limited and fail to reach gaps or other suitable sites, then the disturbance regime may have little or no steady-state impact on local species richness. Sheil and Burslem (2003) also criticized our conclusion (Hubbell et al. 1999), noting that a hump-shaped curve is observed if one graphs the total number of tree species against the percentage of tree species that are pioneers. However, this argument is circular because as the percentage of pioneers becomes large, the percentage of nonpioneers and the total number of species must fall. A hump-shaped curve will be observed for any arbitrary subset of species plotted in this way.

The diversity-stability hypothesis

The BCI forest has exhibited considerable dynamism since the Forest Dynamics Project began (table 15.2), and some parts of the BCI plot have experienced greater change in species richness and composition than others. If a connection between diversity and stability is an important causal factor maintaining tree diversity in the BCI forest, then quadrats containing more species at the start of the study should show less change in species richness and relative species abundance over the quarter-century since then. However, there is no correlation between initial species richness and subsequent change in species richness on any spatial scale in the BCI plot. The same is true of changes in Fisher's α, a diversity index which is sensitive to relative species abundance (Hubbell 2006).

The life-history tradeoff hypothesis

There is a very clear axis of niche differentiation among BCI species in relation to light requirements (Hubbell and Foster 1992), and there is a very well-understood physiological basis for this tradeoff in terms of the leaf economics spectrum (Reich et al. 1997, McGill et al. 2006). There is one unusual BCI species that seems to have high shade tolerance, but which can grow very rapidly like a pioneer when exposed to high light. Perhaps this explains why it is an extremely abundant tree in the BCI forest (Dalling et al. 2001). However, most BCI species are shade-tolerant and do not exhibit such extreme plasticity. In fact, the large number of similarly shade-tolerant species creates a problem for the light-partitioning argument in explaining their coexistence. One would expect finer partitioning of a limiting resource when it is abundant, not when it is scarce. Moreover, there is no evident limiting similarity in shade tolerance. Finally, there seems to be nothing in this hypothesis to predict how many shade-tolerant or shade-intolerant species will coexist. A far simpler argument is that most BCI species have experienced shady environments over their evolutionary history, and as a result they have evolved shade tolerance, irrespective of the number of other species evolving similar traits. This niche-convergence argument runs counter to the standard niche-differentiation argument. However, I have shown theoretically that ecological equivalence can evolve easily under selective regimes that should be commonplace in nature (Hubbell 2006, 2008b). In these models, ecological equivalence leads to long-term nonequilibrium coexistence because niche convergence evolves much faster than competitive exclusion happens, but only when dispersal/recruitment limitation is strong and pervasive among species in the community (see *the dispersal/recruitment limitation hypothesis*, below).

The nutrient ratio or R* hypothesis

This is the iconic niche-assembly theory in plant community ecology. We have recently analyzed variation in soil nutrients across the BCI plot (John et al. 2007) and in relation to R* theory (Hubbell 2009; Hubbell et al., unpublished). Although 70% of BCI species show significant associations with one or more soil nutrients (John et al. 2007), many species show very similar associations. R* theory predicts that local species diversity should increase as the spatial variance in local nutrient ratios increases. However, we found no correlation between local species richness and spatial variance in the ratios of six major macronutrients: N/P, Ca/K, and Mg/Mn (Hubbell 2009; Hubbell et al., unpublished), on any spatial scale in the plot. R* theory also predicts a sequence of species replacements as one traverses nutrient-ratio gradients, but this was not observed either. Instead, most BCI species, including rare species, are nutrient generalists with almost completely overlapping nutrient niches, and there is no evidence of limiting similarity. Of the 187 species abundant enough to test, in 155 species the intersection of their niche breadths was less than 95% of the union of their niche breaths on these nutrient gradients, and in 139 species it was less than 99% (Hubbell 2009). Therefore, R* theory has very little predictive power with respect to species diversity in the BCI tree community.

The reason R* theory does not work well may be that BCI trees violate several key assumptions of the theory. First, the theory assumes independence in the uptake kinetics of each essential nutrient. We now know this assumption is false. For example, calcium facilitates the uptake of many macro- and micronutrients (Epstein and Bloom 2005). Second, R* theory assumes that there is a fixed R* value for each species-nutrient combination, but this also is not true. Plants up-regulate transport enzymes with different substrate affinities and maximum uptake rates depending on the ambient concentrations of nutrients in the soil. Plants also change their root-hair mass and their mycorrhizal associates as a function of soil nutrient concentrations. The result of this plasticity is that plants can maintain relatively constant tissue-nutrient stoichiometry in the face of a wide range of environmental nutrient concentrations. This plasticity provides a simple explanation for why so many BCI tree species are nutrient generalists. Over their evolutionary history, many species have been chronically exposed to locally heterogeneous nutrient environments and have evolved the flexibility to cope with them. This is an example of convergent niche evolution, which I argue is commonplace (Hubbell 2006).

The drought tolerance hypothesis

This hypothesis is relatively new, and research to test it is continuing. There is a very strong rainfall and seasonality gradient across the Isthmus of Panama, with 4000 mm annual rainfall and a two-month dry season on the Caribbean side, to 1200 mm annual rainfall and a five-month dry season on the Pacific side. Englebrecht et al. (2007) demonstrate that the relative density of a number of tree species across the isthmus can be explained in large part by the relative tolerance of their seedlings to drought stress. Drought tolerance also explains some of the species distribution patterns in the BCI plot. The hydrology of the plot is correlated with topography: the slopes are more mesic than the central plateau, which dries out sooner during the four-month dry season. Drought-sensitive species are more often associated with the slopes, and more drought-tolerant species are found on the plateau. However, the hypothesis does not explain how many drought-tolerant or drought-sensitive species to expect in each habitat, and there seems to be no limiting similarity in drought tolerance. Moreover, hydrology seems to be acting simply as a mass environmental filter sensu Weiher and Keddy (2001). The drought-resistant species grow on the plateau but also occur on the slopes, growing right alongside the drought-sensitive species.

The storage hypothesis

We have observed enormous interannual variation in seed rain and seedling germination for the 30 species most abundant in our 20,000 one-m² seedling plots (Hubbell and Comita, unpublished data). This variation is of the sort that Chesson and Warner (1981) said would generate a recruitment advantage for rare species (Wright et al. 2005b). This would seem to be a strong candidate for maintaining BCI tree diversity, but we have reservations about this conclusion based on uncertainties about whether the seedling recruitment fluctuations are sufficiently strong and persistent to produce a distinct rare-species advantage in adult tree populations. This is problematic because the distinctness of year-class cohorts of seedlings is rapidly lost due to increasing within-cohort heterogeneity in seedling growth and survival rates over time. Year classes of seedlings and saplings blur together into indistinguishable size classes in less than five years, but most BCI tree species do not reach reproductive size for many decades. Moreover, the size at reproduction is probabilistic, and it occurs at smaller sizes if trees are in more favorable sites (Wright et al. 2005a). Wills et al. (2007) report evidence of local frequency dependence in survival. At larger spatial scales across the entire tree community, however, there is no evidence that growth rates of individual trees are enhanced by local species richness (figure 15.2). More

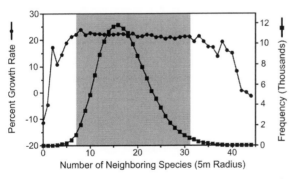

Figure 15.2. Influence of neighborhood species richness (<5 m from focal plant) on mean percent stem growth in DBH (circles) over all species and size classes over a decade (1985–95) in the BCI forest. Squares are the sample size of number of plants of a given neighborhood species richness.

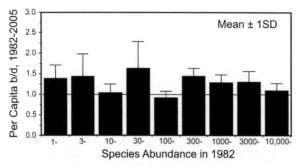

Figure 15.3. Lack of evidence of frequency dependence at the whole-plot level in the BCI tree community over the entire 23 years of the study. Bars are mean per-capita birth rate/death rate (b/d) ratio ± one standard deviation for species binned into half-log base-10 intervals of abundance. In all abundance categories, the confidence limits for b/d are not significantly different from unity; that is, rare species did not enjoy a birth rate advantage over common species for the past quarter-century.

critically, there is no evidence thus far at the whole-plot level of any frequency dependence or rare-species advantage (figure 15.3).

The source-sink hypothesis

How many of the species in the BCI plot are sink populations maintained by immigration from population centers elsewhere? In the absence of manipulative experiments, it may be very difficult to answer this question definitively. Some species that have gone extinct in the plot are weedy species characteristic of pasture edges and other human disturbance. Populations of these species in the plot were probably maintained by immigration from pastures on the eastern end of BCI that have since returned to forest (Hubbell and Foster 1986b). However, many other rare, non-weedy species in the plot are also reproducing poorly, possibly because of Allee effects. The fact that relatively fewer dioecious than hermaphroditic species are rare (Hubbell, unpublished data) provides circumstantial evidence to support this hypothesis, because hermaphroditic species can self-fertilize when no mates are available, whereas dioecious species cannot do so unless they are facultative apomicts.

The dispersal/recruitment limitation hypothesis

The seed rain/seedling germination study, which has now been running continuously for 21 years with biweekly censuses of seed traps, has demonstrated that BCI tree species are strongly dispersal- and recruitment-limited

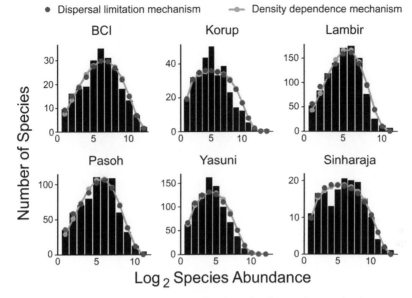

Figure 15.4. Preston-style relative tree species abundance distributions for six 50-ha plots across the new- and old-world tropics, showing the equally good fit of the two current versions of neutral theory: dispersal limitation (circles) and symmetric density dependence (line). The bars are the observed number of species in each log base-two abundance category. The sites are Barro Colorado Island, Panama; Korup National Park, Cameroon, West Africa; Lambir Hills National Park, Sarawak, Malaysian Borneo; Pasoh Research Forest, peninsular Malaysia; Yasuni National Park, Amazonian Ecuador; and Sinharaja Forest Reserve, Sri Lanka. After Volkov et al. (2005).

(Hubbell et al. 1999, Dalling et al. 2002, Muller-Landau et al. 2005, Jones et al. 2005, Hardesty et al. 2006). The most important implications of chronic dispersal and recruitment limitation are that these processes provide a powerful mechanism for long-term nonequilibrium coexistence of tropical trees (Hurtt and Pacala 1995) and set the selective stage for convergent niche adaptation to the most commonly experienced habitats (Hubbell 2006).

Dispersal limitation is pervasive in the BCI tree community, even among the pioneer tree species that are good dispersers (Dalling et al. 2002, Jones et al. 2005). Pioneer species are the only guild of BCI tree species with a soil seed bank. However, when one germinates seeds from soil in recently created light gaps, one finds that seeds of many pioneer species are absent from sites that would be suitable for them (Dalling et al. 1998). Of the more than 200 species whose seeds were collected in traps during the first decade of the study, only 12 had at least one seed in more than 50% of the 200 traps, whereas half of the species deposited seeds in five or fewer traps in 10 years

(Hubbell et al. 1999). Experimental studies with seedlings transplanted to sites where they did not currently occur exhibited survival rates matching or exceeding those planted in sites where they did occur, for all species tested (Svenning et al. 2006).

The symmetric neutrality hypothesis

Data on static (snapshot) species-abundance distributions from BCI and five other tropical forests are remarkably well fit by both of the two current versions of neutral theory (figure 15.4). The dispersal-limitation version is the original generalization of island biogeography theory (Hubbell 2001, Volkov et al. 2003). The second version incorporates symmetric density dependence which, like the storage hypothesis, confers a per-capita fecundity advantage on rare species relative to common species (Volkov et al. 2005). Symmetric density dependence means that every species experiences the same density dependence when at the same abundance. However, a more challenging test of this hypothesis is whether it describes the changes in BCI tree species abundance over time.

Confronting Both Versions of Neutral Theory with Dynamic Data

Can we use the BCI dynamic data to distinguish between the alternative versions of neutral theory? We have considerable power to test these theories because the BCI forest has been remarkable dynamic over the past quarter-century. One-third (91 species, or 31.9%) of all 285 BCI species with two or more individuals in 1982 have changed by more than 50%—and over half (157 species, 55.1%) by more than 10%—in total abundance in just 23 years (table 15.2).

These large changes in abundance are not restricted to rare species, but have also occurred in common species. The two versions of neutral theory yield very different predictions for how the BCI tree community will change. Under the dispersal-limitation version, there is no taxonomic stability, and the community is expected to slowly drift in composition. Under the symmetric density-dependent version, however, there is some taxonomic stability due to rare-species advantage—a frequency dependence that tends to maintain a given taxonomic composition. Thus, in a sense, symmetric density-dependence can be a stand-in for all niche-assembly theories, the only difference being that we assume species are stochastically fluctuating with a Gaussian central tendency around their niche-determined carrying capacities. Note that the storage hypothesis is also a symmetric neutral model by this definition.

To make this test quantitative, let us characterize the loss of similarity in forest composition over time by regressing the log of abundances of species at time $t + \tau$ on their log abundances at previous time t, where τ is a time lag of some number of census intervals—for example, five years. We can then follow the decay in similarity over time by plotting the decrease in the coefficient of determination, R^2, of this relationship (figures 15.5a and 15.5b). What pattern of decay in the R^2 of species abundances do we expect? Under the hypothesis of dispersal assembly and symmetric neutrality, given the observed mortality rate we expect an almost perfectly linear decay in R^2 for at least 300 years (Azaele et al. 2006; figure 15.5c). Under the niche-assembly hypothesis, however, each species' population is stochastically varying around a carrying capacity that is determined by its realized niche under competition from all the other niche-differentiated species in the equilibrium community. In this case the expected decay is curvilinear, decreasing in slope with time to a minimum plateau value at long time lags—a plateau that reflects the underlying stability of community composition (figure 15.5c; Hubbell 2008c). To obtain the expected R^2 decay curve under demographic stochasticity around carrying capacities, I randomly sampled the observed 25-year distribution of intrinsic rates of increase (r) of the BCI tree species. These observed r values are almost perfectly normally distributed with a mean of zero. I then performed 100 stochastic simulations of the community using randomly sampled r values, computed population sizes for each species at successive census times, and calculated the community R^2 values for different time lags.

The results were dramatic and unambiguous: the dispersal-limitation version of neutral theory fits the dynamic BCI data much better that the density-dependent version, which is the symmetric surrogate for niche-assembly theories (figure 15.5d). The observed R^2 decay in community similarity of the BCI tree community is almost perfectly linear. Even more remarkably, the line in figure 15.5d is not a regression line. It is the decay curve predicted in advance by the dispersal-limitation neutral theory from parameters estimated not from the dynamic data, but only from the static relative abundance data available in 1982 at the beginning of the study.

These results do not "prove" the symmetric neutral hypothesis. Indeed, there is abundant evidence that the assumption of per-capita demographic equivalence of BCI tree species is false (Condit et al. 1995). What these results do say, however, is that the patterns of both relative species abundance and community dynamics in the BCI plot can be modeled quite well by the average behavior of species in the BCI forest. In so doing, neutral theory provides a benchmark against which we can test which species deviate from average species demography, and by how much.

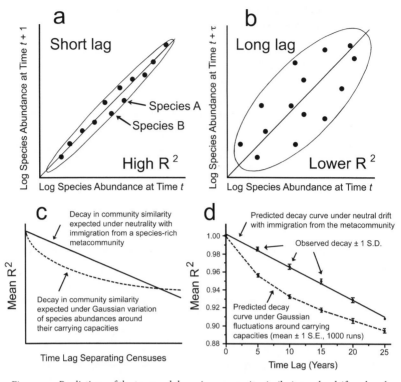

Figure 15.5. Predictions of the temporal decay in community similarity under drift and under stabilizing density dependence. (a, b: The decay in community similarity is measured by the decay in R^2 of the regression of the abundances of species in the community at time $t + \tau$ on themselves at time t as a function of time lag τ. (c) A straight-line decay curve is expected with neutral drift under immigration from a species-rich metacommunity. A —curvilinear decay curve is expected under stochastic Gaussian fluctuations of each species in an equilibrium community around fixed carrying capacities. (d) Observed autocorrelative decay is given by the dots with error bars ± 1 S. D. The straight line curve is the prediction from the dispersal-limitation version of neutral theory with parameters estimated from the 1982 static relative abundance data. The curved line below is the output from the density dependence model of stochastic fluctuations around carrying capacities parameterized from the observed normal distribution of intrinsic rates of increase of BCI tree species over the interval from 1982 through 2005 (after Hubbell 2008b).

Summary

My conclusions have been shaped by the following observations and theory. (1.) Density dependence, although strong, is very local and does not regulate BCI tree species on the scale of a 50-ha plot. (2.) Many tree species appear to have very broad, overlapping niches with little to no evidence of limiting similarity for key resources such as light, water, and nutrients. (3.) Dispersal and recruitment limitation are chronic, affecting all BCI species, and probably have been so over evolutionary time spans. Strong and chronic dispersal and recruitment limitation potentially explains the coexistence problem for large numbers of ecologically equivalent or nearly equivalent species. (4.) In hyper-diverse communities, selection for niche character displacement from pairwise species interactions is very weak and nondirectional in the niche hypervolume. This is because species are rare and rarely encounter one another, and also because biotic neighborhoods are almost never the same from one tree to the next of the same species (Hubbell 2006). (5.) The number of species in guilds of ecologically equivalent species is therefore not determined by pairwise competitive niche differentiation within guilds, but by the biogeographic extent and the persistence over evolutionary time of shared selective regimes. (6.) And, last but not least, neutral theory as an approximation is supported by the remarkably good fits of its predictions to observed static and dynamic patterns of tree species abundance and diversity in tropical forests, including BCI.

The Importance of Place-Based Science

I began this paper by asserting the importance of place-based science for developing a deep understanding of the natural history of ecological communities and ecosystems, and for confronting general theories for how they work. Understanding the BCI forest in the depth as we now do has taken many scientists and assistants working together for quite a long time to gather enormous quantities of data on plant demography and distribution relative to environmental conditions, and on dynamics of community composition. Long-term, place-based research is not easy to do. It required dogged commitment, the continued support of granting agencies both public and private, and institutional encouragement and support—in my case, from the Smithsonian Tropical Research Institute.

My own thinking about complex ecological processes has changed profoundly as a result of the discoveries made on the BCI Forest Dynamics Project (Hubbell 2008). The lack of consistent spatial associations of particular species gave me my first twinges of doubt that niche divergence played a necessary role in coexistence. My skepticism of naïve and facile acceptance of niche-assembly paradigms, at least as applied to hyper-diverse tropical tree communities, has grown with evidence that tropical trees on BCI have very broad and overlapping niches. These doubts stimulated my first attempts to model dispersal-assembly processes, and further developments of neutral theory have gone hand-in-hand with the collection of data on plant demography and community dynamics that we could use to test it.

It is sometimes said that everyone is a prisoner of their experience and the systems they study. I like to think instead that field experience is liberating. It forces one to come to grips with the match between abstract hypotheses and the reality on the ground. This leads to creative modification of theory, and development of new theory, to explore the implications of processes seen in the field for additional attributes of the system. Had I studied species-poor communities at high latitudes, I might have become so impressed by niche differentiation and the strength of pairwise species interactions that I might never have seriously considered the remarkably rich potential consequences of chance and history for tropical forest diversity.

Acknowledgments

I thank Mary Price, Ian Billick, and Patricia Adair Gowaty for reading and commenting on drafts of the manuscript. I thank the National Science Foundation, the John D. and Catherine T. MacArthur Foundation, the A. K. Mellon Foundation, the Pew Charitable Trusts, the Guggenheim Foundation, the Celara Foundation, the Smithsonian Tropical Research Institute, and the more than 100 collaborators and field assistants who have supported or worked on the BCI Forest Dynamics project. I particularly thank Robin Foster, Rick Condit, Joe Wright, Rolando Perez, and Salomon Aguilar for their long-term collaboration on the BCI project. I thank Ira Rubinoff and Peter Ashton for major efforts in launching and financing the Center for Tropical Forest Science, and Elizabeth Losos and Stuart Davies for their direction of CTFS.

References

Adler, F. R, and H. C. Muller-Landau. 2005. When do localized natural enemies increase species richness? *Ecology Letters* 8:438–47.

Ahumada, J. A., S. P. Hubbell, R. Condit, and R. B. Foster. 2004. Long-term tree survival in a neotropical forest: The influence of local biotic neighborhood. Pages 408–32 *in* E. Losos, and E. G. Leigh, Jr., eds., *Forest Diversity and Dynamism: Findings from a Network of Large-Scale Tropical Forest Plots.* University of Chicago Press, Chicago.

Augspurger, C. K. 1984. Pathogen mortality of tropical tree seedlings: Experimental studies of the effects of dispersal distance, seedling density and light conditions. *Oecologia* 61:211–17.

Azaele, S., S. Pigolotti, J. R. Banavar, and A. Maritan. 2006. Dynamical evolution of ecosystems. *Nature* 144:926–28.

Chave, J., H. C. Muller-Landau, and S. A. Levin. 2002. Comparing classical community models: Theoretical consequences for patterns of diversity. *American Naturalist* 159:1–23.

Chave, J., R. Condit, H. C. Muller-Landau, S. C. Thomas, P. S. Ashton, S. Bunyavejchemin, L. L. Co, H. S. Dattaraja, S. J. Davies, S. Esufali, C. E. N. Evango, K. J. Feeley, R. B. Foster, N. Gunatilleke, S. Gunatilleke, P. Hall, T. B. Hart, C. Hernandez, S. P. Hubbell, A. Itoh,

S. Kiratiprayoon, J. V. LaFrankie, S. L. de Lao, J.-R. Makana, Md. N. Supardi Noor, A. R. Kassim, C. Samper, R. Sukumar, H. S. Suresh, S. Tan, J. Thompson, M. D. C. Tongco, R. Valencia, M. Vallejo, G. Villa, T. Yamakura, J. K. Zimmerman, and E. C. Losos. 2008. Assessing evidence for a pervasive alteration in tropical tree communities. *PLoS Biology* 6(3): 1–8.

Chazdon, R., R. K. Colwell, and J. S. Denslow. 1999. Tropical tree richness and resource-based niches. *Science* 285:1459a–60a.

Chesson, P. L., and R. R. Warner. 1981. Environmental variability promotes coexistence in lottery competitive systems. *American Naturalist* 117:923–43.

Comita, L., S. Aguilar, S. Lao, R. Pérez, S. Lao, and S. P. Hubbell. 2007. Patterns of woody plant species abundance and diversity in the seedling layer of a tropical forest. *Journal of Vegetation Science* 18:161–72.

Condit, R. 1998. *Tropical Forest Census Plots.* Springer, Berlin.

Condit, R., P. S. Ashton, P. Baker, S. Bunyavejohewin, S. Gunatileke, N. Gunatilleke, S. P. Hubbell, R. B. Foster, A. Itoh, J. V. LaFrankie, H. S. Lee, E. Losos, N, Manokaran, R. Sukumar, and T. Yamakura. 2000. Spatial patterns in the distribution of tropical tree species. *Science* 288:1414–18.

Condit, R., P. S. Ashton, S. Bunyavejchewin, H. S. Dattaraja, S. Davies, S. Esufali, C. Ewango, R. Foster, I. A. U. N. Gunatilleke, C. V. S. Gunatilleke, P. Hall, K. E. Harms, T. Hart, C. Hernandez, S. P. Hubbell, A. Itoh, S. Kiratiprayoon, J. LaFrankie, S. Loo de Lao, J.-R. Makana, Md. Nur Supardi Noor, A. R. Kassim, S. Russo, R. Sukumar, C. Samper, H. S. Suresh, S. Tan, S. Thomas, R. Valencia, M. Vallejo, G. Villa, and T. Zillio. 2006. The importance of demographic niches to tree diversity. *Science* 313:98–101.

Condit, R., S. P. Hubbell, and R. B. Foster. 1995. Mortality rates of 205 neotropical tree and shrub species and the impact of a severe drought. *Ecological Monographs* 65:419–39.

Condit, R., N. Pitman, E. G. Leigh, Jr., J. Chave, J. Terborgh, R. B. Foster, P. Nuñez V., S. Aguilar, R. Valencia, G. Villa, H. C. Muller-Landau, E. Losos, and S. P. Hubbell. 2002. Beta diversity in tropical forest trees. *Science* 288:1414–18.

Connell, J. H. 1971. On the role of natural enemies in preventing competitive exclusion in some marine animals and in rain forest trees. Pages 298–312 *in* P. J. den Boer, and G. R. Gradwell, eds., *Dynamics of Populations.* Centre for Agricultural Publishing and Documentation, Wageningen, the Netherlands.

———.1978. Diversity in tropical rain forests and coral reefs. *Science* 199:1302–10.

Croat, T. 1978. *The Flora of Barro Colorado Island.* Stanford University Press, Palo Alto, CA.

Dalling, J. W., S. P. Hubbell, and K. Silvera. 1998. Seed dispersal, seedling emergence and gap partitioning in gap-dependent tropical tree species. *Journal of Ecology* 86:674–89.

Dalling, J. W., H. C. Muller-Landau, H. C., S. J. Wright, and S. P. Hubbell. 2002. Role of dispersal in recruitment limitation in six pioneer species. *Journal of Ecology* 90:714–27.

Dalling, J. W., K. Winter, J. D. Nason, S. P. Hubbell, D. Murawski, and J. L. Hamrick. 2001. The unusual life history of *Alseis blackiana:* A shade-persistent pioneer tree? *Ecology* 82:933–45.

Denslow, J. S. 1980. Gap partitioning among tropical rainforest trees. *Biotropica* 12 (suppl.):47–55.

———. 1987. Tropical rainforest gaps and tree species diversity. *Annual Review of Ecology and Systematics* 18:431–51.

Engelbrecht, B. M. J., L. Comita, R. Condit, T. A. Kursar, M. T. Tyree, and S. P. Hubbell. 2007. Drought sensitivity shapes species distribution patterns in tropical forests. *Nature* 447:80–82.

Epstein, E., and A. J. Bloom. 2005. *Mineral Nutrition of Plants.* Sinauer Associates, Sunderland.

Gilbert G. S., S. P. Hubbell, and R. B. Foster. 1995. Density and distance-to-adult effects of a canker disease of trees in a moist tropical forest. *Oecologia* 98:100–108.

Gilbert, G. S., K. E. Harms, D. N. Hamill, and S. P. Hubbell. 2001. Effects of seedling size, El Niño drought, seedling density and distances to nearest conspecific adult on 6-year survival of *Ocotea whitei* seedlings in Panama. *Oecologia* 127: 509–16.

Grubb, P. J. 1977. The maintenance of species-richness in plant communities: The importance of the regeneration niche in plant ecology. *Biological Review* 52:107–45.

Hardesty, B. D., S. P. Hubbell, and E. Bermingham 2006. Genetic evidence of frequent long-distance recruitment in a vertebrate-dispersed tree. *Ecology Letters* 9:516–25.

Harms, K. E., R. Condit, S. P. Hubbell, and R. B. Foster. 2001. Habitat associations of trees and shrubs in a 50-ha neotropical forest plot. *Journal of Ecology* 89:947–59.

Harms, K. E., J. S. Wright, O. Calderón, A. Hernández, and E. A. Herre. 2000. Pervasive density-dependent recruitment enhances seedling diversity in tropical forests. *Nature* 404:493–95.

Hubbell, S. P. 1979. Tree dispersion, abundance and diversity in a tropical dry forest. *Science* 203:1299–1309.

———. 1999. Tropical tree species richness and resource-based niches. *Science* 285:1459–61.

———. 2001. *The Unified Neutral Theory of Biodiversity and Biogeography*. Princeton University Press, Princeton, NJ.

———. 2006. Neutral theory in ecology and the evolution of ecological equivalence. *Ecology* 87:1387–98.

———. 2008. Approaching tropical forest complexity, and ecological complexity in general, from the perspective of symmetric neutral theory. Pages 143–59 *in* W. Carson and S. Schnitzer, eds., *Tropical Forest Community Ecology*, Wiley-Blackwell, New York.

———. 2009. Neutral theory and the theory of island biogeography. *In* J. Losos and R. E. Ricklefs, eds., *The Theory of Island Biogeography at 40: Impacts and Prospects*. Princeton Universiity Press, Princeton, NJ.

Hubbell, S. P., J. A. Ahumada, R. Condit, and R. B. Foster. 2001. Local neighbourhood effects on long-term survival of individual trees in a neotropical forest. *Ecological Research* 16:859–75.

Hubbell, S. P., and R. B. Foster. 1986a. Biology, chance and history and the structure of tropical rain forest tree communities. Pages 314–29 *in* J. M. Diamond and T. J. Case, eds., *Community Ecology*. Harper and Row. New York.

———. 1986b. Commonness and rarity in a neotropical forest: Implications for tropical tree conservation. Pages 205–31 *in* M. Soulé, ed., *Conservation Biology: Science of Scarcity and Diversity*. Sinauer Associates, Sunderland.

———. 1986c. Canopy gaps and the dynamics of a neotropical forest. Pages 77–95 *in* M. Crawley, ed., *Plant Ecology*. Blackwell, Oxford, UK.

———. 1992. Short-term dynamics of a neotropical forest: Why ecological research matters to tropical conservation and management. *Oikos* 63: 48–61.

Hubbell, S. P., R. B. Foster, S. O'Brien, B. Wechsler, R. Condit, K. Harms, S. J. Wright, and S. Loo de Lau. 1999. Light gaps, recruitment limitation and tree diversity in a neotropical forest. *Science* 283:554–57.

Hubbell, S. P., and J. Lake. 2003. The neutral theory of biogeography and biodiversity, and beyond. Pages 45–63 *in* T. Blackburn, and K. Gaston, ed., *Macroecology: Concepts and Consequences*. Blackwell, Oxford, UK.

Hurtt, G. C., and S. W. Pacala. 1995. The consequences of recruitment limitation: Reconciling chance, history, and competitive differences between plants. *Journal of Theoretical Biology* 176:1–12.

Janzen, D. 1970. Herbivores and the number of tree species in tropical forests. *American Naturalist* 104:501–28.

John, R., J. Dalling, K. E. Harms, J. B. Yavit, R. F. Stallard, M. Mirabello, S. P. Hubbell, R. Valencia, H. Navarrete, M. Vallejo, and R. B. Foster. 2007. Soil nutrients influence spatial distributions of tropical tree species. *Proceedings of the National Academy of Sciences (USA)* 104:864–69.

Jones, F. A., J. Chen, G.-J .Weng, and S. P. Hubbell. 2005. A genetic evaluation of long distance and directed dispersal in the neotropical tree, *Jacaranda copaia* (Bignoniaceae). *American Naturalist* 166:543–55.

La Frankie, J. V., P. S. Ashton, G. B. Chuyong, L. Co, R. Condit, S. J. Davies, R. Foster, S. P. Hubbell, D. Kenfack, D. Lagunzad, E. C. Losos, N. S. Nor, S. Tan, D. W. Thomas, R. Valencia, and G. Villa. 2006. Contrasting structure and composition of the understory in species-rich tropical rain forests. *Ecology* 87:2298–2305.

Lewontin, R. 1974. *The Genetic Basis of Evolutionary Change*. Columbia University Press, New York.

Loreau, M., S. Naeem, and P. Inchausti, eds. 2002. *Biodiversity and Ecosystem Functioning: Synthesis and Perspectives*. Oxford University Press, Oxford.

MacArthur, R. H. 1955. Fluctuations of animal populations and a measure of community stability. *Ecology* 36:533–36.

May, R. M. 1973. *Stability and Complexity in Model Ecosystems*. Princeton University Press, Princeton, NJ.

McGill, B. J., B. Enquist, E. Weiher, and M. Westoby. 2006. Rebuilding community ecology from functional traits. *Trends in Ecology and Evolution* 21:178–85.

Metz, M. R., L. S. Comita, Y.-Y. Chen, N. Nordent, R. Condit, S. P. Hubbell, I.-F. Sun, N. S. bin Md. Noor, and S. J. Wright. 2008. Temporal and spatial variability in seedling dynamics: A cross-site comparison in four lowland tropical forests. *Journal of Tropical Ecology* 24:9–18.

Muller-Landau, H. C., R. S. Condit, J. Chave, S. C. Thomans, S. A. Boman, S. Bunyavejchewin, S. Davies, R. Foster, S. Gunatilleke, N. Gunatilleke, K. E. Harms, T. Hart, S. P. Hubbell, A. Itoh, A. R. Kassim, J. V. LaFrankie, H. S. Lee, E. Losos, J.-R. Makama, T. Ojubo, R. Sukumar, I.-F. Sun, N. Supardi N. N., S. Tan, J. Thompson, R. Valencia, G. V. Muñoz, C. Wills, T. Yamakura, G. Chuyong, H. S. Dattaraja, S. Esufali, P. Hall, C. Hernandez, D. Kenfack, S. Kiratiprayoon, J. S. Suresh, D. Thomas, M. I. Vallego, and P. Ashton. 2006. Testing metabolic ecology theory for allometric scaling of tree size, growth and mortality in tropical forests. *Ecology Letters* 9:575–88.

Muller-Landau, H. C., J. W. Dalling, K. E. Harms, S. J. Wright, R. Condit, S. P. Hubbell, and R. B. Foster. 2004. Janzen-Connell effects upon seed and seedling mortality: Disentangling the effects of dispersal patterns, habitat preferences, and density dependence. Pages 340–62 *in* E. Losos, E. G. Leigh, Jr., and R. Condit, eds., *Forest Diversity and Dynamism: Findings from a Network of Large-Scale Tropical Forest Plots*. University of Chicago Press, Chicago.

Orians, G. H. 1982. The influence of tree-falls in tropical forests in tree species richness. *Tropical Ecology* 23:255–79.

Pulliam, H. R. 1988. Sources, sinks, and population regulation. *The American Naturalist* 132:652–61.

Reich, P. B., M. B. Walters, and D. S. Ellsworth. 1997. From tropics to tundra: Global convergence in plant functioning. *Proceedings of the National Academy of Sciences (USA)* 94:3730–34.

Ricklefs, R. E. 1977. Environmental heterogeneity and plant species diversity: A hypothesis. *The American Naturalist* 111:376–81.

Schnitzer, S. A., and W. P. Carson. 2000. Treefall gaps and the maintenance of species diversity in a tropical forest. *Ecology* 82:913–19.

———. 2001. Have we missed the forest because of the trees? *Trends in Ecology and Evolution* 15:367–77.

Sheil, D., and D. F. R. P. Burslem. 2003. Disturbing hypotheses in tropical forests. *Trends in Ecology and Evolution* 18:18–26.

Silvertown, J. M., E. Dodd, J. J. G. Gowing, and J. O. Mountford. 1999. Hydrologically defined niches reveal a basis for species richness in plant communities. *Nature* 400:61–63.

Svenning, J. C., B. M. J. Englebrecht, D. A. Kinner, T. A. Kursar, R. F. Stallard, and S. J. Wright. 2006. The relative roles of environment, history and local dispersal in controlling the distributions of common tree and shrub species in a tropical forest landscape, Panama. *Journal of Tropical Ecology* 22:575–86.

Tilman, D. 1982. *Resource Competition and Community Structure*. Princeton University Press, Princeton, NJ.

———. 1988. *Plant Strategies and the Dynamics and Structure of Plant Communities*. Princeton University Press, Princeton, NJ.

———. 1994. Competition and biodiversity in spatially structured habitats. *Ecology* 75: 2–16.

Tilman, D., and J. A. Downing. 1994. Biodiversity and stability in grasslands. *Nature* 367:363–65.

Uriarte, M., S. P. Hubbell, R. John, R. Condit, and C. D. Canham. 2004. Neighbourhood effects on sapling growth and survival in a neotropical forest and the ecological equivalence hypothesis. Pages 89–106 *in* D. Burslem, ed. *Biological Interactions in Tropical Forests*, BES Symposium, Blackwell Scientific, Oxford, UK.

Volkov, I., J. R. Banavar, S. P. Hubbell, and A. Maritan. 2003. Neutral theory and the relative abundance of species in ecology. *Nature* 424:1035–37.

Volkov, I., J. R. Banavar, F.-L. He, S. P. Hubbell, and A. Maritan. 2005. Density dependence explains tree species abundance and diversity in tropical forests. *Nature* 438:658–61.

Volkov, I., J. R. Banavar, S. P. Hubbell, and A. Maritan. 2007. Patterns of relative species abundance in rain forests and coral reefs. *Nature* 450:45–49.

———. 2009. Inferring species interactions in tropical forests. *Proceedings of the National Academy of Sciences* 106:13854–59.

Weiher, E., and P. Keddy. 2001. *Ecological Assembly Rules: Perspectives, Advances, Retreats*. Cambridge University Press, Cambridge.

Wills, C. 1996. Safety in diversity. *New Scientist* 149:38.

Wills, C., and D. R. Green. 1995. A genetic herd-immunity model for the maintenance of MHC polymorphism. *Immunological Reviews* 143:263–92.

Wills, C., K. E. Harms, R. Condit, D. King, J. Thompson, F. L. He, H. Muller-Landau, P. Ashton, E. Losos, L. Comita, S. Hubbell, J. LaFrankie, S. Bunyavejchewin, H. S. Dattaraja, S. Davies, S. Esufali, R. Foster, R. John, S. Kiratiprayoon, S. Loo de Lau, M. Massa, C. Nath, Md. Nur Supardi Noor, A. R. Kassim, R. Sukumar, H. S. Suresh, I.-F. Sun, S. Tan, T. Yamakura, and J. Zimmerman. 2006. Non-random processes maintain diversity in tropical Forests. *Science* 311:527–31.

Wright, S. J. 2002. Plant diversity in tropical forests: A review of mechanisms of species coexistence. *Oecologia* 130:1–14.

Wright, S. J., M. A. Jaramillo, J Pavon G., R. Condit, S. P. Hubbell, and R. B. Foster. 2005a. Reproductive size thresholds in tropical trees: Variation among individuals, species, and forests. *Journal of Tropical Ecology* 21:307–15.

Wright, S. J., H. C. Muller-Landau, O. Calterón, and A. Hernandéz. 2005b. Annual and spatial variation in seedfall and seedling recruitment in a neotropical forest. *Ecology* 86:848–60.

Part 5

BUILDING THE CAPACITY FOR PLACE-BASED RESEARCH

Ian Billick and Mary Price

Some might dispute Krebs's assertion (chapter 13) that empirical research has contributed most of the conceptual leaps in ecology over the past 50 years, but we think few would dispute the central role of place-based research in advancing ecological understanding. Given that ecological understanding is a desirable goal, and that ecology of place is a means of reaching it, it is logical to ask how we can facilitate place-based research. Doing so means building capacity for (1) the accumulation of diverse types of information about a place; (2) the retention of that information in a form that scientists can access and use; and (3) the recruitment and training of people who use place-based research methods effectively to understand nature. The chapters in this part discuss the means by which scientific infrastructure can build these capacities.

Sustained observation of single systems over time figures heavily in all of the narratives to date in one form or another, because that is how the knowledge of place accumulates. It takes time, as Estes (chapter 8) remarks, "to know a place well." Hence, perhaps the single most important way to facilitate place-based research is to invest in making places available for research over the long term. Paine et al. (chapter 11), for example, could not have worked for more than 30 years at Tatoosh Island without the Makah Nation's support and permission, nor could Armitage (chapter 7) have worked in the East River Valley of Colorado without ongoing permission from the United States Forest Service and private landowners.

Time may be necessary for building the understanding of place, but it is not sufficient by itself for rapid progress in scientific understanding. Aigner and Koehler (chapter 16) point out that knowledge accrues in an accelerating fashion through time. Once a core of knowledge exists, scientists can use it to answer new questions. In other words, knowledge begets more knowledge.

Aigner and Koehler outline ways in which field stations can facilitate this self-catalytic process. The process is "seeded" when researchers build an initial knowledge base. Field stations attract researchers by providing research infrastructure: accessible research sites, lodging, laboratory space, power, communications, research equipment, and databases. Although the simple capacity of a field station affects how fast knowledge accumulates, Aigner and Koehler argue that field stations can facilitate faster progress if they develop their potential as "model ecosystems" that are particularly suitable for addressing particular questions—a theme introduced by Kingsland (chapter 3). By systematically investing in infrastructure and databases that support those research themes, field stations can promote the cooperative effort that builds knowledge rapidly (see chapters in part 3). Aigner and Koehler illustrate these points by describing how the McLaughlin Reserve, a new field station in northern California, has succeeded in just a few years to crystallize research around a few core areas—plant ecological genetics, spatial ecology, geology, and human land use—that take advantage of its unique spatial mosaic of chemically distinct soils.

As the amount and diversity of information about a system grows, however, it becomes more and more difficult to retain and synthesize that information. When research teams work together on a conceptually focused problem, they can manage and use information efficiently. They use standard protocols for collecting data from planned experiments, record that information in a form that they can use, and know the relevance of planned and unplanned observations for the problem at hand. As time goes on, there is turnover of researchers, and research programs evolve and diversify conceptually. The relevance of prior information to current problems becomes ever more obscure, and it can easily be forgotten unless it is archived in a retrievable form.

To be sure, the knowledge gathered by researchers is preserved in the form of scientific publications, the traditional final product of scientific research. However, publication retains and makes available only a small amount of the knowledge gained. Much of the knowledge that underpins a study is not easily published; scientists typically try to keep articles short, and do so by including only the information that is directly relevant to the paper's main point. The rest of the knowledge remains in researchers' heads or personal data files, to be resurrected only if it becomes relevant for another publication or comes up through conversations in research-group meetings or at the dinner table. In these forms, information is easily lost. Preventing the erosion of the accumulated information about a place—its intellectual capital—is just as important a function for field stations as maintaining their brick-and-mortar facilities.

Billick (chapter 17) uses the Rocky Mountain Biological Laboratory (RMBL) to illustrate challenges of capturing, preserving, and making available the conceptually diverse knowledge that accumulates at field stations. The RMBL has particular research strengths in several fields, such as plant-animal interactions, butterfly evolution, marmot population ecology and behavior (Armitage, chapter 7), and stream ecology (Peckarsky et al., chapter 9), but also has facilitated research on such things as spider taxonomy, nutrient flows through ecosystems, climate change, bird disease, burying beetle behavior, and avalanche dynamics (to name just a few topics). Although the "information age" has provided new tools for storing and providing access to information, Billick argues that the diversity of information generated by place-based research presents a unique bioinformatics challenge. Traditional bioinformatics focuses on information that is conceptually and structurally uniform (consider gene sequence data, for example) but that comes from diverse locations or taxa. In contrast, data at field stations not only are conceptually diverse but take on a variety of forms: research plans, photos, datasets, publications, collections, and field notebooks. The only thread that stitches place-based information together is the place itself. The challenge to developing a place-based database is that of devising a way to make conceptually diverse information accessible to researchers so that they can integrate it in novel ways. An even more daunting challenge is devising a means of integrating place-based databases across sites so that the benefits of cross-site comparison (Pulliam and Waser, chapter 4; Rundel, chapter 5) can be fully realized. Although some initiatives for cross-site data-sharing already exist, they do not take advantage of the full array of information available from each site because site-specific databases have not captured that information.

None of this investment in physical and information infrastructure will lead to ecological understanding without people to gather information, interpret it, and build local understanding from it. In fact, a good case can be made that the greatest hurdle to progress in ecological understanding is our limited human capacity for place-based research and for knowing how to use the information gained to manage the planet wisely.

Our strategy for building human capacity needs to be two-pronged. We need to recruit and train scientists to design and carry out place-based research. But, as we have noted in chapter 1, it is unlikely that we can rigorously train and employ enough professional scientists to do the job given the challenges presented by the Earth's ecological diversity. We therefore need a much larger workforce. As Curtin (chapter 12) illustrates, there is a vast untapped reservoir of natural history knowledge among the residents of each place on Earth. If this knowledge is channeled properly, it can be deployed to develop local understanding and solve local problems. Feinsinger et al. (chapter 18) describe approaches they have developed and implemented in Latin America to involve local communities and citizen-scientists in the process of scientific inquiry and environmental problem-solving. Their programs concentrate on empowering nonscientists, such as schoolchildren and park guards, to pose research questions, answer those questions through well-designed field research, analyze and reflect on the results, and apply the knowledge gained to improve understanding of the local ecosystem and, in many cases, its conservation and management. The only distinction between these "inquiry cycles" and the formal place-based research described elsewhere in this book is that the inquiry cycles need not be linked explicitly to general ecological concepts and theories, although those are often implicit in the genesis of the research question and the discussion of results. Rather, the goal is local understanding.

The methods of Feinsinger et al. work well in rural settings, where people enjoy everyday contact with the ecosystems they depend on. Here, the challenge is to empower them with methods of inquiry. In our increasingly urbanized world, however, many people grow up with little exposure to nature (Waller and Flader, chapter 3), which is the most powerful stimulus to curiosity, motivation, and personal knowledge. In rural places, people naturally develop these capacities as children; in urban places, the capacities must be fostered deliberately. Hence, to recruit citizen-scientists in urban places, there must be investment in outdoor-education programs that motivate children by affording them the opportunity to experience and observe nature directly, as well as in science-education programs that teach them inquiry skills.

Mobilizing an army of citizen-scientists requires a corps of professional scientists to train them. Hence, building human capacity also requires that we train scientists how to do place-based research and how to teach others to do it. The first step is to awaken their motivation and knowledge through outdoor education programs. The motivational power of direct experience with nature is clear: 69 % of 95 professional ecologists surveyed by the Ecological Society of America (Ecological Society of America 2008) credit their career choice to their having been exposed to nature during childhood or high school. A second step is to improve ecology curricula in secondary schools and universities, something that Aldo Leopold thought a lot about. Leopold was highly critical of traditional university curricula that deemphasized direct experience with nature and instead had "students memorizing the names of the bumps on the bones of a cat. It is important, of course, to study bones, otherwise we should never comprehend the evolutionary process by which animals came into existence. But why memorize the bumps?" (Leopold 1966). He felt that the objective of a liberal education in wildlife is "to teach the student to see the land, to understand what he sees, and enjoy what he understands" (Leopold 1942). To that end, Waller and Flader (chapter 3) note that Leopold not only took his classes into the field at every opportunity, but was also revising his pioneering text (Leopold 1933) at the time of his death to include a series of place-based exercises in which students could learn to see the seemingly innocuous features of the landscape that provide clues to the history and destiny of the people and organisms that inhabit that land. The efficacy of Leopold's philosophy is borne out by the fact that the second most frequent stimulus to career choice discovered in the Ecological Society of America's survey (2008) was university- or postgraduate-level coursework (31 percent) that involved independent study. Structuring curricula to emphasize individual field study and prepare students to "see the land" is an important means for recruiting them to the field sciences.

The essays in this book chart a course for understanding the complex and diverse ecosystems of this planet—the ecology of place—and the chapters in part 5 indicate ways in which we can increase our capacity for place-based science. Building such capacity will require sustained investment by society and by individuals: investment in places where long-term research can be done; in the physical infrastructure that supports the accumulation, management, and use of place-based information; and in the training and support of generations of scientists and citizen-scientists. Most important of all, it requires commitment of the passion, time, and energy of individual investigators to the pursuit of ecological understanding. It is clear that how quickly we progress in solving the pressing environmental problems we humans face will depend on the size of these investments.

Ecological research has long been grossly underfunded; it was so even in Aldo Leopold's day of massive public works spending to combat the Great Depression. "Half a dozen New Deal Bureaus are spending a score of millions on wildlife work, but not a red penny for research," Leopold complained. "They [come] to some research Unit whose total budget would not pay their office boys and say: 'Please give us the facts on which to build our program.' Naturally we can't." (Leopold 1937) What he wrote 70 years ago unfortunately rings all too true today, when the Nature Conservancy, a private nonprofit organization, spends approximately as much on overhead as the National Science Foundation does on all of environmental biology in the United States. If we want to understand the ecosystems upon which our economies and qualities of life depend, we must dramatically increase our investment in ecological research and education.

References

Ecological Society of America. 2008. (http://www.esa.org/education/ecologists_profile/ EcologistsProfileDirectory/index.php).

Leopold, A. 1933. *Game Management*. Charles Scribner's Sons, New York.

———. 1937. Research program. *Transactions of the Second North American Wildlife Conference:* 104–107. Cited on page 175 *in* J. N. Newton. 2006. *Aldo Leopold's Odyssey: Rediscovering the Author of A Sand County Almanac*. Island Press, Washington, DC.

———. 1942. The Role of Wildlife in a Liberal Education. *Transactions of the Seventh North American Wildlife Conference:* 485–489. Cited on Pages 302–3 *in* S. L. Flader, and J. B. Callicott, eds. (1991), *The River of the Mother of God and Other Essays by Aldo Leopold*. University of Wisconsin Press, Madison.

———. 1966. Natural history. Pages 188–95 in *A Sand County Alamanac with Other Essays on Conservation from Round River*. Oxford University Press, New York.

16 *The Model Ecosystem as a Paradigm of Place-Based Research*

THE INTERSECTION OF GEOLOGY, ECOLOGY, AND ECONOMICS AT THE MCLAUGHLIN RESERVE

Paul A. Aigner and Catherine E. Koehler

You can, if you like, among the existential boulders on the hillsides of life, set up an experiment that calibrates the sync between paradigm and practice.

Figure 16.0. A poem by David Robertson (2006), from *Walks in the World #19, McLaughlin Natural Reserve*, describing an intersection of science and aesthetics. Photo courtesy of David Robertson.

Abstract

From the study of molecules to whole organisms, advances in biology have been facilitated by widespread reliance on model organisms—species that are intensively studied with the expectation that such study will yield broad insights about biological processes. We argue that model ecosystems can similarly facilitate the study of ecology and evolutionary biology, and that biological field stations can increase their value to science by developing facilities, infrastructure, data collections, and management policies that support a model ecosystem approach. The McLaughlin Reserve is an example of a relatively new field station that is deliberately developing its potential as a model ecosystem for research in plant ecological genetics and spatial ecology—areas of study for which the natural- and human-caused edaphic heterogeneity of the reserve make it particularly well suited. We discuss strategies of field station development and management that have been designed to increase the convenience and value of the reserve for conducting the types of experiments (e.g., reciprocal transplants and diversity manipulations) traditionally associated with these areas of study, and also to encourage a more holistic view of the ecosystem that emphasizes how humans interact with it, and that attracts creative endeavors in the arts and humanities.

Introduction

Of the more than 1.7 million described species on earth, a certain few will be familiar to any student of biology. Although these model species—the vinegar fly *Drosophila melanogaster*, the nematode *Caenorhabditis elegans*, the diminutive mustard *Arabidopsis thaliana*, the brewer's yeast *Saccharomyces cerevisiae*, the zebrafish *Danio rerio*, various pathogenic bacteria, and some others—account for only a tiny fraction of the diversity of life, they form much of the foundation of our understanding of how biological organization at the organismal level is determined by molecular, cellular, and developmental processes. Are there similar model systems in the study of ecology? Many biologists would undoubtedly recognize a few locations that have been the source of important insights, landmark experiments, or long-term ecological research—perhaps the Galápagos Islands for studies of adaptation and character divergence (see Grant and Grant, chapter 6), or Hubbard Brook Experimental Forest for studies of chemical cycling; but most would be hard-pressed to produce a short list of sites that account for the bulk of our understanding of biological process at the population, community, and ecosystem level. There are some obvious reasons why an emphasis on model systems is a more practical approach for studying organisms than ecosystems, not the least of which is that organisms can be replicated in laboratories across the world and ecosystems cannot. Nevertheless, this volume is in essence a recognition that, despite its limitations, some aspects of the model system approach have been productive in understanding life at these higher levels of organization.

What is the conceptual basis for applying the model system approach to studies of populations, communities, and ecosystems? To answer this question it is instructive to consider why model organisms have been so important to the study of biology at the molecular, cellular, and organismal levels. A model species is adopted for its suitability in addressing a particular set of biological questions, and it is intensively studied with the expectation that this will yield insights that are broadly applicable to other organisms. Such inference is possible because all species are related by common ancestry, and we expect that studies of model species will be most fruitful when the genetic, cellular, and developmental processes under investigation have been conserved through evolution. When this assumption is met, as in the use of *Drosophila* to unravel the chromosomal basis of inheritance, reliance on model organisms can be spectacularly successful. When it is not, model organisms can be blamed for slowing progress in a field. Thus, some critics have argued that the failure of medical science to develop a comprehensive cure for many cancers has been caused by overreliance on a model system (the house mouse) that is too distantly related to humans (Rangarajan and Weinberg 2003). Lack of evolutionary conservatism (i.e., the evolution of novel features) can confound the model system approach, but it does not invalidate it. The best model organisms are known not primarily because they are in some way average or representative of a group of species, but instead because they possess unique features that simplify the study of particular questions. *Drosophila* was suited for the study of inheritance because it possesses remarkably few chromosomes (Sturtevant 1965), and zebrafish are suited to the study of development because they have uniquely transparent embryos and eggs (Dawid 2004). The underlying rationale for the model system approach is that science can advance more rapidly when investigators choose study systems that are particularly amenable to the question at hand, and when they work cooperatively, learning from previous experiences and building upon results.

Populations, communities, and ecosystems are not strictly related by common descent, and perhaps the only undisputed law in ecology is that populations, communities, and ecosystems are universally marked by strong variation in space and time. Ecologists at times seem so overwhelmed by this variation that they despair of uncovering general principles (see Pulliam and Waser, chapter 4). We suspect that if polled, ecologists would cite this variation as a primary reason why model systems have not been as prominent in ecology as in organismal biology. But while there may not be "representative" ecosystems, all ecosystems are shaped by common geological, climatological, ecological, and evolutionary processes, and particular classes of ecosystems (e.g., alpine, Mediterranean, desert) show convergent properties. Certain ecosystems may have features that make them particularly well suited to study some of the processes, and the study of any ecosystem proceeds more efficiently as accumulated knowledge of the system grows.

Even without common descent, much of the rationale used for relying on model organisms can be applied to ecosystems. Thus, model ecosystems can be defined by two properties similar to those that define model organisms: (1) they are especially suited to the study of particular ecological or evolutionary questions, both because of their biological attributes and because they are convenient places to work, and (2) they simplify and accelerate the scientific process by providing an ever-increasing base of knowledge about the system.

To the extent that model ecosystems have been employed in the study of ecology, they are frequently associated with biological field stations. While model organisms must be amenable to survival and reproduction in the laboratory, model ecosystems must be suited to function as laboratories. Field stations, because they supply the on-site infrastructure and facilities for ecosystem studies, are an essential element of the model ecosystem approach. A site may have the ideal biological conditions for addressing a particular question, but without infrastructure it may have limited value for research. Conversely, a model ecosystem approach can probably enhance the success of a field station (if research productivity is taken as a metric of success), and we suspect that those field stations that cultivate a model ecosystem approach contribute disproportionately to the advancement of ecology. That is, field stations are likely to be most successful when they have clearly identified the scientific questions they are uniquely suited to study and have systematically developed the infrastructure, data collections, and policies to support those studies.

The process by which a particular species comes to be recognized as a model organism is inherently idiosyncratic (e.g., Somerville and Koornneef 2002), and to some degree this also may be true of model ecosystems. In many cases, the value of a model system for addressing a particular question is only apparent in retrospect, and some model systems (including ecosystems) may become attractive simply because they accumulate a body of research that serves as the foundation for additional studies. Furthermore, although the suitability of a particular system for studying a given question may be predictable over the short term, the features that will be important for future utility are often difficult to anticipate. Brewer's yeast has the smallest genome of any eukaryotic model organism, which made it an ideal choice as the first eukaryote to be completely sequenced. But the simplicity of *S. cerevisiae*, which has attracted researchers for more than 100 years, may ultimately limit its utility as we develop more sophisticated tools to work with complex organisms (Fields and Johnston 2005). Despite these sources of uncertainty, the development of model systems is not without planning, and model ecosystems are particularly amenable to planning. Model ecosystems, because of their fixed geographic extent, have small communities of researchers compared to model organisms, and this simplifies the planning effort.

Perhaps the most obvious examples of the success of a model ecosystem approach come from field stations in which the planning has been continuous and deliberate. Field stations such as Hubbard Brook Experimental Forest and Coweeta Hydrologic Lab (see Pulliam and Waser, chapter 4) were designed specifically as model systems for the study of chemical cycling at the watershed scale. The founders of these sites recognized that a model system for the study of hydrology and chemical cycling required the protection of a large watershed comprised of many smaller ones that could be individually manipulated, and the installation of basic equipment (e.g., weirs) to monitor water flow and chemical fluxes. Notable successes are also evident at field stations where the model ecosystem approach was relatively unplanned. Thus, the Rocky Mountain Biological Laboratory became a model system for the study of floral evolution driven by pollinator-mediated selection only after the number of related investigations at the site reached a critical mass. Similarly, Jasper Ridge Biological Preserve emerged as a model system for studies of ecosystem impacts of climate change when experiments initiated by just two investigators yielded insights that have since attracted a community of researchers to that field station.

A planned effort to develop a model ecosystem is underway at the Donald and Sylvia McLaughlin Reserve, a relatively young field station located in a geologically heterogeneous landscape. The development of this landscape into a model ecosystem for studies of ecological genetics and spatial ecology has been deliberate since the reserve's inception. Our purpose here is to discuss the natural attributes of the reserve that make it particularly suited to serve as a model ecosystem, and to outline a strategy to develop its research potential, particularly from the perspective of providing appropriate infrastructure, facilities, and management policies. Our thesis is that the fields of ecology and evolutionary biology would benefit from greater emphasis on model ecosystems, and that biological field stations can increase their value to science by promoting and developing their sites as model ecosystems. This process should be guided by the natural attributes and research potential of the sites, and depending on these attributes, it may not require extensive resources or elaborate facilities.

The Serpentine Syndrome

The McLaughlin Reserve is one of 36 sites within the University of California's Natural Reserve System (NRS). The goal of the NRS is to provide protected research and teaching sites that represent the diversity of ecosystems present in California, and McLaughlin's niche in this system is that of a serpentine mosaic ecosystem. The unique value of the McLaughlin Reserve as a research site stems from its location in California's inner north-coast range at the intersection of three major geologic formations—one sedimentary, one volcanic, and one ultramafic. This geologic condition strongly influences the economic history of the region, the origins of the reserve, and the suitability of the location as a model ecosystem.

Geology

Ultramafic rocks have long been recognized as originating in the earth's upper mantle, and are so named for their high content of magnesium and iron and low content of silica relative to the lighter rocks that form in the crust (mafic is a contraction of magnesium-ferric). Ultramafic rocks occur worldwide at the earth's surface either in igneous forms, such as peridotite or dunite, or more commonly in hydrated forms, especially serpentinite. Although technically the term serpentine refers to a group of minerals resulting from the hydration of ultramafics, which in various combinations comprise the rock serpentinite, it popularly refers to a spectrum of ultramafic rocks including, for example, peridotite, serpentinite, and partly serpentinized peridotite, as well as the geologic formations composed of these rocks.

Serpentine is usually found in a distinctively ordered sequence of rocks called an ophiolite, in which serpentine, the basal layer of the sequence, is overlain by a layer of mafic gabbro followed by sheeted dikes, pillow basalts, and a cap of marine sediments. Ophiolites were a source of controversy to geologists for nearly 70 years after they were first described in the early part of the twentieth century (Coleman and Jove 1992, Moores 2003). Although the ultramafic components of the ophiolites were recognized as having an origin in the mantle, their method of emplacement in the upper crust puzzled geologists. In the early twentieth century, prominent hypotheses had ophiolites representing the vertical intrusion of either molten or partially fused mantle material, but this view was complicated by the empirical work of Bowen and Tuttle (1949), which showed that serpentine magmas could not exist at temperatures typical in the earth's crust. Starting in the early 1960s, the study of ophiolites became central to the emergence of a revolutionary and unifying idea in geology: the theory of plate tectonics. With the recognition of mid-ocean spreading centers, it became clear that ophiolites represent cross sections of ocean crust and upper mantle that have been accreted onto continental margins during subduction of dense oceanic crust beneath lighter continental crust. The newfound theory of plate tectonics shed light on why ultramafics are distributed in narrow belts along the present and (what we now recognize as) past margins of continents.

Once exposed at the earth's surface, serpentine weathers into soils that are often high in heavy metals such as nickel, chromium, and cobalt; are low in nutrients such as potassium and phosphorous; have low water-retention capacity;and possess an extremely low ratio of calcium to magnesium. These factors, especially the low Ca:Mg ratio, render serpentine soils hostile to most plants (Brooks 1987). Consequently, serpentine soils around the world support vegetation that is sparse and stunted relative to adjacent non-serpentine soils, and that includes many serpentine-endemic species and serpentine-tolerant ecotypes.

The serpentine economic syndrome

The unique chemical composition and infertility of serpentine soils has produced a peculiar syndrome of human land use (Kruckeberg 1984). We are aware of no comprehensive studies of the use of serpentine landscapes by indigenous people, but archaeological work in the vicinity of the McLaughlin Reserve suggests that serpentine areas were used lightly by Native Americans relative to nearby non-serpentine areas, with most occupation apparently transient and for hunting. Certainly, serpentine areas offered few resources to attract European settlers in North America. Productivity of serpentine soils is usually low enough to preclude or limit such economic uses as forestry, agriculture, and grazing. Historically in California, mining has been the primary economic activity in serpentine landscapes. The heavy metal concentration of serpentine has resulted in economic deposits of chromium and nickel in some areas, and mercury deposits are often associated with the faulting that occurs along the boundary of serpentine formations. Other minerals that have been mined from serpentine include talc, soapstone, asbestos, and magnesite. More recently in California, the open, desert-like vegetation and the lack of demand for competing use has made serpentine landscapes a popular place for off-road vehicle recreation. Also, outcrops of serpentine in coastal California near dense population centers are increasingly under pressure from urban development.

Serpentine and the edaphic ecotype

The edaphic conditions that render serpentine landscapes relatively unattractive for most economic land uses have excited botanists who strive to understand evolutionary processes and mechanisms of adaptation. Because of the close association between plants and soil, there should be strong selection for plants to evolve tolerance to extreme soil conditions (Rajakaruna 2004). It is not surprising, then, that some of the textbook examples of ecological genetics involve plant adaptation to extreme edaphic conditions

(Kruckeberg 1951, 1954, 1967; Antonovics et al. 1971; Macnair and Christie 1983).

Kruckeberg (1951, 1954, 1967) conducted pioneering studies of serpentine ecotypes in species that appear "indifferent" to serpentine (i.e., species whose populations are found both on and off serpentine). Seed collected from 39 species growing on and off serpentine in California and Washington were reciprocally transplanted on serpentine and non-serpentine soils, variously in the greenhouse and in the field. More than two-thirds of these species showed evidence for differentiation into edaphic ecotypes (i.e., when grown in serpentine soil, seed collected from serpentine soil outperformed seed collected from non-serpentine soil). Kruckeberg gave special attention to *Achillea borealis*, a species that was central to the classic experiments of Clausen, Keck, and Hiesey, which had explored the occurrence of distinct climatic ecotypes (Clausen et al. 1948). Kruckeberg showed that within a single climatic race of *Achillea borealis* there occur at least two edaphic races. His analysis encouraged a more holistic view of the ecotype (i.e., one that considers multiple environmental factors).

Kruckeberg (1954) also investigated the problem of serpentine endemism. In reciprocal transplants of edaphic ecotypes, he found that serpentine ecotypes always performed as well or (usually) better when grown on non-serpentine soils. Why then are serpentine endemics—or serpentine-tolerant ecotypes, for that matter—mostly restricted to serpentine soils? Kruckeberg postulated that the evolution of serpentine tolerance involves a trade-off in the loss of competitive ability, and demonstrated that this was the case in a serpentine endemic species of the jewelflower, *Streptanthus*.

Serpentine mosaics and patterns of species diversity

In the same way that serpentine drives the differentiation of edaphic ecotypes, the physiognomy and species composition of serpentine plant communities can contrast sharply with adjacent communities growing on non-serpentine soils. Whittaker pioneered the quantitative study of vegetation in the 1950s and 1960s (Whittaker 1954, 1960), extending the "ecotype" concept from species to entire plant communities. Whittaker's view of plant communities mirrored Kruckeberg's holistic view of the ecotype. Whittaker rejected the strict typological view of the vegetation climax and coined the term *coenocline* to describe variation in the structure and composition of entire plant communities in response to climatic and edaphic gradients. It is not surprising that he focused his studies of vegetation on the mosaic landscapes of serpentine and non-serpentine parent materials that occur in the Siskiyou Mountains of Oregon and California. Here, transitions from non-serpentine parent materials, such as diorite and gabbro, to serpentine are reliably indicated by striking ecotones where dense, closed-canopy evergreen forests of Douglas fir and Port Orford cedar give way to open stands of stunted pines and other conifers.

Whittaker's (1960) landmark study of the vegetation of the Siskiyou mountains is particularly impressive because, in an era preceding the widespread use of multivariate statistics, he analyzed the effect of multiple environmental factors on vegetation. This approach led Whittaker to what is perhaps his most enduring insight: the hierarchical nature of species diversity. Whittaker classified his samples along two climatic gradients: a regional gradient from coastal to inland sites, and a local topographic moisture gradient from mesic to xeric sites. Total species diversity increased from the coastal to the inland sites, and this was in part because the vegetation of inland sites showed more differentiation along local moisture gradients. This result led Whittaker to propose that the total species diversity of a region (gamma diversity) is a product of the diversity of a particular stand or community (alpha diversity) and the change in community composition along complex environmental gradients (beta diversity). Whittaker's hierarchical framework is still adhered to by ecologists, and it is central to studies of scale-dependence in the relationship between productivity and species richness (Chase and Leibold 2002).

Hard Hats and Birkenstocks: How Miners and Ecologists Found Common Ground at the McLaughlin Reserve

As in other serpentine-dominated areas, mining has been the predominant human activity that has impacted the McLaughlin Reserve. Beginning in the 1860s and continuing for more than 100 years, as many as five independent mercury mines operated on land that ultimately was incorporated into the reserve. The Knoxville-Reddington mine was the largest of these; in the late nineteenth century it supported a resident population of more than 300 (Praetzellis and Praetzellis 1985). However, the unique geology of McLaughlin proved even more lucrative for mining than any of these early miners could have imagined. In the 1970s, geologists with Homestake Mining Company developed a new model predicting that gold deposition should be associated with hot spring activity. In the late Pliocene and Pleistocene, volcanic activity along a major fault separating serpentine from sedimentary formations had produced such hot springs at the McLaughlin Reserve, and by 1980 Homestake had discovered at this site the first, and to date the only, economic gold deposit in the coast range of California (Tosdal et al. 1993).

Obtaining the permits necessary to open a new mine in California in the 1980s was no trivial undertaking, and Homestake was under intense scrutiny as they laid out their proposal for the McLaughlin Mine. The

California Surface Mining and Reclamation Act of 1975 requires every new mine to develop a closure and reclamation plan that provides for the "protection and subsequent beneficial use of the mined and reclaimed land." Homestake hired a local county planner, Ray Krauss, as its environmental manager, and Krauss developed a visionary and innovative plan for post-mining land use at McLaughlin. Recognizing that the geology and ecosystem of the McLaughlin mine region were of tremendous scientific interest, Krauss proposed that upon closure the mine property be converted into a field station for environmental research.

Although the original closure plan did not identify an entity to manage the research facility, geologists from the University of California (UC) at Davis had been drawn to the site before it was discovered by Homestake, and geologists and ecologists from Davis continued to visit the area after Homestake began its operation. UC Davis, which already managed two nearby sites as part of the statewide NRS, quickly emerged as a likely partner. The Donald and Sylvia McLaughlin Reserve was formally established in 1993, and until 2002 it operated as an enclave within the Homestake gold mining operation and property. This enclave was affectionately dubbed the "Birkenstock zone" by early users, to contrast it with areas of active mining where research required close coordination with Homestake to avoid unplanned encounters with heavy equipment, and in some cases also required the donning of safety gear such as hardhats and steel-toed boots. During this period a temporary field station was created in a Homestake warehouse, the university was granted access to much of the Homestake property, and the reserve began to capture the imagination of researchers who saw its patchy landscape as an ideal place to study local adaptation and the effect of edaphic heterogeneity on the ecological processes that controlled species diversity. In 2002 the McLaughlin Mine was decommissioned, and a new agreement initiated the transfer of ownership of more than 400 hectares and a 550-square-meter administration building from Homestake to the university. The university was also granted management rights and access to an additional 2,400 hectares of Homestake property. Of the more than 2,800 hectares comprising the reserve, about 20% was directly impacted by the McLaughlin Mine while the remaining 80% could be considered natural (i.e., it has a much less distinct signature of past human land use). Also in 2002, the university hired a full-time resident director and provided funding to convert the administration building into a permanent field station.

Building a Serpentine Mosaic Model Ecosystem

Geological and biological heterogeneity at the McLaughlin Reserve is evident at several spatial scales (figure 16.1). At the scale of kilometers, the

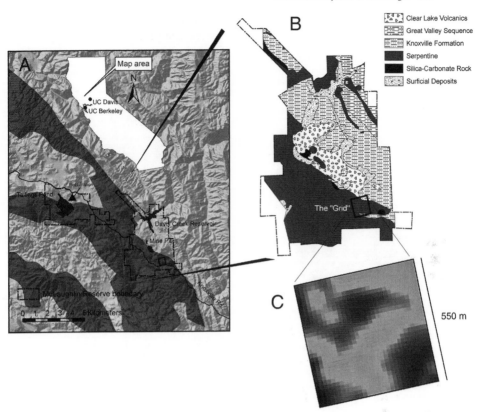

Figure 16.1. Multiple scales of environmental heterogeneity at the McLaughlin Reserve: (A) regional distribution of serpentine (dark shading) around the reserve; (B) geologic heterogeneity in the vicinity of the Grid (sedimentary formations include the Knoxville Formation and the Great Valley Sequence; silica-carbonate rock forms in association with past hot spring activity); (C) fine-scale variation in the soil Ca:Mg ratio on the Grid (the degree of shading is inversely proportional to the log of Ca:Mg, so dark areas indicate strong serpentine influence).

Stony Creek Fault divides the reserve almost equally into two major units. The northern unit is dominated by soils derived from sandstone and shale; the southern is dominated by soils derived from ultramafic rocks, particularly serpentine. The fault's complex tectonics, combined with later volcanic activity, has resulted in the mixing of substrates in the vicinity of the fault, producing multiple transitions from serpentine to non-serpentine at the scale of hundreds or tens of meters. Subsequent bedrock weathering and soil transport have produced variation in the degree of serpentine influence at the meter and sub-meter scale. The distribution of vegetation types is strongly governed by this geologic mosaic, with oak woodlands found on mudstones, chamise (*Adenostoma fasciculatum*) chaparral on sandstones, dense mixed chaparral on some sedimentary and volcanic substrates, and

Figure 16.2. An edaphic ecotone on the McLaughlin Reserve Grid. The non-serpentine substrate in the upper third of the photo supports blue oak (*Quercus douglasii*) woodland, which
is a common vegetation type throughout the foothills of California. The serpentine substrate
in the lower part of the photo supports a sparse chaparral community consisting of serpentine
endemic shrub species such as leather oak (*Quercus durata*), Jepson's ceanothus (*Ceanothus
jepsonii*), and silk tassel bush (*Garrya congdonii*), as well as more common shrub species such
as toyon (*Adenostoma fasciculatum*), that are indifferent to serpentine. Photograph by P. A.
Aigner, 2006.

sparse serpentine chaparral and cypress (*Cupressus* spp.) woodlands on serpentine substrates (figure 16.2). Annual grasslands dominated by old-world
species occur on various non-serpentine substrates, whereas native grasslands and forblands are found in serpentine areas with deeper soils.

  Starting in the 1990s, the earliest studies at the McLaughlin Reserve,
spearheaded by Susan Harrison and her lab group at UC Davis, began to
address questions about how species diversity is shaped by ecological and
evolutionary processes across the reserve's edaphic mosaic. These studies
examined how soil and other factors such as habitat patchiness, grazing,
roads, and fire interact in their effects on native and exotic species diversity
(Harrison 1997, 1999a, 1999b; Safford and Harrison 2001, 2004; Gelbard and
Harrison 2003), how plants tolerate edaphic stress gradients through local
adaptation and phenotypic plasticity (Jurjavcic et al. 2002), and how the size
and isolation of serpentine outcrops influence the ecological processes that
affect the persistence of serpentine endemics (Wolf et al. 2000, Wolf 2001,
Wolf and Harrison 2001). In 1999 the idea that McLaughlin would serve as
a model ecosystem was formalized when six investigators from UC Davis
were awarded two large grants from the Mellon and Packard Foundations

for work on "Ecological and Evolutionary Responses of Plants to Habitat Mosaics." These workers argued that ecologists and evolutionary biologists need a model ecosystem in which to understand how biological diversity is shaped and maintained by a heterogeneous environment, and that the opportunity to develop such a system was available at the McLaughlin Reserve. The centerpiece of this project, located squarely on the Stony Creek Fault, is a carefully surveyed 500 × 550–meter grid that straddles some of the most striking gradients in geologic, soil, and plant diversity found at McLaughlin (figures 16.1 and 16.2). Soils and plant diversity are characterized in detail at 50-meter intervals across the grid, with even more fine-scaled surveys occurring in six 50 × 100-meter "gridlets" where the same variables are measured every ten meters. Data from the grid are archived and made available to all users of the reserve.

These grants spawned a new wave of research on the ecological and evolutionary consequences of edaphic heterogeneity, extending the themes that were pioneered by Kruckeberg and Whittaker. Ecological geneticists at McLaughlin combined molecular marker techniques, selection analysis, and classical reciprocal transplant experiments to investigate the tension between natural selection and gene flow that may result in the evolution of edaphic ecotypes (Sambatti and Rice 2006, 2007; Wright et al. 2006b; Wright and Stanton 2007), and to understand the relative importance of local adaptation versus phenotypic plasticity in conferring serpentine tolerance (Murren et al. 2006). A search for the mechanistic basis of serpentine tolerance proceeded on two fronts: one focused on ecological interactions, the other on plant physiology and genetics. Ecological studies have shown that intraspecific facilitation, mychorrizae, soil microbes, and pathogens may all play a role in mediating serpentine tolerance (Batten et al. 2006, 2008; Espeland and Rice 2007; Springer 2007; Springer et al. 2007; Schechter and Bruns 2008). The physiology and genetics of serpentine tolerance are only beginning to be investigated (Brady et al. 2005), but this is an area in which the use of model organisms and model ecosystems may intersect. A large-scale screening of *Arabidopsis thaliana* mutants has revealed that loss of function of a single gene, CAX1, which codes for a calcium-proton antiporter in the tonoplast that maintains cytoplasmic Ca^{2+} homeostasis, produces many of the traits associated with serpentine tolerance in other species (Bradshaw 2005). It is notable that the Brassicaceae (mustard family), primarily the genera *Streptanthus* and *Arabis*, is disproportionately represented in the serpentine endemic flora of California relative to the California endemic flora as a whole (Safford et al. 2005). Currently, investigations are underway to understand the genetic basis of serpentine adaptation in three genera—*Streptanthus, Collinsia,* and *Mimulus*—with serpentine ecotypes at the McLaughlin Reserve.

Community ecologists have studied spatial patterns of species occurrence and diversity at McLaughlin using three distinct modeling frameworks: one focused on the ecological niches of individual species (Baack et al. 2006, Moore and Elmendorf 2006, Wright et al. 2006a, Elmendorf and Moore 2008), the second using statistical methods to predict the shape of species-area relationships (Green et al. 2003, Harte et al. 2005), and the third partitioning causal factors into two categories: local (e.g., competition) and regional (e.g., long-distance dispersal, environmental heterogeneity, climate, and biogeographic history). Work in this latter area, following the hierarchical diversity framework established by Whittaker, has been particularly fruitful. Plant ecologists have long recognized that processes controlling species diversity operate at different spatial scales. Species diversity is usually related to the productivity of a community, and this pattern has been found to be scale-dependent. At regional scales, diversity is usually positively related to productivity, but when diversity is compared among field plots the relationship may be neutral, unimodal, or negative. The invasibility of communities shows similar patterns: native and exotic species diversity are often positively related at large scales but negatively related at small scales. Recent work at McLaughlin has shown that spatial heterogeneity of the environment—in this case, soil—does explain this scale transition in the pattern of association between native and exotic diversity (Davies *et al.* 2005), but does not explain the transition in the productivity-diversity relationship. That is, more productive regions are not necessarily more heterogeneous than less productive ones (Harrison et al. 2006). An alternative explanation for the scale transition in the productivity-diversity relationship is that strong regional patterns in diversity are determined primarily by earth's climatic and evolutionary history, which has been biased towards warm and productive conditions (Harrison and Grace 2007). If local communities are limited by propagule input from regional species pools, such that they do not receive or cannot absorb all of the diversity in the region, then diversity should have less dependence on productivity at the local scale (Harrison et al. 2006). Indeed, work at McLaughlin has shown that local diversity within serpentine islands (regions) is largely dependent on patterns of propagule input from regional species pools (Freestone and Harrison 2006, Freestone and Inouye 2006).

Facilitating Research in a Model Ecosystem

The research themes of a field station will largely depend on the natural attributes of the site, but its potential as a model ecosystem will also depend on carefully developed facilities, infrastructure, and policies. Just as successful model organisms must be convenient to work with, model ecosystems must be convenient places to work. A model ecosystem will usually have easy access; places for investigators to meet their basic needs for food, water, and shelter; infrastructure and facilities to support necessary experimental work; well-managed data collections; and research and land management policies that allow experimentation but protect the integrity of the ecosystem and its future value for research. When Homestake decommissioned its mine in 2002, the university began to develop the facilities, infrastructure, and management policies that would establish and maintain the serpentine mosaic landscape as a model ecosystem.

Clearly, a field station with overnight facilities and other indoor space contributes tremendously to the attractiveness of a particular location as a field site. These basic needs were quickly met as the university took over management of the property. The reserve was fortunate to inherit a large office building from Homestake Mining Company, so the first stage of facilities development has focused on transforming this office space into a comfortable field station to support basic researcher needs. Following several years of remodeling, the current facilities include bunkrooms, bathrooms with showers, a kitchen, a large dining and recreation area, a classroom, and a laboratory—all conveniently packaged in a drab brown prefabricated metal box-like structure. The facilities at McLaughlin reflect the industrial history of the site; what they lack in charm, they make up for in utility.

Once the minimum need to provide basic amenities to long-term visiting researchers was met, emphasis shifted to providing additional facilities that specifically enhanced the research potential of the serpentine mosaic ecosystem. Many of the experimental techniques employed at McLaughlin require the growing of plants in the greenhouse, followed by planting of seedlings in field plots. To enable investigators to conduct both steps of this process at the reserve, near-term development will include a small research-grade greenhouse specifically designed for controlled crosses and short-term rearing of seedlings.

Probably no form of infrastructure has more impact on research and management at the reserve than roads. Most prospective investigators who visit the reserve for the first time looking for research sites come with specific requests to find particular species, plant communities, or ecological conditions. Almost universally, their second criterion in selecting a study site is that it must be easily accessible year-round; few projects are undertaken beyond 200 meters from a road. The reserve inherited an extensive network of unpaved roads created by past mining and ranching, and the university has maintained as many of these as has been practical. Somewhat ironically, reclamation needs forced Homestake to construct a road through the "grid" shortly after it was established. Although this disturbance caused considerable alarm among the early grid investigators, the long-term effect has most certainly been to increase research productivity at the site. Manipulative experiments have occurred that probably never would have been attempted without the convenience of hauling research equipment by automobile.

Planning and management of roads to support a model serpentine mosaic ecosystem presents distinct challenges. Most obvious is the engineering challenge of maintaining miles of unpaved roads through a rainy season when research activity is at its peak. But as serious as this problem is, it is far less important than that of managing the insidious threat that roads may pose to the ecosystem itself.

Roads are increasingly recognized as being primary dispersal corridors for invasive animals and plants (Gelbard and Harrison 2003), and the roads on the McLaughlin Reserve are no exception. Maintaining the native diversity of the serpentine ecosystem is the first land management priority at the reserve. Serpentine substrates support a large number of endemic species with distributions that include only a few northern California counties. They are also refuges for many native species of grasses and forbs that have been largely replaced on more benign substrates by non-native invasive species, which for the most part do not tolerate serpentine. Perhaps the greatest threat to the integrity of McLaughlin's serpentine mosaic ecosystem is the recent (circa early 1990s) introduction of barbed goatgrass (*Aegilops triuncialis*), which is recognized as a noxious weed of agricultural and wild lands in California. Barbed goatgrass is unique among the suite of non-native invasive species at McLaughlin in its tolerance of extreme serpentine conditions. It is rapidly spreading through serpentine grassland and chaparral communities, and in some grassland sites it has completely replaced the native flora. Not too surprisingly, dispersal and establishment of barbed goatgrass appears to be greatly facilitated by roads and by researcher foot traffic.

The challenges of managing a biological invasion in a research environment are emblematic of a larger policy issue at McLaughlin: the tension between facilitating research—particularly manipulative experiments—and maintaining the long-term integrity and research value of the serpentine mosaic ecosystem. The spread of barbed goatgrass is, of course, of scientific interest in itself, and it has spurred new lines of investigation in the serpentine mosaic ecosystem (Batten et al. 2005, 2006; Meimberg et al. 2006). But as the invasion spreads, we are increasingly faced with difficult decisions about whether to restrict researcher access to infested areas and whether to curtail ongoing eradication efforts to allow continued research.

Managing research to preserve the ecosystem is not limited to addressing the unintended consequences of researcher activity. Proposed experiments that directly or indirectly change natural patterns of species distributions, gene flow, and population genetic structure are perhaps the most common source of this tension. Common gardens, reciprocal transplants, and manipulations of diversity through species additions and removals are all fundamental tools of ecological genetics and spatial ecology. The classical reciprocal transplant experiment is the primary method for understanding whether populations have differentiated in response to natural selection within their local environment. In the serpentine mosaic ecosystem, many investigators have tested the hypothesis that particular species have evolved ecotypes adapted to local edaphic conditions by moving genotypes found off serpentine onto serpentine, and vice versa. Similarly, the introduction of novel species into a research plot is a method to test directly whether species diversity is controlled by local processes such as interspecific competition within plots, or by regional processes such as propagule input from a regional species pool. All of these techniques have the potential to introduce non-local species or genotypes into a natural population. This has a number of potential consequences, both for the conservation of regional species and genetic diversity and for the preservation of the reserve's future research value. For example, if distinct ecotypes are maintained by strong selection, then introduction of non-local genotypes should impose a genetic load (reduction in mean fitness) on a local population (Montalvo and Ellstrand 2001). A large transplant into a small local population could in theory drive the population extinct, although empirical evidence for such an effect is lacking. Even if natural selection acts similarly in two environments, introduction of non-local genotypes can have detrimental effects if previously isolated populations have independently evolved coadapted gene complexes.

Adopting a zero-tolerance policy for artificial gene flow might be the ideal strategy for preserving the ecological integrity and evolutionary potential of the serpentine mosaic ecosytem, but such a constraint would severely limit the current research value of the reserve. Besides, such a policy would be ludicrous, given the degree to which researchers unintentionally cause gene flow by dispersing propagules on their clothing or research equipment—and also given the degree to which past human activities may already have altered the "natural" distribution of genotypes. We instead work with all investigators to find creative ways to answer their research questions while at the same time minimizing the impact of their research on the integrity of the ecosystem and on the future research value of the reserve. For example, in animal-pollinated species, most pollen is likely to be transferred to nearest neighbors, so most pollen exported by non-local transplants can be intercepted by collecting all fruits from neighboring local plants before they mature. Aside from minimizing the biological impact of transplants, we can further safeguard the reserve's future research potential by keeping careful maps and records of which populations have been subjected to manipulations.

Humans in the Ecosystem and Alternative Paradigms

The view that it is the relatively pristine nature of the edaphic mosaic that makes the McLaughlin Reserve a model system for studies of ecological genetics and spatial ecology ignores the persistent role of humans in the ecosystem. "Artificial" gene flow that is intentionally or accidentally introduced by experimenters is only the most recent example in a long history of human impact on the ecosystem. Edaphic ecotones on the reserve not only are hot spots of microevolution but also tend to be associated with the most economically important mineral deposits, such as gold and mercury. Areas of transition between serpentine and other substrates have received far more impact than large areas with more homogeneous geology and vegetation; all five historic mercury mines and the one gold mine on the reserve were located in such transition zones.

Researchers at the reserve have adopted various approaches to incorporating the effect of humans on the ecosystem into their own work. By far the most common approach in studies of ecological genetics and spatial ecology is to ignore the role of humans by selecting study organisms and sites that presumably have been negligibly affected by human activities. A contrasting approach has been to study explicitly how human activities such as grazing, mining, and the introduction of invasive species interact with properties of the ecosystem to affect its function. For example, a series of studies at the reserve has shown that grazing and other human disturbances interact with the degree of serpentine influence in the soil to affect community invasibility (Safford and Harrison 2001, Williamson and Harrison 2002, Harrison et al. 2003). Yet another research approach that embraces human presence is to use the vast areas disturbed by 20 years of industrial mining as a blank experimental canvas for studies of spatial ecology. To this end, Homestake Mining Company has cooperated with a group of academic ecologists in the revegetation of more than 100 hectares of mine waste rock that was capped with topsoil. Native species are being planted in an experimental design that will test how the spatial organization of woody plant species diversity affects ecosystem processes from pollination and herbivory to nutrient cycling.

The most novel approaches to studying humans in the ecosystem have come from outside the biological sciences. Academics in disciplines such as the arts and humanities have shown that the model ecosystem concept can extend beyond the traditional realm of biology as a means for organizing inquiry and creative endeavor. The reserve has hosted a number of classes and events for programs, such as UC Davis's "Nature and Culture" and "Artists in the Bioregion," that integrate art, nature, and science. "Nature and Culture" is a truly multidisciplinary undergraduate program that seeks to understand and expand the limits of traditional modes of inquiry and expression such as science, art, writing, and spirituality. "Artists in the Bioregion" appoints artists- and writers-in-residence who are expected to do work that responds to the people and places of the two major watersheds straddled by the McLaughlin Reserve: those of Cache and Putah Creeks. Interestingly, the creative endeavors of participants in these groups have been focused as much on the serpentine mosaic ecosystem as have the efforts of scientific investigators. Participants in these programs quickly realize that an appreciation of the McLaughlin Reserve requires effort and experience. The landscape presents a stark aspect upon first introduction. On serpentine, the chaparral grows thinly on rocky slopes, with scant shade provided by a smattering of sinuous ghostlike pines that rise over the shrubs. Off of serpentine, the chaparral may grow so thickly and monotonously over vast expanses that it is impenetrable to virtually all animals except rodents and birds. Intruding into this natural austerity, the McLaughlin gold mine has stripped a broad promontory of its native contour and vegetation and replaced it with abrupt terracing and engineered non-native grasslands. Finally, the field station itself and the supporting infrastructure retain the industrial flavor of a modern mining operation. For artists and scientists alike, a sense of the reserve's beauty is enhanced by an understanding of the geologic, biotic, and human processes that have shaped it. Non-science groups have enriched our appreciation for the serpentine mosaic ecosystem by focusing on the interaction of human and natural forces in the evolution of the landscape, and by producing works that manipulate and enhance the emotional and aesthetic response that the landscape draws from its human occupants.

The model ecosystem concept has served the McLaughlin Reserve well as a paradigm for planning and management, but it is not the only model for biological field station development. After all, only a tiny fraction of the important research questions in ecology can ever be answered by relying on a few model ecosystems. The foundation of ecology and evolutionary biology is rooted in (sometimes undirected) natural history observations made across ecosystems. Often, the important questions only become apparent after we have worked in a wide variety of ecosystems. The role of field stations in supporting such exploratory work is paramount, traditional, and should not be sacrificed for the model system approach. Indeed, no incoming graduate student would be turned away from McLaughlin for having an insufficiently developed research question, and research at the reserve is by no means entirely focused on the serpentine mosaic ecosystem. A number of grassland investigators seek specifically to avoid serpentine because, although serpentine grasslands may provide insight to the condition of California grasslands before European settlement, they are not representative of the current state of grasslands in the California floristic province. Other investigators, such as those studying the social behavior of animals, have been indifferent to the edaphic conditions of their study sites.

Biological field stations are by default centers of place-based inquiry, but not all should or will become model ecosystems. The natural features of some field stations may be less obviously suited for a particular research focus which thus develops only gradually or not at all. Other field stations may be constrained by preservation missions that preclude some of the developments or intensive experimentation that would be necessary to create a model ecosystem. The research and management that occurs at any biological field station will sooner or later contribute to that station's particular sense of place. In some cases, as in that of the McLaughlin Reserve, the sense of place drives the research and management, and can in itself become a powerful investigative force.

References

Antonovics, J., A. D. Bradshaw, and R. G. Turner. 1971. Heavy metal tolerance in plants. *Advances in Ecological Research* 7:1–85.

Baack, E. J., N. C. Emery, and M. L. Stanton. 2006. Ecological factors limiting the distribution of *Gilia tricolor* in a California grassland mosaic. *Ecology* 87:2736–45.

Batten, K. M., K. M. Scow, K. F. Davies, and S. P. Harrison. 2006. Two invasive plants alter soil microbial community composition in serpentine grasslands. *Biological Invasions* 8: 217–30.

Batten, K. M., K. M. Scow, and E. K. Espeland. 2008. Soil microbial community associated with an invasive grass differentially impacts native plant performance. *Microbial Ecology* 55:220–28.

Batten, K. M., J. Six, K. M. Scow, and M. C. Rillig. 2005. Plant invasion of native grassland on serpentine soils has no major effects upon selected physical and biological properties. *Soil Biology and Biochemistry* 37:2277–82.

Bowen, N. L. and O. F. Tuttle. 1949. The system MgO-FeO-SiO$_2$. *Geological Society of America Bulletin* 60:439–60.

Bradshaw, H. D., Jr. 2005. Mutations in CAX1 produce phenotypes characteristic of plants tolerant to serpentine soils. *New Phytologist* 167:81–88.

Brady, K. U., A. R. Kruckeberg, and H. D. Bradshaw, Jr. 2005. Evolutionary ecology of plant adaptation to serpentine soils. *Annual Review of Ecology, Evolution, and Systematics* 36: 243–66.

Brooks, R. R. 1987. *Serpentine and its Vegetation: A Multidisciplinary Approach* (Ecology, Phytogeography and Physiology Series #1). Dioscorides Press, Portland, OR.

Chase, J. M., and M. A. Leibold. 2002. Spatial scale dictates the productivity-biodiversity relationship. *Nature* 416:427–30.

Clausen, J., D. D. Keck, and W. M. Hiesey. 1948. *Experimental Studies on the Nature of Species. III: Environmental Responses of Climatic Races of* Achillea. Carnegie Institution of Washington, Washington, DC.

Coleman, R. G., and C. Jove. 1992. Geological origin of serpentinites. Pages 1–17 *in* A. J. M. Baker, J. Proctor, and R. D. Reeves, eds., *The Vegetation of Ultramafic (Serpentine) Soils.*

Proceedings of the First International Conference on Serpentine Ecology, University of California, Davis, 19–22 June 1991. Intercept, Andover, UK.

Davies, K. F., P. Chesson, S. Harrison, B. D. Inouye, B. A. Melbourne, and K. J. Rice. 2005. Spatial heterogeneity explains the scale dependence of the native-exotic diversity relationship. *Ecology* 86:1602–10.

Dawid, I. B. 2004. Developmental biology of zebrafish. *Annals of the New York Academy of Sciences* 1038:88–93.

Elmendorf, S. C., and K. A. Moore. 2008. Use of community-composition data to predict the fecundity and abundance of species. *Conservation Biology* 22:1523–32.

Espeland, E. K., and K. J. Rice. 2007. Facilitation across stress gradients: The importance of local adaptation. *Ecology* 88:2404–9.

Fields, S., and M. Johnston. 2005. Whither model organism research? *Science* 307:1885–86.

Freestone, A. L., and S. Harrison. 2006. Regional enrichment of local assemblages is robust to variation in local productivity, abiotic gradients, and heterogeneity. *Ecology Letters* 9:95–102.

Freestone, A. L., and B. D. Inouye. 2006. Dispersal limitation and environmental heterogeneity shape scale-dependent diversity patterns in plant communities. *Ecology* 87:2425–32.

Gelbard, J. L., and S. Harrison. 2003. Roadless habitats as refuges for native grasslands: Interactions with soil, aspect, and grazing. *Ecological Applications* 13:404–15.

Green, J. L., J. Harte, and A. Ostling. 2003. Species richness, endemism, and abundance patterns: Tests of two fractal models in a serpentine grassland. *Ecology Letters* 6:919–28.

Harrison, S. 1997. How natural habitat patchiness affects the diversity of Californian serpentine chaparral. *Ecology* 78:1898–1906.

———. 1999a. Local and regional diversity in a patchy landscape: Native, alien, and endemic herbs on serpentine soils. *Ecology* 80:70–80.

———. 1999b. Native and alien species diversity at the local and regional scales in a grazed California grassland. *Oecologia* 121:99–106.

Harrison, S., K. F. Davies, H. D. Safford, and J. H. Viers. 2006. Beta diversity and the scale-dependence of the productivity-diversity relationship: A test in the Californian serpentine flora. *Journal of Ecology* 94:110–17.

Harrison, S., B. D. Inouye, and H. D. Safford. 2003. Ecological heterogeneity in the effects of grazing and fire on grassland diversity. *Conservation Biology* 17:837–45.

Harrison, S., and J. B. Grace. 2007. Biogeographic affinity helps explain productivity-richness relationships at regional and local scales. *American Naturalist* 170:S5–15.

Harte, J., E. Conlisk, A. Ostling, J. L. Green, and A. B. Smith. 2005. A theory of spatial structure in ecological communities at multiple spatial scales. *Ecological Monographs* 75:179–97.

Jurjavcic, N. L., S. Harrison, and A. T. Wolf. 2002. Abiotic stress, competition, and the distribution of the native annual grass *Vulpia microstachys* in a mosaic environment. *Oecologia* 130:555–62.

Kruckeberg, A. R. 1951. Intraspecific variability in the response of certain native plant species to serpentine soil. *American Journal of Botany* 38:408–19.

———. 1954. The ecology of serpentine soils. III. Plant species in relation to serpentine soils. *Ecology* 35:267–74.

———. 1967. Ecotypic response to ultramafic soils by some plant species of northwestern United States. *Brittonia* 19:133–51.

———. 1984. *California Serpentines: Flora, Vegetation, Geology, Soils, and Management Problems.* University of California Press, Berkeley.

Macnair, M. R., and P. Christie. 1983. Reproductive isolation as a pleiotropic effect of copper tolerance in *Mimulus guttatus*. *Heredity* 50:295–302.

Meimberg, H., J. Hammond, C. Jorgensen, T. Park, K. J. Rice, and J. K. McKay. 2006. Molecular evidence for an extreme genetic bottleneck during the introduction of barbed goatgrass, *Aegilops triuncialis*, to California. *Biological Invasions* 8:1355–66.

Montalvo, A. M., and N. C. Ellstrand. 2001. Nonlocal transplantation and outbreeding depression in the subshrub *Lotus scoparius* (Fabaceae). *American Journal of Botany* 88:258–69.

Moore, K. A., and S. C. Elmendorf. 2006. Propagule vs. niche limitation: Untangling the mechanisms behind plant species' distributions. *Ecology Letters* 9:797–804.

Moores, E. M. 2003. A personal history of the ophiolite concept. Pages 17–29 *in* Y. Dilek and S. Newcomb, eds., *Ophiolite Concept and the Evolution of Geologic Thought*. Geological Society of America Special Paper 373, Boulder, CO.

Murren, C. J., L. Douglass, A. Gibson, and M. R. Dudash. 2006. Individual and combined effects of Ca/Mg ratio and water on trait expression in *Mimulus guttatus*. *Ecology* 87:2591–2602.

Praetzellis, M., and A. Praetzelllis. 1985. Historical archaeology in the Knoxville and Morgan Valley areas, Lake and Napa counties, California. Report for Davy McKee Corporation and Homestake Mining Company.

Rajakaruna, N. 2004. The edaphic factor in the origin of plant species. *International Geology Review* 46:471–78.

Rangarajan, A., and R. A. Weinberg. 2003. Comparative biology of mouse versus human cells: modelling human cancer in mice. *Nature Reviews Genetics* 3:952–59.

Safford, H. D., and S. Harrison. 2004. Fire effects on plant diversity in serpentine versus sandstone chaparral. *Ecology* 85:539–48.

———. 2001. Grazing and substrate interact to affect native vs. exotic diversity in roadside grasslands. *Ecological Applications* 11:1112–22.

Safford, H. D.., J. H. Viers, and S. P. Harrison. 2005. Serpentine endemism in the California flora: A database of serpentine affinity. *Madroño* 52:222–57.

Sambatti, J. B. M., and K. J. Rice. 2006. Local adaptation, patterns of selection, and gene flow in the Californian serpentine sunflower (*Helianthus exilis*). *Evolution* 60:696–710.

———. 2007. Functional ecology of ecotypic differentiation in the Californian serpentine sunflower (*Helianthus exilis*). *New Phytologist* 175:107–19.

Schechter, S. P., and T. D. Bruns. 2008. Serpentine and non-serpentine ecotypes of *Collinsia sparsiflora* associate with distinct arbuscular mycorrhizal fungal assemblages. *Molecular Ecology* 17:3198–3210.

Somerville, C., and M. Koornneef. 2002. A fortunate choice: The history of *Arabidopsis* as a model plant. *Nature Reviews Genetics* 3:883–89.

Springer, Y. P. 2007. Clinal resistance structure and pathogen local adaptation in a serpentine flax-flax rust interaction. *Evolution* 61:1812–22.

Springer, Y. P., B. A. Hardcastle, and G. S. Gilbert. 2007. Soil calcium and plant disease in serpentine ecosystems: A test of the pathogen refuge hypothesis. *Oecologia* 151:10–21.

Sturtevant, A. H. 1965. *A History of Genetics*. Harper and Row, New York.

Tosdal, R. M., D. A. Enderlin, G. C. Nelson, and N. J. Lehrman. 1993. Overview of the McLaughlin precious metal deposit, Napa and Yolo counties, northern California. Pages 312–29 *in* J. J. Rytuba, ed., *Active Geothermal Systems and Gold-Mercury Deposits in the Sonoma-Clear Lake Volcanic Fields*. California Society of Economic Geologists, Littleton, CO.

Whittaker, R. H. 1954. The vegetational responses to serpentine soils. *Ecology* 35:275–88.

————. 1960. Vegetation of the Siskiyou Mountains, Oregon and California. *Ecological Monographs* 30:279–338.

Williamson, J. N., and S. Harrison. 2002. Biotic and abiotic limits to the spread of exotic revegetation species in oak woodland and serpentine habitats. *Ecological Applications* 12:40–51.

Wolf, A. T. 2001. Conservation of endemic plants in serpentine landscapes. *Biological Conservation* 100:35–44.

Wolf, A. T., and S. P. Harrison. 2001. Effects of habitat size and patch isolation on reproductive success of the serpentine morning glory. *Conservation Biology* 15:111–21.

Wolf, A. T., S. P. Harrison, and J. L. Hamrick. 2000. Influence of habitat patchiness on genetic diversity and spatial structure of a serpentine endemic plant. *Conservation Biology* 14:454–63.

Wright, J. W., K. F. Davies, J. A. Lau, A. C. McCall, and J. K. McKay. 2006a. Experimental verification of ecological niche modeling in a heterogeneous environment. *Ecology* 87:2433–39.

Wright, J., and M. L. Stanton. 2007. *Collinsia sparsiflora* in serpentine and nonserpentine habitats: Using F2 hybrids to detect the potential role of selection in ecotypic differentiation. *New Phytologist* 173:354–66.

Wright, J., M. L. Stanton, and R. Scherson. 2006b. Local adaptation to serpentine and nonserpentine soils in *Collinsia sparsiflora*. *Evolutionary Ecology Research* 8:1–21.

17 *Managing Place-Based Data*

THE ROCKY MOUNTAIN BIOLOGICAL LABORATORY AS A CASE STUDY

Ian Billick

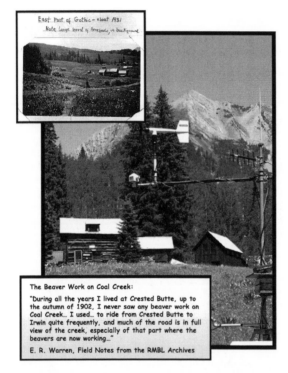

The Beaver Work on Coal Creek:

"During all the years I lived at Crested Butte, up to the autumn of 1902, I never saw any beaver work on Coal Creek... I used... to ride from Crested Butte to Irwin quite frequently, and much of the road is in full view of the creek, especially of that part where the beavers are now working..."

E. R. Warren, Field Notes from the RMBL Archives

Figure 17.0. Modern automated weather stations and archived historical photographs and journals contribute to the Rocky Mountain Biological Laboratory's place-based data management system. Image by David Larson, 2009.

Abstract

The concept of ecology of place emphasizes the importance of integrating diverse types of information for interpreting and generalizing scientific studies. Using the Rocky Mountain Biological Laboratory as a case study, I discuss how we archive important datasets, collect and manage spatial data, and encourage scientists to submit data. This approach involves managing a wide variety of data from a single location, rather than a more traditional approach of archiving the same type of data across multiple locations. Additionally, I outline four decision points that field stations can consider when implementing a place-based database management system. These decision points will affect the costs of the system's installation and maintenance, as well as the type of information available to scientists.

Introduction

One of the many things that drew me into field research was the opportunity to see new and unexpected things in the field. I have a vivid memory, for example, of noticing yellow jackets picking off queens of the ant *Formica obscuripes* as the queens sat on bushes above colonies waiting for visiting males to inseminate them. At a different site our research group noticed the use of treehopper honeydew by different wasps when we experimentally excluded tending ants. While my research was focused on understanding ant-insect mutualisms, seeing how the wasp community interacted with the ants and treehoppers provided a fascinating glimpse of the context of this interaction. Certainly test-tube science conducted in a laboratory has its advantages, but opportunities to observe the richness of ecological interactions in the field is not one of them.

In similar fashion, one of the joys and benefits of working at a field station is that of sharing such observations with other scientists. Not only are you surrounded by people who are excited by stories that in other contexts might get you an odd look, but you also interact with careful observers who spend a great deal of time in the field. These scientists accumulate odd facts such as where to find the fungus that hijacks the reproductive machinery of plants, or the best time of year to look for the slave-making ants in action. And field stations are important for the dissemination of such information because if you don't hear it directly from a scientist, you often won't hear about it any other way; much of this natural history information is never published.

While sharing such natural history observations over a meal or a campfire is fun, the observations are too important to be retained solely through oral history. They can set off a light bulb in a scientist's head, providing the clue that explains some results or serves as the basis for a new extramurally funded research project. During my time as executive director of the Rocky Mountain Biological Laboratory (RMBL, or "lab"), we have made it a priority to capture the rich knowledge base that has accumulated since the lab's founding in 1928. To that end, we have worked on building a place-based database management system (DBMS) that can handle the diverse types of data associated with a single place.

Database management systems allow scientists to store, find, and extract information. They are a means of ensuring that future scientists can pull new insights from old data. Typically they are not just compilations of unrelated data sets; rather, they involve related pieces of information along with the tools needed to put them together in interesting ways.

A number of DBMSs already exist for field-collected data. Herbaria, for example, are systems for managing data associated with plant specimens. Originally non-digital, many herbaria have begun to digitize the information within their collections. The U. S. Geological Survey provides a DBMS for the North American Breeding Bird Surveys. The National Atmospheric Deposition Program manages information collected at more than 200 precipitation-monitoring sites throughout North America.

Place-based database management systems are different from the typical DBMS associated with field data. The latter are conceptually organized around specific, narrow types of data (e.g., plant collections, counts of breeding birds, precipitation records) that are obtained from diverse locations. In contrast, place-based systems are geographically narrow (e.g., the upper Gunnison Basin of Colorado in the case of the RMBL), but conceptually and taxonomically diverse (e.g., genetics, physiology, behavior, and ecology). Consequently a major challenge for such a place-based system is that of handling the wide range of data types that can be generated in a single region, including all those interesting natural history observations that one just stumbles upon in the field.

As with any DBMS, a place-based system provides for the archiving and accessing of data. The archiving function is critical for field data because these usually are difficult to obtain and perhaps impossible to replicate. Some studies may only have a relatively short window of time in which to work because of seasonal or funding constraints. Additionally, repeating long-term studies—for example, thirty years of phenology data or of finch measurements—is impractical. Finally, because ecological systems are dynamic, and because many ecological studies involve understanding how characteristics of ecosystems change through time, historical measurements cannot be repeated. Given the range of questions that ecologists and evolutionary biologists have yet to answer, the limited resources to support such science, and the significant investment in existing research, in terms of both money and time, it is important to maximize the scientific value of field work by ensuring future access to the data.

Ultimately the greatest value of a place-based DBMS may be to facilitate the integration of diverse types of data about a specific place, which can range from formal field data to photographs to natural history anecdotes. Such information can be important for asking questions, designing experiments, interpreting results, and ultimately for developing general insights. Historically this process of accumulating site-specific information has largely happened through long-term individual experience with a study site, through oral history, or through published information. For example, a combination of anecdotal observations with historical data on population densities of the western tiger salamander (Whiteman and Wissinger 2005) led to an extramurally funded research program examining natural causes of amphibian population cycles. In similar fashion, 25 years of data on marmot emergence and first sightings of migratory birds allowed RMBL researchers to examine how plants, and migrating and hibernating animals, are responding to climate change (Inouye et al. 2000). A place-based DBMS provides the formal infrastructure for storing such information and making it accessible. While a place-based system will not replace individual experience and oral tradition, it can help minimize the loss of information over time and facilitate its synthesis.

In this chapter I use the RMBL's efforts as a case study of a place-based DBMS. Specifically I talk about the general goals and specific objectives we have established for data management, the various types of data that flow into our system, the challenges associated with encouraging scientists to contribute data, and the costs of establishing and maintaining a system. Because "place" is an important tool for integrating field data, I discuss how the RMBL generates and manages spatial data and uses location as the "glue" to hold together different pieces of information. Indeed, many of the goals, challenges, and tools associated with our system are a direct result of its place-based nature. I also discuss some general recommendations for establishing a place-based DBMS.

Case Study

A history of RMBL's data-management efforts

The RMBL is situated in the Elk Mountains of west central Colorado at an altitude of 2900 m. Founded in 1928 as a private nonprofit corporation with no institutional affiliation, the RMBL provides research and educational opportunities in the biological sciences. Approximately 160 scientists and students use the facility, generating about 10,000 user-days each year. As the research emphases have changed through the years, the lab has hosted a diverse range of research programs, including but not limited to work on pollination biology, butterfly ecology and evolution, mutualisms, stream ecology, physiological ecology, and marmot social behavior.

This scientific activity has been accompanied by a high level of scientific productivity. Each year approximately 35 scientific papers are published based on fieldwork at the RMBL; and the lab's scientists have produced more than 1,300 peer-reviewed publications. Because many researchers have returned to the lab for years, some of these studies have been conducted over the course of decades (e.g., Calder et al. 1983, Armitage 1991, Inouye et al. 2000, Whiteman and Wissinger 2001, Saleska et al. 2001). This level of productivity is particularly impressive given that the RMBL provides little or no direct financial support of research; for the most part scientists pay all of their expenses associated with working at the lab.

Although a considerable body of work exists in the published literature, much of the information has only a precarious existence. Some data sets have never been published. For example, the late Dr. Robert Enders trapped small mammals on a grid for many years; we have been able to find only a small amount of data associated with that project. After more than 40 years of collecting in the vicinity, Dr. Harriett Barclay left only two single-page publications and 499 pressed specimens in the RMBL herbarium. She left behind neither field notes nor any synthesis of that botanical work. Even the published data sets are typically only partly documented. Most papers provide only summaries of data, not the raw data itself. Critical information needed to extend such published work, such as exact locations, is often missing.

Recognizing the importance of such historical information, and prompted by the National Science Foundation, the lab has taken steps to manage its intellectual resources. While individual scientists originally bore the brunt of archiving and providing access to important datasets, the RMBL has recognized that management of data is just as important to the infrastructure of a field station as construction and maintenance of laboratory space and living facilities. The RMBL's early efforts with a formal data management system focused on compiling a list of peer-reviewed publications conducted through the lab. In 2003 it established an searchable online database providing open access to citations of peer-reviewed papers, student papers, and theses. In the late 1990s the lab had developed a form to document scientific data sets; use of the form was not mandatory and participation was only haphazard. But in 2005 the form was updated, and it can now be accessed through the RMBL's Web site. Since 2006 the lab has also provided a searchable database for its herbarium.

We are still only in the beginning stages of developing a system that does full justice to the rich history of scientific work conducted at the RMBL. The rest of this chapter reviews the lab's efforts, and discusses how we have upgraded our system. The discussion focuses on the nuts and bolts of establishing a place-based DBMS, rather than on its uses; our system is still in development, and until scientists have access to a wide range of information associated with the RMBL, it will be difficult to know how the system will be used in the future.

Goals of RMBL's DBMS

The goals of RMBL's DBMS are to help people ask new questions and/or ask old questions better. We also have a number of immediate objectives, one of which is to provide scientists access to what is known about their sites. As several chapters in this volume make clear, the importance of knowledge about the environmental context of a field project cannot be overstated. For example, scientists may use weather data to interpret results (Grant and Grant, chapter 6; Armitage, chapter 7) or use historical disturbance information in deciding where to conduct a study (Estes, chapter 8).

One way the lab makes information accessible is by helping scientists precisely measure the locations of their study plots and/or study organisms, and helping to manage that spatial information. By explicitly embedding their studies within the greater landscape, scientists can take full advantage of available information in order to design their studies better and interpret their results.

The second and perhaps most important objective is to ensure that data collected are appropriately archived and available for future scientists. Once a researcher dies or stops actively working with a data set, the information is typically lost forever. Minimizing this kind of loss is the lab's highest current priority. Part of archiving datasets is ensuring that they are documented clearly. Unless someone who has had nothing to do with collecting the data can understand how they were collected and what the variables mean, the data will be useless.

Range of datasets available at the RMBL

It is easy to underestimate the variety of scientifically useful information available at a field station (see box 17.1). While the primary focus of data management is often on data collected as part of a scientific study, there are many other types of information that can also be useful. For example, when a scientist initiates research at the lab, she or he files a fairly detailed research plan. This plan can be used to identify potential datasets that may not have made it to publication. Scientists also map their study sites. When locations are known precisely, future scientists can extend the work done at the location and develop long-term data sets, or they can interpret their results in light of historical information about the site. As part of a study, scientists may place a voucher specimen in the lab collections or take photographs related to the study. Scientific publications and dissertations are another type of information. Publications, or student papers from the lab's coursework program, may contain anecdotal observations in addition to data. Each of

BOX 17.1. Diverse types of information can be useful for place-based research. Here are listed some types of data found at the RMBL, with notes on their potential use and file format.

Research plans
- used to interpret datasets and to understand lines of research thought that have not made it into the literature
- stored as text files

Maps of research site locations
- used to find research sites years later and extend upon earlier work
- stored as position coordinates in a coordinate system

Data sets collected by scientists in the field
- used and published to allow the verification of other published analyses as well as the initiation of new analyses
- stored as text fields or within software packages such as Excel or Systat

Scientific publications
- used as the traditional means by which scientific knowledge is stored and transmitted
- stored as articles in paper and/or digital journals

Photographs (of which the RMBL has approximately 4,000)
- used to document long-term changes in ecosystems (e.g., Curtin, chapter 12).
- stored as both hard copies and digital copies

Biological collections
- used to help understand distributions and the timing of species invasions
- stored as physical specimens and written collection data, much of which has been digitized and is available online

Human disturbances
- used for experimental design
- stored in a GIS database

Data from sensors, including weather stations, HOBOs, and stream gauges, which collect information about the environment
- used to provide important context about the abiotic environment
- stored typically in digital form

these pieces of information can be valuable on their own, but linked together they create the potential for interesting follow-up studies.

Scientists are not the only source of important data. Indeed, information about the abiotic and biotic environment that is collected as part of a monitoring program is essential for some scientific studies. The RMBL maintains a weather station on behalf of the U. S. Environmental Protection Agency, and many researchers have used the information about intra- and interannual weather variation to help interpret experimental findings. The lab also maintains stream gauging stations that provide information on stream discharge, pH, conductivity, temperature, and dissolved oxygen.

One data set that has received the greatest attention and use at the RMBL is generated by Billy Barr, the RMBL's business manager and year-round resident. Barr has collected more than 30 years of information on snowfall, avalanches, and the first spring appearances of the local fauna. This information collection has not been question-driven; rather, the focus has been on documenting annual variation. Because snow pack and snowmelt data are important drivers in the subalpine system around the RMBL, this dataset has been extremely useful. The snowfall data have been used to interpret ant population dynamics, plant phenology patterns, and results of climate change experiments. The avalanche data have also been used in a number of different contexts. For example, they have been used to examine the extent to which power laws describe the frequency and size of avalanches (Birkeland and Landry 2002) and to help a scientist design experiments on fireweed (*Epilobium angustifolium*), a plant typically found in avalanche paths. The date-of-first-appearance data have been used to make inferences about long-term responses of organisms to climate change (Inouye et al. 2000).

Some important data may not be obviously scientific. To manage its infrastructure, the RMBL maintains a number of spatially explicit data sets. For example, we map our utilities and buildings. In the simplest sense, this information enables us to track research sites so as to minimize conflicts with infrastructure access and maintenance. Also, historical building sites and infrastructure disturbance may alter soil characteristics and plant community dynamics; understanding the extent to which a site has been previously disturbed may be important to the interpretation of some studies. Moreover, the lab has delineated wetlands and mapped the 30- and 300-year avalanche zones, and this information can also serve as important biological references.

We have also reached beyond the lab to acquire and compile spatially explicit information that is available through local agencies. We have obtained information on roads, trails, precipitation, wetlands, and vegetation from the local county, and the Colorado Division of Wildlife has provided coverages (geographic data sets) on wildlife (see box 17.2). Some of the more

---

BOX 17.2. Elements of the Rocky Mountain Biological Laboratory's geo-database. Here I summarize the GIS datasets available through the RMBL's data management system, along with information on the source and precision of the data, and the number of datasets.

1. Local government data include roads, property boundaries, wetlands, soil maps, vegetation
   - source: Gunnison County
   - precision: highly variable
   - number: 23 datasets

2. Digital elevation models describe the elevation of the area around Gothic
   - source: U.S. Geographical Survey
   - precision: 10 meters and 30 meters
   - number: two separate coverages of different resolution

Orthorectified photos: aerial photos for which absolute location has been established
   - source: USDA, RMBL, Gunnison County
   - precision: one to ten meters
   - number: three different sets of photos of the region around Gothic

Wildlife distributions: data includes mule deer winter range
   - source: Colorado Division of Wildlife
   - precision: highly variable
   - number: 128 coverages for 32 species

RMBL-generated datasets: locations of research sites and infrastructure data
   - source: RMBL and RMBL scientists
   - precision: five millimeters to one meter
   - number: approximately 200 coverages

---

useful spatial data sets have been digital elevation models (DEMs) and orthorectified aerial photos. The DEMs are digital maps indicating the known elevation of some points and providing information on study site slope and aspect. Orthorectified photos are aerial photos that have been corrected for location. If study sites are mapped they can be overlaid onto the orthorectified photos and the DEMs, and information about the context within which a study has been conducted can be presented visually.

Generating spatially explicit data

One of the challenges of integrating such a wide range of data is that of establishing the relative positions of various pieces of information. This requires mapping study sites in a common coordinate system. The RMBL provides scientists access to three different grades of measuring equipment: handheld mapping grade global positioning system (GPS) units, a survey-grade GPS unit, and a survey-grade conventional unit. These units have different levels of accuracy, price, and ease of use (see box 17.3). The handheld mapping-grade units are the easiest to use, but they are dependent upon satellite coverage and offer only sub-meter accuracy.

The survey-grade GPS unit also depends upon satellite coverage, but it offers centimeter accuracy. This increased accuracy comes at a cost. The equipment is more complicated to use, and it requires the establishment of a localization grid. This is a set of monuments—bronze caps on top of a reinforcing bar embedded in concrete and sunk six feet into the ground—for which the relative positions are known. The locations of these monuments are typically determined using traditional, non-GPS survey techniques and then wedded to a GPS coordinate system. Future measurements of location start by setting the survey-grade unit up on the monuments, which presumably have not moved, and matching the GPS coordinates to the local grid. The ability to wed the relative locations of known monuments to the absolute locations provided by GPS satellites reduces error. Measurements can then be taken within the vertical and horizontal boundaries of the grid to a precision of millimeters.

Conventional survey units offer even higher accuracy. The RMBL uses a survey-grade conventional unit with a potential accuracy of three to five millimeters. This survey unit works by establishing a baseline between two points in space, and then using a compass and laser to measure an object's distance from one of the points and the angle of its position relative to the baseline. This conventional unit is more complicated to use than the survey-grade GPS unit. It requires three units to be working in the field with a line of sight between two of the units and the central unit, whereas the survey-grade GPS instrument involves two units that need only to be in radio contact with each other. An additional drawback of the conventional unit is that it only establishes relative positions. If two locations are measured using separate baselines for each, and one baseline has not been surveyed relative to the other, it is impossible to know where the two locations are relative to each other. This problem is eliminated if the conventional unit is being used within a localization grid that has been wedded to a GPS coordinate system. Positions measured with a conventional unit from known monuments can easily be entered into such a system. The conventional unit, however, does offer the advantage of not requiring satellite coverage. Consequently it is useful in situations in which the horizon is significantly obscured—for example, if mountains are in the way and satellite coverage is problematic.

BOX 17.3. For survey equipment, different levels of precision and accuracy come at different costs. A survey-grade conventional unit measures the location of a point relative to two other points; unless additional work is done to embed those relative points into a projection system, the location information cannot be managed well in a geographical information system. The other units do generate location information within a projection system, however, and this facilitates management of the information within a GIS. Here is a list of precision thresholds and examples of available survey units, along with the associated costs and the pros and cons of each unit.

10 meter accuracy
- consumer-grade GPS units
- cost: $100–300
- cheap and easy to use, no staff required, data available in projection system
- very low resolution, requires satellite coverage

Submeter accuracy
- mapping-grade GPS unit
- cost: $4000
- minimal staff required, data available in projection system
- requires post-processing and satellite coverage

10–20 millimeter accuracy
- survey-grade GPS unit
- cost: $35,000
- great accuracy/precision, data available in projection system
- expensive, requires satellite coverage and staff support

5–10 millimeter accuracy
- survey-grade conventional unit
- cost: $10,000
- great accuracy/precision, does not require satellite coverage
- requires line of sight between three units, establishes relative locations, requires three individuals to operate

The RMBL actively uses all three of these types of measuring equipment. The mapping-grade GPS unit is commonly used to map the location of study sites in which sub-meter accuracy is sufficient. One group has used it to map marmot dens, and another has used it to map the locations of mist nets. We used the conventional survey-grade unit to establish the localization grid, and it has proven useful in getting measurements on locations on the side of Gothic Mountain, which sits directly to the west of the field station. We have used the survey-grade GPS unit to map streams and wetlands, study sites, and some of the station's infrastructure. Many of the study sites are on the order of 1 × 1 m in size. They include long-term demographic plots of the plant *Ipomopsis aggregata* as well as 35-year-old plant phenology plots. Measurements of these plots to an accuracy of millimeters will allow future scientists to return to them decades later without the need to maintain phys-ical markers at the corners. If substantial differences occur in densities of *I. aggregata*, or in the relative abundance of flowering plants in the phenology plots, scientists will know the changes are real and not simply the result of an error in locating the plots.

Managing spatially explicit data

The RMBL manages spatial information using the geographical information system (GIS) software package ArcSDE. While ArcSDE is more complicated and expensive than ArcGIS, a program designed to run on an individual machine, it allows geographic information to be managed over a local area network. This makes it fairly simple for multiple users to access the same in-formation while avoiding the problems that can rise with multiple versions of the same data set. The lab uses ArcSDE to maintain a core set of coverages on a server that can be accessed from work stations in a computer lab, and to maintain a geo-database that maintains spatial data in either vector or raster form, with additional information associated with each individual point. For example, if you measure a plant's location, any additional information you collect, such as species, height, and number of flowers, will be stored in conjunction with that point in the geo-database.

Managing vaguely spatial data

Not all of the RMBL's information can be managed with GIS; a considerable amount of it "vaguely spatial," meaning that either its exact location (e.g., a GPS coordinate) is not known, or that it does not make sense to speak of it as having an exact location. For example, plant collectors typically describe the locations of plants collected for the herbarium, but for older specimens there are no GPS coordinates; instead, there is a description of the collecting locale. In some instances the description is rather specific; in others it may simply refer to a general area. Regardless of how good the description is, if there is no GPS coordinate the data cannot be reasonably managed through GIS. Scientific publications represent another example of vaguely spatial information. Scientific articles based on research conducted at the lab are commonly based upon work done in multiple areas or over a large region. It does not necessarily make sense to assign them a specific location. Consequently this information is best managed outside a GIS environment.

To manage such nonspatial information, The lab uses MySQL, an open-source database that facilitates the archiving and retrieval of data. While maintaining a MySQL database requires expertise, the maintenance cost, both in terms of time and money, is considerably less than for commercial applications such as Oracle. Currently the lab maintains a number of separate databases. For example, we have one database specifically for scientific publications and another for our herbarium collection. Ultimately, however, the lab is moving towards a system that will integrate all the information, including both spatial and vaguely spatial data.

Managing information flow for a place-based DBMS

A critical piece of any DBMS is managing the flow of information from creation to storage to access and manipulation. The RMBL currently collects all information, except historical information that does not yet exist in digital form, through a Web site, which enables scientists to provide information without being onsite. Porting this information directly from the website to a MySQL server requires a greater level of staff expertise, since webmasters do not necessarily have database experience.

Ensuring that datasets are available to future scientists is not trivial. They must be saved in a file format (e.g., text format) that will be readily accessible even after the programs that created the files are long obsolete. Additionally, it is important that files are backed up so that they do not disappear in case of a hard drive failure or a fire.

Perhaps the most difficult step in archiving data is ensuring that the files are appropriately documented through comprehensive metadata. Metadata are data about data, and they are necessary to maintain the usefulness of data. Without information on what the data represent and how they were collected, it is typically impossible for individuals not involved in the actual study to use the data. The RMBL's experience is that it is very difficult to design a metadata form that is easy to use, that can be passively collected through a website without review, and that adequately documents a data file. We have found that for the most part, metadata and data submissions must be reviewed by an experienced scientist, and additional questions must be asked in order to ensure that the data files can by interpreted by future generations. Such a quality-control step, critical in many cases, adds substantially to the cost of maintaining a DBMS.

Additionally, the lab has found it necessary to establish different protocols for metadata associated with research plans, research sites, spatial data, vaguely spatial data, photographs, and papers. Because a place-based DBMS must manage diverse types of datasets, diverse types of metadata information are needed. Consequently we have submission forms that are unique to each data type. Establishing these submission protocols involved both a senior-level scientist familiar with the content and purpose of the different types of information, and a database expert familiar with data management techniques and with ways to facilitate search and retrieval. Once the necessary protocols were established, however, it took significantly less staff time to maintain those systems.

The flip side of making certain the lab does a good job of receiving information is making certain it does a good job of providing information. Ideally, relevant data sets can be easily found and relationships between disparate data sets can be established. Because of the lab's current focus on archiving, we have not spent much time in developing and testing systems for data retrieval. Currently, scientists can search our publication, herbarium, research sites, and research data bases through our website. Eventually we will provide other data bases, such as photo and research plan data bases. We will also implement tools that allow individuals to search across the databases. However, more work will be needed to fully integrate the different types of information—for example, to make explicit the linkages between a research plan, the geographic coordinates of research sites, a data set, and a publication based upon that work.

Encouraging participation in the DBMS

One of the greatest barriers to achieving the full potential of a place-based DMBS is lack of participation. Most scientists have established methods that they use for entering, backing up, and accessing their data. Participation in an institutional DBMS involves providing metadata, and may also involve converting data sets to a common format. Consequently it requires a nontrivial amount of time. If scientists are already comfortable with their own methods for data storage, there may be no perceived benefit that justifies the costs of participation in a larger DMBS. Scientists may also hesitate to participate because of concerns about control over data sets. Given the significant investment typically involved in a field study, scientists reasonably want time to process and publish their data before these become generally available.

The RMBL deals with this problem by minimizing the costs of participation, providing tangible benefits, and in some instances requiring participation. One way to minimize the cost of participation is to take advantage of information that scientists are already submitting. Whereas some institutions collect metadata in the absence of actual data sets, so that other scientists can know what type of data resources exist, the lab relies upon research plans to capture that information. It has long required that scientists submit research plans describing their studies in order to minimize conflicts between research projects and also to determine whether the projects are appropriate for the lab. By making research plans available through a DBMS, the lab can eliminate the need for scientists to submit metadata before they are ready to submit actual data sets. Not only does use of existing information minimize the amount of additional work required, but integration of the information in research plans, maps of research sites, data sets, and publications is often more useful for maintaining the utility of data sets than the information scientists provide on standard metadata forms.

Access to GPS equipment has also facilitated participation in the DBMS, because it is a benefit of participating. Since use of the Lab's systems requires a staff person to process the data, the lab has more control over that data than it has over data that scientists collected individually. The lab also needs basic information about the data set, such as contact information, in order to process it. This information can be minimally elaborated and turned into metadata. The lab can dramatically cut down on the metadata scientists must provide because it can standardize the methods used to collect spatially explicit data, and can directly attach metadata information to data sets without requiring input from scientists. For example, spatially explicit metadata require information on the projection and coordinate systems within which GPS information is collected. Scientists at the RMBL do not need to provide this information because technicians attach those metadata to their datasets.

The lab has also tried to encourage participation in the DBMS by putting the greatest possible control over datasets into the hands of the scientists. Current trends are clearly towards open sharing of data. The National Science Foundation requires sharing of data collected with public funds, and prominent journals are quickly headed in the direction of requiring scientists to provide access to the datasets upon which their published work is based. However, some scientists are clearly worried that by they will lose control over their data. At some point, even at retirement or death, they will lose those concerns. But by then it may be impossible to recover the data or metadata. Thus, the lab's philosophy has been that it is more important to get data archived and documented than to push scientists to make them publicly available. Access to an archived dataset can always be provided in the future, but getting a dataset archived and documented may be impossible if too much time passes.

Establishing a DBMS

Establishing RMBL's DBMS has proven significantly more expensive and time-consuming than maintaining it. The lab has relied primarily upon a three-person team to manage the data capture through its Web page, to establish the GIS and GPS systems, to organize the MySQL server, and to bridge the gulf between field biologists and the experts in databases, GPS, GIS, and Web page development. The systems administrator, who had a strong database and Web page background but no experience in field biology, was responsible for Web page and MySQL development. The GPS/GIS technician, who had a Master's degree in biology with extensive field experience and some basic training in GIS, oversaw installation of the GPS and the GIS systems. As the team's third member, I had extensive field research experience and was responsible for overall project management.

The GPS/GIS technician's field biology experience proved more important to the success of the installation of the GIS and GPS systems than any prior GPS/GIS training. She was able to pick up the technical aspects of the systems fairly quickly. After taking classes through ESRI, the GIS software company, and going through two full-day training sessions for the conventional unit and the survey-grade GPS unit (which was included as part of an installation package), she was able to manage both the GPS and GIS aspects of the DBMS, translate the technology into a format that biologists found accessible, and ensure that biological questions were driving its application.

The systems administrator's technical expertise was invaluable in developing an overall DBMS of professional quality. While he focused primarily on the MySQL server, his understanding of data management was extremely useful in informing development of the ArcSDE system and in managing data integration between the GPS, GIS, Web page interface, and MySQL database. Additionally, his familiarity with best practices for data management (e.g., security protocols, backup procedures, and minimizing of errors during data capture) was invaluable. To minimize costs, all three members of the team worked together to design efficient and easy-to-use Web interfaces (a task that must be ongoing). The Web interfaces that the lab currently uses reflect a familiarity with Web site development and an understanding of how biologists work and how they structure their data sets.

Decision Points

Cost is a major concern to any institution. While installation of a place-based DBMS can easily run into hundreds of thousands of dollars, some well-defined objectives can be achieved more cheaply. In this section I discuss four decision points that will determine both the costs of a place-based DBMS and what it will be able to achieve.

Managing spatial data on a single computer or serving it to distributed users

The cost of a software system to serve geographic information can be reduced by several thousand dollars by maintaining coverages on a single machine as opposed to serving them to multiple users, either over the Internet or over a local area network. In terms of GIS software, this involves choosing among ArcGIS (a single machine), ArcSDE (serving information across a local area network, or LAN), or ArcIMS (serving information over the web). Serving information over a LAN or the Web makes it much easier for multiple users to access and edit coverages. Additionally, at least for GIS software, there is an assumption that software designed to serve information to multiple users will be handling more information and larger files. Consequently the software designed to work on single machines has limitations on the sizes of files that can be easily accessed or the amount of information that can be stored.

Serving information to multiple users does require more staff expertise. For the RMBL, installation of ArcSDE involved a staff person taking a five-day course and having significant staff expertise with data bases. Once ArcSDE was installed, the maintenance required significantly less expertise. As data sharing over the Web becomes even more ubiquitous, however, the difference in costs between this system and one using a single machine will shrink.

Meter or centimeter precision?

The degree of precision to which location will be measured will have a large effect on start-up costs, needed staff expertise, and long-term maintenance costs. If the goal is to be able to return to one-meter plots with sub-centimeter precision, then the high-end units are absolutely necessary. However, if the primary goal is to be able to put study plots within a larger ecological context, then sub-meter precision will most likely suffice. While the mapping-grade GPS unit offers only sub-meter accuracy, much of the contextual spatial information available at the Lab is on a coarser scale. For example, the digital elevation model (DEM) available for the area, which provides information on elevation and aspect, is accurate on a 10-meter scale. The ortho-rectified photos (aerial photos overlaid on a GPS coordinate system) are on a one-meter scale. Consequently the high-end units constitute overkill if the primary objective is integration of study plots with existing, externally derived coverages.

Avoiding the high-end units will save quite a bit of money. The mapping-grade units are not only cheaper to purchase (see box 17.3), but easier to operate. Scientists can easily learn how to operate them, significantly reducing the staff needed to support them. The equipment does not require a full-time position, nor does it require someone with a survey background. However, a trained staff person will need to be present when it is operated. The precision of the mapping-grade units is increasing, though it is unlikely that they will provide the precision of the high-end units anytime soon.

Archiving only, or analysis too?

The third decision is about whether to focus on archiving or analysis. For example, a scientist might want to use GIS software to develop a model predicting the location of marmot dens based on coverages such as aspect, elevation, and vegetation. While the RMBL provides all of the information, we do not help scientists use the GIS software to conduct the actual analyses. Indeed, the lab has reduced the amount of staffing the GIS systems require by not providing institutional support for the extensive analytical capabilities of the software. Considerably less expertise and staff time is needed to keep our coverages current and to help scientists access the materials over the local area network or to provide the coverage for scientists to take to their home institutions.

In similar fashion, the lab has minimized maintenance costs associated with the MySQL database by focusing primarily on archiving and accessing data. Again, because we are not providing extensive analytical tools or using the full range of tools available within MySQL to query data, we can keep maintenance costs relatively low. Eventually, however, the lab will move beyond the initial stage of archiving and accessing material, both for the MySQL and the ArcSDE databases. We anticipate that significantly more staff time will then be needed to maintain the DBMS.

Customized or off-the-shelf management systems?

The final decision concerns the degree of control exerted over the flow of information. For example, the Knowledge Network for Biocomplexity (KNB) has developed a methodology that allows it to accept data sets along with the documentation needed for their use. Because the tools are readily available on the Web, scientists can obtain them directly; no staff is needed. The alternative approach, which involves developing those protocols internally, is considerably more expensive. Customization requires having staff with programming, data base, and Web-server skills. It also means managing the data so as to establish backup protocols and ensure future access.

There are some substantial benefits to customization, however. For one, as mentioned above, it can reduce the burden on the scientists by building information into the metadata systems. More importantly, it allows the capture and integration of information from sites used by multiple scientists (box 17.1). In contrast, the KNB is focused on archiving actual data sets, and similarly limited data base tools have been developed for some of the other types of information as well (e.g., the herbarium management system known as Specify). But there are a number of other data resources, such as research plans and research sites, for which off-the-shelf database management tools are unavailable. Additionally, because field-based DBMSs have traditionally focused on working with similar types of data from geographically dispersed areas, the tools simply have not been developed that allow integration of heterogeneous data collected from one location.

Conclusion

Place-based DBMSs provide a formal process for embedding scientific studies within a larger ecological context. Ecologists have long gained important insights from contextual information, whether it takes the form of data collected by other scientists or anecdotal information transmitted orally. However, the rapid development of both geographical information and global positioning systems has created powerful new tools for making such contextual data available.

In developing its DBMS, the RMBL has focused on the word "place." The time and location of ecological phenomena are important; many ecological interactions are localized and ecological conditions vary across space. Tools such as GIS software allow scientists to link disparate and seemingly unrelated pieces of information; they are the glue that cements individual studies in the larger landscape.

Because deep understanding of a single ecosystem is built by integrating diverse types of contextual information, the sort of relational database described in this chapter provides the foundation for the ecology of place approach. However, as several chapters in this book articulate (e.g., Pulliam and Waser, chapter 4; Rundel, chapter 5), comparison of several well-studied places is a powerful route to ecological understanding. Hence, we also need a means of integrating data from multiple places. Several such data management initiatives are underway. The Long-Term Ecological Research (LTER) Network includes 26 intensively-studied sites. While each site has its own research emphasis, the sites also host research on core research areas common to the entire network, such as patterns and controls of primary productivity (Ceballos et al. 2002). Participation in the LTER Network requires that sites actively manage data. The way they do this varies substantially, and tends not to include place-based contextual information. However, management by the LTER network of data sets on core ecological processes at multiple sites facilitates cross-site comparison.

Another initiative that is providing opportunities for cross-site comparisons is the National Science Foundation–sponsored National Center for Ecological Analysis and Synthesis (NCEAS), one of whose primary objectives is to use existing data resources for ecological synthesis and analysis. By hosting workshops, NCEAS allows scientists the opportunity to integrate data across field sites. Along with the LTER network, the Organization of Biological Field Stations, and the Knowledge Network for Biocomplexity, it makes data management a central part of such synthetic efforts. This means that there are a substantial number of databases associated with field data.

Most such NSF-sponsored data management has focused on either cataloging or documenting datasets. There has been little emphasis on tracking the less traditional types of data discussed in this chapter. However, such efforts to make field data available are a strong foundation for the expansion of site-specific database systems to include contextual data of the sort RMBL is incorporating into its place-based system, and also for the use of such systems for interesting cross-site syntheses. It will take more time for bioinformatics to revolutionize field biology the way it has revolutionized disciplines like molecular biology. But the tools for data management are reaching the point where they allow scientists to pursue new types of questions.

References

Armitage, K. B. 1991. Social and population dynamics of yellow-bellied marmots: Results from long-term research. *Annual Review of Ecology and Systematics* 22:379–407.

Birkeland K. W., and C. C. Landry. 2002. Power-laws and snow avalanches. *Geophysical Research Letters* 29:1554.

Calder, W. A. III, N. M. Waser, S. M. Hiebert, D. W. Inouye, and S. M. Miller. 1983. Site-fidelity, longevity, and population dynamics of broadtailed hummingbirds: A ten-year study. *Oecologia* 56:359–64.

Ceballos, G. C., M. Goodchild, R. Goodman, M. L. Guerinot, F. Harris, A. Hastings, A. Karp, D. T. Kingsbury, L. Krishtalka, J. N. Levitt, L. McDade, B. Paschke, H. M. Swain, W. T. Sommers, L. Wallace, F. Wagner, and E. A. Zimmer. 2002. *Long-Term Ecological Research Program: Twenty-Year Review.* National Science Foundation. http://intranet.lternet.edu/archives/documents/reports/20_yr_review/.

Inouye, D. W., B. Barr, K. B. Armitage, and B. D. Inouye. 2000. Climate change is affecting altitudinal migrants and hibernating species. *Proceedings of the National Academy of Sciences (USA)* 97:1630–33.

Saleska, S., M. Shaw, M. Fischer, J. Dunne, M. Holman, C. Still, and J. Harte. 2001. Plant community composition mediates both large transient decline and predicted long-term recovery of soil carbon under climate warming. *Global Biogeochemical Cycles Biology* 16:1055.

Whiteman, H. H., and S. A. Wissinger. 2005. Population cycles in tiger salamanders: The importance of long-term research for amphibian conservation. Pages 177–84 *in* M. J. Lanoo, ed., *Status and Conservation of Amphibians in North America*. University of California Press, Berkeley.

18 Local People, Scientific Inquiry, and the Ecology and Conservation of Place in Latin America

Peter Feinsinger, Samara Álvarez, Geovana Carreño, Edmundo Rivera, Rosa Leny Cuéllar, Andrew Noss, Félix Daza, Mireiza Figuera, Edith Lanz, Lainet García, Maikel Cañizares, Aylin Alegre, and Alejandra Roldán

Figure 18.0. A community member, with baby, of Molleturo (El Cajas, Ecuador), developing questions. Photograph by Peter Feinsinger, 2004.

Abstract

In rural landscapes across Latin America, local people are learning to apply firsthand scientific inquiry to answer their own questions about the ecology of place. Their investigations follow the scientific method of the basic or applied inquiry cycle. In "community inquiry," local residents such as subsistence farmers or fishermen propose their own concerns, develop research questions to address them, design and carry out the complete investigations to answer the questions, reflect on the results, and finally apply the reflections to their decision-making. In "park guard inquiry," park rangers and other personnel of protected areas follow a similar sequence to answer questions that arise from observations they have made during their considerable time in the field. In "schoolyard ecology," children perform hands-on scientific investigations on the ecology of the place most accessible to them—the schoolyard—and their teachers integrate these into the formal education curriculum. "Visitor inquiry" induces visitors to protected areas and other venues to engage in active instead of passive learning, and to relate what they've experienced to their actions and attitudes about the ecology and conservation of their own place. Examples from Bolivia, Cuba, Ecuador, and Venezuela illustrate these four approaches. Whether place-based inquiry takes root in a given landscape often depends on whether or not it is continually encouraged, but never controlled, by a local facilitator who might or might not have formal training in ecology.

Introduction: The "Four Cs"

If science is to help in biological conservation, it must be a much more inclusive and widespread science than we know now.   —*Allan Y. Cooperrider (1996)*

"*Para conservar, hay que conocer, comprender y cuestionar.*" The slogan of "the four Cs" in Spanish means simply that effective conservation (*conservar*) requires that we know intimately (*conocer*) the nature of the local ecosystem, understand (*comprender*) the ecological processes and interactions occurring in it, and constantly refine that understanding by asking questions (*cuestionar*). To whom does "we" refer? To ecologists and conservation biologists who grew up and trained elsewhere, such as most of this chapter's authors? Yes, sometimes. To those people who grew up in the landscape in question and have the greatest stake in its destiny? Yes, absolutely.

Throughout Latin America, local people of all ages are now applying the powerful tool of firsthand scientific inquiry to investigate the ecology and conservation of place: their place. By posing and answering their own research questions, schoolchildren, subsistence farmers and fishermen, park rangers, and many others are learning much they didn't know before, gaining understanding of ecological processes and human effects on them, honing their critical thinking skills, and in some cases applying lessons learned from their investigations to better manage and conserve their surroundings. This chapter describes the current status of this continually evolving approach, and illustrates it with examples from five quite different rural landscapes.

Scientific Investigation, the Inquiry Cycles, and the Fifth C

First-hand scientific inquiry begins with a question about one's surroundings that can be answered through studies carried out in those same sur-

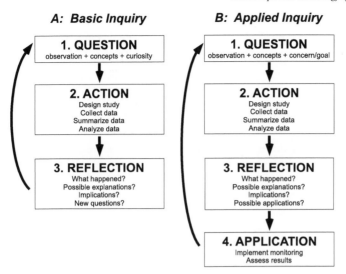

A: Basic Inquiry

1. QUESTION
observation + concepts + curiosity

↓

2. ACTION
Design study
Collect data
Summarize data
Analyze data

↓

3. REFLECTION
What happened?
Possible explanations?
Implications?
New questions?

B: Applied Inquiry

1. QUESTION
observation + concepts + concern/goal

↓

2. ACTION
Design study
Collect data
Summarize data
Analyze data

↓

3. REFLECTION
What happened?
Possible explanations?
Implications?
Possible applications?

↓

4. APPLICATION
Implement monitoring
Assess results

Figure 18.1. Crucial elements of the (A) basic and (B) applied inquiry cycles. Both cycles start with a question and continue through the action and reflection stages. The applied cycle (B) contains an additional stage: application. The basic and applied cycles loop back to new questions. Cycles vary in the motivation for and source of the question, and in the primary output of reflection.

roundings. Observations, prior knowledge, general concepts, and either simple curiosity or explicit concerns or management goals combine to generate the question. Simple curiosity leads to a basic study ("pure science"): the answer increases knowledge and understanding. An explicit concern or management goal leads to an applied study ("applied science"); the answer to the question explicitly informs decision-making. Of course, in real life, research questions present a continuum from "pure" to "applied." Furthermore, a basic study might turn up previously unrecognized threats or management possibilities, while an applied study always adds to knowledge and understanding. Whether an investigation is basic or applied, its quality and usefulness depend on certain key elements: common sense, intellectual honesty, logical reasoning, humility and, where ecology or conservation of place is involved, awareness of the fifth C: *todo cambia*,[1] everything changes. The results from the study carried out in its particular place and time don't necessarily apply to other places or times.

Imagine yourself a subsistence farmer or fisherman, a park ranger, a schoolchild, or the trained scientist (larval or adult) you probably are if you're reading this chapter. You're about to undertake either a basic investigation, following the basic inquiry cycle that figure 18.1a summarizes, or an applied one following the applied inquiry cycle shown in figure 18.1b.[2] For your study to be useful and doable, the question should be articulated so as to satisfy four criteria (Feinsinger 2001, 2004; Arango et al. 2005). It should:

1. be answerable by means of collecting new information;
2. be comparative, basing the comparison on the general concept, concern, or goal;
3. be seductive ("sexy") in two senses: (a) the answer isn't obvious beforehand, and (b) the effort or expense required to get the answer won't overwhelm its usefulness or allure; and
4. be phrased in clear, understandable, jargon-free language.

Clearly, criterion 3 depends on who you are. A question that's seductive to the schoolchild might not be so to the farmer or trained scientist, and vice versa. Criterion 4 also depends on who you are, but less so. Whoever you are—schoolchild or trained investigator—you should be able to phrase your research questions in terms the subsistence farmer and park ranger can understand. If you're unable to do so, most likely your question isn't well thought out. Perhaps it isn't even valid.

Criteria 1 and 2 apply equally to all research questions regardless of whose they are. If your question clearly indicates what you'll compare, then it complies with criterion 2. If it indicates clearly what you'll observe (measure) in each entity to be compared, then it complies with criterion 1. As a trained ecologist you'd probably label "what's being compared" more formally—as the design factor, treatment factor, or input variable—and for statistical analysis, convert it into the independent variable. You'd probably label "what's being measured" as the response variable or output variable, and for statistical analysis convert that into the dependent variable. What's to be compared and what's to be measured in the studies led off by the following questions?

- *Schoolchild*: What kinds of plants grow in the middle of the footpath, at its edge, and at an arm's length from the edge?
- *Ecology graduate student*: How do species diversity and composition of invertebrate prey vary among rocky intertidal zones with and without generalist predators?

Once your question is ready, it's time for you to take action to answer it (figure 18.1). First, you carefully design exactly how the study will be carried out (see box 18.1 below). Next, you collect the data, whether this takes 45 seconds or 45 years. Finally, you compile, process, analyze, and present the results in tables, graphs, or drawings.

The third step of any inquiry, reflection (figure 18.1) is in many ways the most critical. Just as the question and the observations and general concepts that generated it end up being the introduction to a research paper, and the action ends up being methods and materials and results, reflection is exactly what a well-written, thoughtful discussion should cover. Whatever your age or background, you must review the precise results of the study done at its particular time and place in accord with the particular design, then move beyond them to speculations on the causes, generalizations, and new proposals. Take pains to avoid the logical trap of confusing "what is"—the data themselves—with "what might be" or "what might have been": the possible but unexamined causes of trends the data display; possible extrapolations to other places, contexts, and times; and possible applications. As you reflect on "what is" and ponder "what might be or might have been," new questions spring to mind. Many of these can be adjusted to meet the four criteria and lead to new inquiries.

An applied study on the conservation of place or a similar theme has one additional step (figure 18.1b). You began the applied inquiry with a concern or goal, not just curiosity. You complete it when your reflection leads to a management guideline—in the broadest sense—that addresses that concern or goal (Feinsinger 2001, 2004). Now you decide to take the step of putting the management guideline into effect. Can you go home now? No; you know enough about the fifth C, "*todo cambia*," to realize that implementing the guideline might have unexpected consequences—if not now, then next year. Each surprise, of course, generates a new concern or goal that leads into a new applied inquiry cycle (figure 18.1b). A conservation biologist or manager recognizes this process as adaptive management (Holling 1978).

Is This Science?

Perhaps you feel uncomfortable discussing the inquiry cycles as if they involved rigorous science and believe that the latter is best accomplished by formal Popperian or hypothetico-deductive (HD) methods (e.g., Underwood 1990, Peters 1991). In strict HD science, the investigator begins by combining induction from observations with deduction from general theories to propose a universal scientific hypothesis—or, more correctly, a universal scientific alternative hypothesis and its complementary null hypothesis (Underwood 1990). Next begins the testing process (not to be confused with hypothesis testing in statistics). From the universal (alternative) hypothesis the investigator deduces a specific prediction of the form: "If the universal hypothesis is true whenever and wherever its conditions are met, and if those conditions are met in a particular time and place under my control, then results of a properly designed study will show such and such a tendency." The investigator and others deduce prediction after prediction, make test after test of the universal (alternative) hypothesis, ad infinitum. If the HD method is followed rigorously, a single study that fails to support its prediction falsifies the universal (alternative) hypothesis.

In the inquiry cycles, then, where are those general hypotheses that relate to fundamental theories? Where is the formal logical structure with alternative and null hypotheses (Underwood 1990)? Where is there scope for Popperian falsification? In fact, with one exception (see below) the basic inquiry cycle (figure 18.1A) parallels the HD method. In simpler terms, the theory or paradigm and the formal scientific (alternative) hypothesis of HD methodology are the "general concepts" and "prior knowledge" that, combined with observations and curiosity, lead to the question that begins the inquiry (Feinsinger 2001; see Wilson and Armesto 2006). Once it complies with the four criteria detailed above and clearly states the particular time and place to which it applies, the question in fact becomes equivalent to the HD prediction. Indeed, as the research questions of the inquiry cycle are nondirectional, they are more realistic and revealing for field studies than are the directional predictions that characterize most HD studies (see examples below, and Lombardi and Hurlbert 2009). An investigator completing an HD study reevaluates the prediction and its progenitor hypothesis in light of the results. An investigator following the inquiry cycle engages in reflection, which—among many other things—explicitly reexamines the general concepts and prior knowledge (i.e., hypothesis) that led to the question, and explores all other possible causes of the observed results, from artifacts of the study design and unforeseen effects of history or complexities of natural history to alternative underlying concepts (i.e., consideration of multiple hypotheses). Because the question is open-ended, this reexamination and exploration are equally complete and rigorous whether the results have confirmed or contravened conventional wisdom. These and other reflections generate new questions that lead off new inquiries, analogous to the HD process of proposing new hypotheses and testing them (see Wilson and Armesto 2006).

The one outstanding difference between inquiry cycles and formal HD methods followed in the physical sciences and sometimes in ecology (see, however, Camus and Lima 1995) is in fact the latter's focus on testing universal scientific hypotheses through a theoretically infinite number of carefully controlled experiments replicated at different times and places. In strict HD methodology, any one study alone is of little intrinsic interest, in part because experimental protocol should have already controlled for any unique features due to extraneous factors that might vary in time and space. In contrast, inquiries generated by local people—and most professional studies in field ecology or conservation, especially the place-based investigations that are the theme of this book—focus on the study per se as it incorporates all the quirks of time and place in which it's carried out. The "universal hypothesis" serves instead as the context for orienting the research question and its two key features: what will be compared and what will be measured.

Later, reflection on the locally based study's results generates new insights that can be generalized, with caution, to what might happen at other times and places.

As numerous authors have pointed out over the years, even if field ecologists and conservation biologists wished to perform strict HD science they couldn't do so unless their study were a rigorously controlled field experiment. Why not? "*Todo cambia.*" So true replication among field studies is impossible even if we yearn for it—which we don't. Even given spatial autocorrelation (Legendre 1993), two places quite close to one another will have unique present-day ecologies that reflect unique biological, geological, biogeographical, and human histories. Successive years in the same place are each unique. Two places or time intervals farther apart are even more distinct, on average. That's the challenge of conservation of place, because it's dangerous to blindly apply lessons learned at one time in one landscape to decision-making for other landscapes or years. That's also the beauty of ecology of place. A certain landscape attracts field ecologists into returning year after year not as a simple testing ground for universal scientific hypotheses but instead as a unique, fascinating *n*-dimensional puzzle in which scientific hypotheses may help them to frame the most enticing questions, and possible alternative explanations for the answers they obtain. We suspect that most field ecologists will prefer to think of their years-long or decades-long studies as having great intrinsic interest instead of just being drops in the infinitely vast ocean of evaluation of a universal scientific (alternative) hypothesis. Flip through the pages of *Ecology, Conservation Biology*, or a similar journal, and you'll find that few if any published field studies rigorously follow HD science.

A schoolchild, park ranger, farmer, or fisherman is even less likely to develop a given inquiry in order to test a universal scientific hypothesis. Again, well-reasoned general concepts—which are analogues of formal scientific hypotheses, but are often derived from local observations and common sense rather than from external sources—help to frame the question, such as possible effects of chemical and organic fertilizers on crop yield and soil fauna, or possible effects of different perturbations on the rate of invasion of exotic plants. The focus, though, is explicitly local. Whether or not patterns in the results also apply to other schoolyards, rivers, forests, watersheds, or continents isn't of primary interest, although it certainly figures in the reflection.

So, do the inquiry cycles (figure 18.1) represent real science? If they're followed attentively, yes. Either the local residents who follow those cycles rigorously are doing science, or very few of us are. The formal HD method is not very relevant to developing a useful understanding of ecological systems where "*todo cambia*" and local idiosyncrasies intrude but are also prime objects of interest. And it certainly doesn't work for local people focused on questions about their immediate surroundings.

Four Modes of Inquiry by Local People

Community inquiry

Members of the local community—farmers, fishermen, loggers, hunters, medicinal plant harvesters—articulate the concerns most significant to them and most amenable to investigation. Next, they formulate research questions to address those concerns, according to the four criteria. They then design studies to answer their questions (box 18.1; steps 1–7); carry out the inquiries as defined by the questions (step 8); compile, summarize, and present the results; reflect; and, in most cases, decide on and implement the management guidelines whose effects and effectiveness they monitor from then on. That is, community inquiry usually follows the applied inquiry cycle (figure 18.1b), which in many ways simply codifies what many local people do in their daily lives.

How do community members reach the point of designing and carrying out their own scientific investigations? A five-day workshop nearly always provides them with tools, self-confidence, and hands-on practice they need to accomplish the studies they propose, whether these will involve a few days or several years of data collection. After the workshop the facilitators' only remaining role is to provide encouragement, moral support, and limited practical advice during the inquiries without ever manipulating, directing, or "upgrading" the studies. The protagonist and owner of each inquiry is the local person or group that generated it. If the goals or concerns that generate the inquiries involve health issues, crops, or potable water instead of focusing directly on the ecology and conservation of place, that's their choice—as long as the questions comply with the four guidelines.

Is community inquiry just a variation on the well-known and increasingly widespread theme of "participatory management?" No; in the participatory approaches, specialists or consultants from outside work together with local stakeholders to develop conservation plans and management guidelines (Berkes and Folke 1998, Margoluis and Salafsky 1998, Forgie et al. 2001, Selman 2004). The "outsiders" are often trained social scientists or other experts from government institutions, aid programs, and local to international nongovernmental organizations (NGOs). Often donor-driven, the participatory process is organized and guided by those outsiders. Scientific investigation is neither sufficient nor always necessary for the participatory process. If planning does require scientific investigations, the specialists provide them, or at least direct them.

BOX 18.1. Designing an inquiry in eight steps. Bracketed phrases are the "official" terms you might use when designing your own study as a trained ecologist or conservation biologist (Feinsinger 2001, 2004).

1. Look again at the research question and make sure it states the scope in space and time that you wish to investigate, and that realistically you can investigate.
2. According to the question, what will you compare? [the design factor and its levels]
3. What is one independent case of what you'll compare? [the response or experimental unit]
4. Draw a scheme to show how those unitary cases will be distributed in space and in time, so that you won't mistake the effects of different soils, elevations, exposures, seasons, or other factors that might confound the effects of what you're comparing.
5. According to the question, what will you observe (measure) and record in each case of what you're comparing? [the response variable(s)]
6. What methods will you use to make measurements in those standard units?
7. What standard unit—for example, parcel, volume of soil, transect, plant, single leaf—will you use so that your data will be comparable from case to case? [the evaluation unit] Will one such measurement suffice for each case, or will you need several to get an accurate idea of what you're measuring for the case as a whole? [subsamples]
8. Get to work!

In contrast, community inquiry focuses strictly on firsthand investigation conceived and carried out by local residents. Possibly, participatory approaches have generated the concerns or goals. Possibly, an autochthonous investigation will contribute to a participatory process later. The inquiry per se, though, should be independent of ideas and significant resources from outside. In fact, one desired outcome of community inquiry is to counter local people's increasing dependence on the ideas, direction, and dollars of well-meaning institutions and persons from outside—in Spanish, to counter *"el asistencialismo"*—by empowering residents with the skills and self-confidence to take charge of their own learning and decision-making in management and conservation.

Park guard inquiry

Park guards already devote themselves to biological conservation. Many rural landscapes in Latin America include protected areas, whose guards often come from local communities. These guards are familiar with local natural and cultural history long before their first day on the job. Usually their stated duty is to patrol the protected area and deal with illegal activities. They're also uniquely positioned to undertake long-term investigations and/or to monitor the results of management guidelines that they've implemented. While on patrol day after day, year after year, they can't help but observe natural history phenomena and possible conservation threats, and are respectively curious or concerned about what they see. All they need is practice at proposing research questions that comply with the four criteria; at applying the inquiry cycle or the applied inquiry cycle; and at designing a complete study that follows the sequence outlined in box 18.1. A hands-on workshop of five to twelve days provides park guards with these intellectual tools, and they leave the workshop with fully designed investigations ready to launch. As in community inquiry, the facilitators then provide them only with encouragement, moral support, practical advice, and help with preparing the talks they will give at scientific meetings or the papers they will write.

Schoolyard ecology

Now consider the local people who don't yet have much experience with the ecology of place: children. When and where might children best learn place-based ecology and scientific inquiry at the same time? When they're at school, and where ecology is most accessible: the schoolyard and its surroundings. In schoolyard ecology, known in Spanish-speaking countries as "la EEPE" (*la enseñanza de ecología en el patio de la escuela*), schoolchildren undertake inquiries during school hours that can be closely integrated with the formal education curriculum (Feinsinger et al. 1997, Arango et al. 2005). Most studies focus on ecological interactions that are accessible in the school-yard—an approach championed a century ago by Anna Botsford Comstock (1911). Most follow the basic inquiry cycle (figure 18.1a), although by middle school, children and teachers might also employ the applied inquiry cycle (figure 18.1b) to address such concerns as soil erosion, effects of fire or pesticides, herbivory in the school garden, or invasion of exotic plants.

An inquiry into schoolyard ecology cannot help but include skills and competencies mandated by the curriculum in natural sciences, mathematics, and language. Nearly any such inquiry can also involve social science themes, including ethics (a required subject in some countries)—if not in the action phase, then in a far-ranging reflection subtly guided by the teacher. Many inquiries explicitly include features of local culture, such as medicinal plants or farming practices. Back in the classroom the young investigators' oral and graphical presentations of the inquiry can include art, speech, drama, and music. A single, imaginatively designed inquiry whose action in the schoolyard lasts 40 minutes or less can include not only a significant hands-on experience with the ecology of place, but also a significant fraction of the year's mandated curriculum. At the same time, though, the children are getting to know, understand, and ask questions about the ecology and culture of place.

Who guides schoolchildren through the process of place-based scientific inquiry? Facilitators from outside? No, their classroom teachers, following their own interaction with facilitators. At what point will teachers feel competent to do this? Once they themselves have discovered not only the wonders of the ecology of their own schoolyards but also their own often impressive capacities to observe, formulate research questions, answer those questions through a well-designed study, and then reflect. How do teachers arrive at those discoveries? By participating in a four- to five-day EEPE workshop in which they alternate between developing those capacities through abundant hands-on practice in the schoolyard, and brainstorming on the means to integrate the inquiry cycles and schoolyard ecology into pedagogical philosophy, daily lesson plans, and the formal education structure.

Visitor inquiry

What about people who often aren't local but are just visiting? Consider a local "green place" or other natural history venue, whether it is a protected area, the village's communal forest, or an urban green space. Visitors might range from local schoolchildren and families to tourists from distant continents. Except for schoolchildren accompanied by teachers, most visitors come to relax, not to spend their time pursuing scientific inquiry. Yet didactic instruments already exist in many green places—self-guided trails with interpretive signs, tours led by a park guide, and visitor centers. Traditionally these instruments have promoted passive learning, as visitors' eyes and ears, if not their brains, absorb the information presented by signs, tour guides, or displays in visitor centers. Now, however, some visitors in Latin America find themselves both observing and thinking about what they observe. They're on, or in, trails, tours, and visitor centers of inquiry.

The great variety of green places, kinds of visitors, and modes of engaging those visitors in inquiries doesn't permit an adequate discussion here. In all cases, though, visitor inquiries use a truncated version of the basic inquiry cycle (figure 18.1a) as the template. The visitors themselves don't formulate the research question. Instead, signs, brochures, or guides present them with it. Whether the question is presented in writing or orally, it must be clearly answerable, explicitly comparative, extraordinarily "sexy" from the visitor's point of view, and absolutely understandable. The action must be capable of yielding clear results in a very short time, perhaps as short as 15 seconds in some cases. If the visitor has come from a different landscape, the suggestions for reflection must take this into account: the inquiry is here, but the visitor's place and its particular ecology are there. Thus, either the signage or the guide must engage visitors in answering a question about the surroundings of the moment, but end up reflecting on their own surroundings.

Could inquiry also take place in other "natural history venues?" Of course. Visitor inquiry includes not only park inquiry, but also urban parks, demonstration farms, central plazas, museums, botanical gardens, and zoos of inquiry. For example, a museum display might present visitors with a question that complies with the four criteria, asks them to observe how a particular trait varies among several different specimens, and suggests points for reflection. Imagine all the Inquiries that schoolchildren could undertake in the reptile house of a zoo. But zoos and museums aren't very relevant to this book's theme, so let us return to the unfettered ecology of place.

How did all this start?

For the most part the approaches described here have developed accidentally, not deliberately. The idea for schoolyard ecology sprang up in 1985 during a graduate course the senior author of this chapter was teaching at the University of Florida (Arango et al. 2005). "Modern" schoolyard ecology took on solid form in 1994 in San Carlos de Bariloche, Argentina—the night before the first-ever workshop exclusively for school teachers. Also in 1994 in Bariloche, late one night two friends from Parque Nacional Nahuel Huapi proposed the idea for visitor inquiry. The idea for community inquiry surfaced in 1996 during a rapid retreat by canoe from the Afro-Ecuadorian community of San Miguel, northwestern Ecuador. Community leaders had just vehemently rejected a well-intentioned proposal for research projects that would be "to their benefit" but which had been thought up by biologists (including the senior author) sitting around a table in Quito. Of the four approaches to local scientific inquiry discussed above, only park guard inquiry (2000), had been developed, a priori, by leaders of Proyecto Kaa-Iya in Bolivia's Gran Chaco (see below).

Park guard inquiry and all other approaches have evolved considerably over the years and continue to evolve, thanks to input from the thousands of people involved. Aside from schoolyard ecology in North America, which before 1994 had already taken a divergent route, the genesis and development of all the approaches has been strictly Latin American. The guiding philosophy is that each approach must be adapted to the ecology and culture of the place involved. The only constants are the inquiry cycles (figure 18.1) and the four guidelines for questions.

Examples

Bolivia: The Coroico region

The municipality of Coroico is the principal town in the Nor Yungas province of La Paz department. It has around 12,000 inhabitants and 93 communities, many with one-room schoolhouses. Most people are descendants of Aymara-speaking immigrants from the Bolivian Altiplano. Coroico's extraordinarily rugged topography, at a median elevation of about 1700 m, was originally covered by moist montane forests. The steeper slopes are still forested. Crops include citrus, shade coffee, legally grown coca, and vegetables.

Schoolyard ecology first came to Coroico in April 2000 by means of an EEPE workshop for local schoolteachers, biologists, and agronomists from many parts of Bolivia. Workshop participants spread the approach among teachers and education authorities throughout Coroico and beyond, reaching even the most isolated one-room schools. Facilitators continued to provide training and backup as the teachers gained confidence and expertise. Today some 40 key teachers from throughout the zone have assumed leadership roles, facilitating training workshops for new teachers and refresher workshops, integrating schoolyard ecology into both the mandated curriculum and local realities, and promoting "scientific congresses" in which schoolchildren present their work.

A few examples of the inquiries that schoolchildren have completed— and presented in their scientific congresses—include the following.

- Which *bichos* (macroinvertebrates) live in the river just below the school, which live in the pond, and which live in both? (Children's reflections led into the possibility of using stream invertebrates as indicators of water quality, and new inquiries pursued that theme.)
- How does production of oranges vary among trees with and without infestations of mistletoe (*jamillo*)?

Parents have become increasingly involved in schoolyard ecology, after initially complaining that their children were spending far too much time "playing in the schoolyard" and too little time sitting obediently at the desks and reciting from memory. Now schoolyard ecology is self-sustaining and highly regarded, especially since the children most exposed to schoolyard ecology are demonstrably outperforming others.

Among Coroico's community inquiries is one on controlling pests of the (legal) coca plant with homemade concoctions such as garlic in water.

Bolivia: Isoso (Gran Chaco)

The Gran Chaco, mostly in Santa Cruz department, is a vast, thorny, and diverse dry forest that's best represented in the nearly uninhabited 3,441,115 ha of the National Park and ANMI (Área Natural de Manejo Integral) Kaa-Iya. A handful of guards patrol the park. The Río Parapetí parallels the park about 30 km to the west, and along the Parapetí are 25 communities collectively called the Isoso. Most of the 10,000 Isoceños, principally from the Guaraní ethnic group, engage in subsistence hunting and fishing, small-scale agriculture, and livestock raising. Since 1997 biologists of the Wildlife Conservation Society and members of the local government, the Capitanía del Alto y Bajo Isoso, have worked with the communities first on sustainable hunting practices and later on aspects of their use of natural resources. Involved in these efforts are trained local parabiologists and hunting monitors, as well as environmental education promoters who circulate among communities. The collaborations have already produced many resources written in the local dialect of Guaraní as well as Spanish (e.g., Combés et al. 1997; Combés 1999; Componente de Educación Ambiental 2000; Guerrero et al. 2000; Bourdy 2001, 2002).

Park guard inquiry arrived in the Isoso in May 2000, along with schoolyard ecology. In 2002, community inquiry was introduced, and in 2003 some communities began to design Trails of Inquiry. Now, around 60% of the 120 teachers in the Isoso practice schoolyard ecology with their students on a regular basis, and another 20% do so sporadically. As in Coroico, key teachers have taken the lead in training their colleagues, supporting one another's activities, and incorporating schoolyard ecology into the mandated curriculum. Many schoolyard inquiries resemble those in Coroico—for example, comparing mammal tracks, *bichos*, or plants in different habitats or microhabitats, or in areas with different intensities of disturbance. Other inquiries involve medicinal plants or other aspects of ethnobotany, such as

- How do the numbers and kinds of medicinal plants vary between the vegetation in the schoolyard and the disturbed forest across the road?

or other themes altogether, such as

- How many and what kinds of seeds occur in horse manure and in goat manure? (Livestock may play major roles in seed dispersal close to communities.)

In three communities, schoolteachers have worked with environmental education promoters and other community members to develop trails of inquiry for schoolchildren and other residents. For example, visitors guided by handmade signs compare the birds or vegetation among different habitats along a disturbance gradient.

Among the community inquiries undertaken in the Isoso, one of the most rapidly completed also had the most striking results. The question was:

- How many vinchucas occur in houses whose interior walls either are or are not plastered? (Vinchucas, or kissing bugs [Heteroptera: Reduviidae], carry Chaga's disease, a serious health problem in the Isoso.)

Members of the Rancho Viejo community examined 15 houses and found slightly more vinchucas in houses with plastered walls. Not only was the finding counterintuitive, but the presence of any live vinchucas at all came as a surprise because a recent government fumigation program had supposedly eliminated them. Even though the local investigators reflected that their sampling had been inadequate, their discovery led to new ideas on how community members could control vinchuca without depending solely on fumigation.

More park guard inquiry takes place in the Isoso and the Kaa-Iya park than at any other Latin American site to date. Some inquiries have already accumulated three or more years of data. These studies include:

- How does the abundance of the corechi (the armadillo *Tolypeutes matacus*) vary between patches of grassland and patches of choroquetal (low forest dominated by *Ruprechtia triflora,* Polygonaceae) in the Kaa-Iya national park? (Corechi flesh is prized as food, its fat as medicine.)
- How does the extraction of colonies of native bees for honey production affect the health and survivorship of nest trees?

Park guards, parabiologists, hunting monitors, and environmental education promoters have initiated dozens more in-depth inquiries. Once each year they present their studies—alongside schoolyard inquiries, community inquiries, and schemes for trails of inquiry—in a congress to which the entire Isoso is invited, along with key people involved in firsthand inquiry from nearby municipalities and from elsewhere in Bolivia. The protagonists have presented several park guard Inquiries at formal scientific congresses, both national and international.

In short, enthusiasm for inquiry pervades the Isoso. Understanding of the ecology and culture of place in all its complexities is expanding rapidly among elementary schoolchildren answering schoolyard questions in a few hours, as well as among park guards answering "professional-level" questions through years of intensive field work.

Cuba

Inquiry first arrived in Cuba in 2002. Schoolyard ecology now takes place in rural communities near protected areas, towns in agricultural zones, and large cities such as Santiago de Cuba and Havana, where "la EEPE" was featured on national television. At this time (early 2009) local efforts are finally becoming self-sustaining.

Visitor inquiry is under development at several protected areas. The symbol of the Varahicacos Ecological Reserve, a tiny protected area surrounded by the intense tourism of the Varadero Peninsula, is the aguacate cimarrón (*Dendrocereus nudifloris*), an endangered giant cactus whose ancient exemplars host myriad epiphytes, vines, invertebrates, and lizards. Entering the reserve, visitors first encounter a full-sized replica: an aguacate cimarrón of concrete. The concrete statue supports algae, mosses, spiders, ants, and snails. Naturalist guides present children from nearby schools with the following question:

- Which plants and animals live on the real aguacate cimarrón, and which live on the cement replica? (Of course children find a far greater biodiversity on the real cactus, but also discover that the cement one isn't as sterile as they'd expected. Reflections involve the contrast between complex natural environments and simplified urban ones, but also lead children to recognize that ecology of place is everywhere, even in the most apparently barren places. The conceptual range and depth of the reflections generated by this inquiry exceed those of almost any other the authors know.)

Visitor inquiry has also been developed at the visitor centers of the Zapata Swamp and Alejandro de Humboldt National Park, among other places.

In 2004, government ministries issued directives for Cuba's protected-area personnel to undertake investigations aimed at developing local management guidelines. Coincidentally, we had just planned the first training workshop on park guard inquiry for early 2005, in the Reserva Ecológica Alturas de Banao. In early 2006 a second workshop took place in the Reserva Ecológica Siboney-Jutici. In the latter, for example, guards from the Humboldt national park, Cuba's largest and least disturbed, developed the following inquiry from real-life management concerns:

- How do the population structure and dynamics of two sought-after species of jutía (large rodents) vary between the border region of the Ojo de Agua sector of the park and the buffer zone alongside—where clandestine hunting occurs—in the wet and dry seasons of two consecutive years?

Personnel of the host reserve proposed another inquiry:

- In the 33 small sand mines now protected within Siboney-Jutici, how does the vegetation's speed of recovery vary between 16 control sites and 17 randomly selected sites which each have three three-tiered artificial perches for birds?

Community inquiry in eastern Cuba began in early 2008 and is just now being expanded.

Ecuador: El Cajas National Park

High in the Andes of southern Ecuador (El Azuay province), El Cajas National Park includes 235 lakes and a striking landscape of cold, wet alpine páramo, wet montane forest, and patches of *Polylepis* (Rosaceae) woodlands. No settlements exist within the park, but about 750 people from nine communities in the buffer zone raise dairy cattle and cultivate potatoes and other highland crops. El Cajas is administered by the local public works administration and provides drinking water for the city of Cuenca. The park straddles a major highway and attracts many visitors: residents of Cuenca, other highland cities, and Guayaquil; trout fishermen; and ecotourists from Europe and North America.

Schoolyard ecology is well established in Cuenca. The El Cajas region has some schoolyard ecology, but other forms of inquiry are more common. Park guard inquiry got off to a promising start in 2003, but due to a frequent rotation of guards, it now remains mostly on hold. Community inquiry in the buffer zone began with a workshop on the theme "inquiry and the management of family farm plots." In less than three and-a-half days, community members had complete designs for highly replicated investigations such as

- How do quality and quantity of milk production differ between cows foraging on native grasses and those foraging on cultivated grasses?
- How does vegetable production vary between truck gardens bordered by alder trees and equivalent gardens without a tree border?

Some months after the workshop, political changes in the park's administration prevented the key facilitators from returning to the villages. Some inquiries didn't survive, but these two and several others have continued with no additional encouragement from outside.

Visitor inquiry began even earlier, with a self-guided trail of inquiry in 2001, two more in 2004, and training workshops for ecotourism guides. Many visitors to El Cajas, though, spend less time out on the trails than they do warming up inside the modern visitor center. There they encounter panels with two complete inquiries, which are interspersed among other well-designed displays. One panel presents color photos of four tree species frequently seen in and around the park, two of them native (including

Polylepis) and two exotic: *Pinus* sp. and *Eucalyptus* sp. Just below these photos, the visitor encounters the question:

- How many kinds of plants do you see growing on the limbs and trunks of the four trees in the photos? (Visitors compare the photos and search for different kinds of plants visible in each.)

Next follows a bit of background information:

- Most of the plants that grow on tree limbs and trunks don't damage the trees. They're called epiphytes. Epiphytes retain water. They're home to many small animals, such as frogs, spiders, and insects. They also help to maintain high humidity in the forest.

Then come four points for reflection:

- On which trees in the photos did you count the greatest number and the most kinds of epiphytes?
- What do you think might happen with the rain that falls on the forest if these epiphytes were removed? (Many visitors to El Cajas illegally harvest epiphytic bromeliads, ferns, and orchids for use as ornamentals at home.)
- Which of these trees most resembles those in your yard or in the city park nearest your house?
- If you had the chance to plant new trees in your neighborhood, which of these would you choose?

Venezuela: Maripa and the Río Caura

The Río Caura of eastern Venezuela drains a watershed of over 45,000 km² and flows north into the Orinoco. The Caura basin includes low mountain ranges, flat-topped tepuís, and extensive lowlands—much still covered by nearly virgin forests or savannas. Stands of moriche palm (*Mauritia flexosa*) occur along the river and also as *morichales* along small, clear-water spring runs that feed into the Caura. Ethnic and cultural diversity is high. Criollos and Afro-Venezuelans predominate in or near towns such as Maripa, not too far upstream from the Orinoco. Between Maripa and the Caura's headwaters, seven indigenous groups live in many small settlements. Throughout the watershed, activities include subsistence and small-scale commercial fishing, hunting, gathering of forest products, and cattle ranching. Many local people are familiar with the ecology of place, but some have recently arrived from quite different landscapes. Favorite spots such as the *morichales* attract increasing numbers of visitors—local families, tourists from large industrial cities, and even a few adventure tourists from overseas.

The lower Caura had its first workshops in schoolyard ecology in 2000. More teacher training took place in March 2005 as part of a ten-day workshop that also included community inquiry and visitor inquiry. Between April and September 2005 alone, some 94 schoolyard inquiries took place just in the five schools close to Maripa, in communities of three different indigenous groups. Other sites with ongoing schoolyard ecology range from the Maripa secondary school to Ye'kwana communities near the Caura headwaters, several days away by canoe. Refresher workshops and numerous follow-ups continue to multiply and strengthen both schoolyard ecology and community inquiry.

The themes of the schoolyard inquiries we know best, those in the five schools near Maripa, concentrate on the biodiversity of *bichos* and plants in different microhabitats of the schoolyard. For example:

- Which and how many *bichos* live underneath fallen trunks or live tree roots, in vegetated patches, and in bare patches of the schoolyard? (This inquiry took place at the school of the community of Colonial, founded just three years earlier by a Piapoco community arriving from the quite different Amazon watershed.)

Other inquiries deal with food preferences among ants, flower visitors, scale insects on native plants and exotic fruit trees, the relative velocities of ants climbing trunks with smooth and rough bark, stream invertebrates, frugivorous birds, soils, shapes of insects, leaf texture, microtopography, fishes, (micro)landscape ecology, and the biota of cement and soil substrates. Teachers ensure that each inquiry addresses a specific list of mandated skills, abilities, and competencies from across the required curriculum.

Community inquiries designed in 2005 and 2006 span concerns ranging from human health, ethnobotany, food crops, and dairy cattle to the threat of unsustainable fishing practices. For example:

- How does the incidence of malaria per household vary with distance from Maripa's open garbage dump? (The concern was that the municipality was about to start a housing project next to the dump.)
- How does fishing success currently vary among lakes fished only with traditional hook-and-pole methods, lakes recently fished with barbasco [a native plant toxin with limited residence time in water], and lakes recently fished with DDT [a technique employed by commercial fishermen who come in high-powered boats by night, dump 55-gallon drums of DDT in lakes and sell the piscine casualties in large cities]? (This inquiry, like most community inquiries, has an extraordinarily strong research design with abundant replication and precise, standardized methodology, all proposed by the subsistence fishermen.)

- In the Ye'kwana community of Santa María Erebata, how does the incidence of acute diarrhea and vomiting vary among children in households using the whitewater river for drinking water and households that switch to the darkwater river, which according to legend is still polluted from the blood of the gigantic serpent that the eagle killed upstream long ago? (Note the extraordinarily daring step of putting to test a taboo. The protagonists proposed that the accepted practice of drinking only from the "clear" stream could well have worse health consequences than drinking from the tannin-rich one.)

The Environmental Brigade for the Conservation of the Río Caura Watershed, a corps of recent graduates of the Maripa high school, not only facilitates schoolyard ecology by providing continuous support and leadership for teachers and students, but also provides extensive support for leaders of the community inquiries.

Conclusions

Do the seeds of scientific inquiry always germinate and thrive in local soils?

These examples represent a very small sample of the landscapes in which local people are applying scientific inquiry to explore the ecology, conservation, and culture of place. Children and teachers in ten additional Latin American nations plus Guyana practice schoolyard ecology, not just in rural landscapes but also in the middle of large cities. Park guards and/or communities in Argentina, Brazil, Chile, Colombia, Paraguay, and other countries have inquiries underway. Trails, tours, and centers of inquiry hook visitors into exploring local ecology in Argentina, Chile, Paraguay, Brazil, and elsewhere.

This doesn't mean that every local initiative takes root and bears fruit. Workshops for community members, park guards, teachers, and designers of inquiries for visitors always sow many seeds. With few exceptions, though, the seeds fail to germinate unless fertilized—or they germinate, but then wither away without producing fruit. What's the fertilizer? Money? No; in fact, the infusion of money is often counterproductive. People? Yes; just one key person, an enthusiastic facilitator who lives in the community and provides encouragement and support without being domineering, can make all the difference. This person might be a schoolteacher or school administrator, a community leader, a young biologist or agronomist, or a park guard or other member of a protected area's staff. Position and formal training don't matter. What matters is the person: his or her personality, grasp of the inquiry process, and sensitivity to the local culture.

Without such a key person backing them up, many teachers soon find themselves overwhelmed by the demands and petty jealousies of a school, especially if they lack colleagues or directors who are also enthusiastic about schoolyard ecology. Without encouragement, many park guards and community members soon lose their self-confidence and either abandon the inquiries they began so eagerly, or else relinquish them to specialists from outside. In fact, the latter often happens when government or NGO personnel pass through and, with the best of intentions, feel they must provide "expert advice"—or, at worst, opine that it's ridiculous for local people without formal training or deep pockets to even think of engaging in scientific inquiry. Without the constructive criticism of a key person, local designers of trails, tours, and centers of inquiry almost always retreat back into standard nature interpretation instead of challenging visitors to question, observe, and reflect. In some cases not even a key local person can counter political or administrative intervention. Sometimes citizen-initiated inquiry is either ignored or perceived as a threat by those already wielding power, including some education authorities. Finally, in some landscapes it is especially difficult to sustain autonomous learning, investigating, thinking and decision-making in the face of the intellectual and financial *asistencialismo* proffered by aid programs, NGOs (see Chapin 2004, Frazier 2006), high-tech environmental education programs, and well-meaning professionals from outside.

Nevertheless, more often than not, some seeds sown during a workshop do germinate and take root. The fortunate combination of enthusiasm and a key local facilitator can lead to the sustained and sustainable use of scientific inquiry by local residents, as some thousands of people throughout Latin America (tens of thousands including schoolchildren) can attest. The fondest wish of this chapter's authors, frequent facilitators ourselves, is that we become obsolete.

The fruits produced

It's difficult to quantify the positive outcomes of firsthand scientific inquiry by local people, or to attribute positive outcomes solely to the introduction of inquiry cycles into rural communities. Here, though, is a sample of observations, in addition to those already made in the previous examples:

- community inquiry: increasing pride and self-confidence with respect to traditional and recently acquired knowledge about the ecology of place
- community inquiry: increasing self-reliance with respect to investigative skills, creative thinking, and reflection

- synergy between community inquiry and increasing legal control over, and responsibility for, natural resources and ancestral lands (especially with regard to indigenous groups in Bolivia)
- park guard inquiry: increasing skills at observation and the investigative process as a whole, including careful, objective, and abundantly replicated data collection following rigorous research design
- park guard inquiry: marked increases in the self-confidence needed to think critically and creatively, instead of just patrolling
- park guard inquiry: papers presented at regional, national, and international meetings, to widespread acclaim
- schoolyard ecology: changes in how science is taught throughout schools, districts, or regions
- schoolyard ecology: improved student performance with respect to mandated skills, abilities, and competencies
- schoolyard ecology: top awards in local, regional, and national science fairs for students presenting schoolyard inquiries
- schoolyard ecology, park guard inquiry, and community inquiry: positive behavioral and attitudinal changes induced by hands-on experiences with the ecology of place during the *action* phase of inquiries.

Positive effects of visitor inquiry can't yet be reported, because most initiatives have only recently been completed or are still in the design phase, and also because it's difficult to evaluate effects on visitors who return home afterward.

The ecology of place and the four Cs

Many residents of Latin America's rural landscapes, including schoolchildren, are exceptionally keen observers and know a great deal about local natural history without ever having heard of the inquiry cycles. Sometimes, though, natural history observations accumulate without a clear understanding of the ecological processes behind them. Unfounded assertions often substitute for understanding. Except when dealing with crop plants, or with native plants that might have medicinal properties, local people rarely make the conscious choice to ask and answer questions themselves, in order to acquire more knowledge and/or make decisions. Sometimes low self-esteem deters them from such endeavors. Most rural residents, especially indigenous people, are fully in favor of biological conservation in theory: ensuring that today's natural resources will still be around for their great-grandchildren. They are often unaware, though, of connections between their own natural history knowledge, their own critical and creative thinking abilities, and conservation in practice.

Firsthand scientific inquiry does more than increase the self-esteem and motivation of those who use it. Let's start with *"conocer."* Through hands-on investigation, schoolchildren acquire a diverse and profound knowledge of the ecology of place. By focusing their observational skills on the subjects of their inquiries, experienced park guards, subsistence farmers, and fishermen also add substantially to their prior knowledge. Even visitors who complete a 50-second inquiry will discover novel features of the ecology of place and better appreciate the ecology of their own place upon returning home.

As far as *"comprender"* goes, a visitor undertaking a 50-second inquiry with no replication isn't likely to leave a protected area's landscape with a vastly greater understanding of the ecological processes taking place there. By applying a rigorous inquiry cycle, though, all local residents, from schoolchildren to their grandparents, cannot help but acquire a much more profound understanding than they had before about the processes responsible for their observations on the ecology of place. They acquire a healthy skepticism about unfounded, simplistic assumptions and a deep appreciation for the complex influences of historical and current factors on what they observe today.

Local people often grasp the inquiry process itself (*cuestionar*) at least as rapidly as do field ecologists or conservation biologists suffering from chronic "hypothetico-deductivitis." Understanding the basic or applied inquiry cycle simply requires common sense, critical thinking ability, and an appreciation for the fifth C (*"todo cambia"*). These requisites are universal in rural communities, whatever the particular culture. What does vary among cultures is the custom of reflecting on the results of a study. In some communities it takes some time before people grasp reflection and feel comfortable doing it; in others, it takes no time at all.

So, what about *"conservar"*? In some instances, inquiries have already led to local conservation initiatives or management guidelines. In many cases, though, conservation of place will come about more gradually—for example, as sustainable farming practices develop, or as children experienced in schoolyard ecology reach adulthood. What's important to recognize is that the tool of scientific inquiry empowers local people of all walks of life to seek their own understanding of the ecology of place, to reflect, and to apply reflections to problem-solving.

The ecology of place and the fifth C

Finally, let's talk about you, assuming that you're an ecologist or conservation biologist. If ecological phenomena were consistent across the globe, you wouldn't be reading a book on the ecology of place, because that would be a non-concept. A rigorous scientific investigation of the ecology of place B wouldn't add anything to what had been learned from investigations of the ecology of place A, other than demonstrating that results were replicable. Physicists don't need to concern themselves with the physics of place in the same way that ecologists do. In contrast, most social scientists aren't at all ashamed to put the sociology or anthropology or history of place at the forefront. Most are fully aware that apart from general concepts and basic patterns, *todo cambia*. It's been said far too often that most ecologists are wannabe physicists who detest being lumped together with social scientists. Sorry; *todo cambia* in ecology, too. Ignore that fact—or, if you prefer, ignore the incontrovertible fact of spatial and temporal variation—at your peril. Instead, glory in its existence and dive into the ecology of place. Feel free to bring along some brand-name scientific alternative hypotheses to help you frame research questions and to suggest reflections afterward. Just don't become infected by "hypothetico-deductivitis" (Willson and Armesto 2006).

Your understanding and wonderment will be even richer, of course, if you can compare your place with other places. Thus, the more people who are investigating the ecology of their places, the richer and more diverse will be your own comparisons. From this selfish perspective alone, you should encourage place-based investigations by park guards, communities, and schoolchildren. The authors hope, though, that your altruistic side will also recognize the value of firsthand inquiry into the ecology of place to local people themselves, and help to facilitate it for that reason too. People everywhere, and the planet, will thank you.

Acknowledgments

So many friends, colleagues, and institutions, local to international, have contributed to these efforts that we cannot even begin to list them all here. Instead we mention just the three international organizations that have supported many initiatives in firsthand inquiry by local people, including some cases described here. These are the Wildlife Conservation Society, the National Audubon Society and—often through one of the other of the first two—the John D. and Catherine T. MacArthur Foundation.

References

Arango, N., M. E. Chaves, and P. Feinsinger. 2005. *Schoolyard Ecology: A Conceptual and Methodological Guide.* National Audubon Society, New York.

Berkes, F., and C. Folke, eds. 1998. *Linking Sociological and Ecological Systems: Management Practices and Social Mechanisms for Bbuilding Resilience.* Cambridge University Press, New York.

Bourdy, G. 2001. *Plantas del Chaco: Guía para el docente*. Proyecta Kaa-Iya, Santa Cruz de la Sierra, Bolivia.

———. 2002. *Plantas del Chaco II: Usos tradicionales Izuceño-Guaraní. Yaijua Vaera Jare Yaikatu Vaera Yaiporu Ñande ïvïra*. Proyecto Kaa-Iya, Santa Cruz de la Sierra, Bolivia.

Camus, P. A., and M. Lima. 1995. El uso de la experimentación en ecología: Fuentes de error y limitaciones. *Revista Chilena de historia natural* 68:19–42.

Chapin, M. 2004. A challenge to conservationists. *World Watch* (Nov-Dec): 17–39.

Combés, I. 1999. *Arakae: Historia de las comunidades Izoceñas*. Proyecto Kaa-Iya, Santa Cruz de la Sierra, Bolivia.

Combés, I., J. Aguirre, A. Noss, J. Ventocilla, and A. M. Saavedra. 1997. *Educación ambiental en el Izozog*. Proyecto Kaa-Iya, Santa Cruz de la Sierra, Bolivia.

Componente de Educación Ambiental. 2000. *Ore, Mbaembimba Kaa Ipo Reto. Nosotros, los animales del monte*. Proyecta Kaa-Iya, Santa Cruz de la Sierra, Bolivia.

Comstock, A. B. 1911 [facsimile edition 1985]. *Handbook of Nature Study*. Cornell University Press, Ithaca, NY.

Cooperrider, A. Y. 1996. Science as a model for ecosystem management: Panacea or problem? *Ecological Applications* 6:736–37.

Cushman, K., ed. 1999. The cycle of inquiry and action: Essential learning communities. *Horace* 15, no. 4 (April–May).

Feinsinger, P. 2001. *Designing Field Studies for Biodiversity Conservation*. Island Press, Washington, DC.

———. 2004. *Diseño de estudios de campo para la conservación de la biodiversidad*. Editorial FAN, Santa Cruz de la Sierra, Bolivia.

Feinsinger, P., L. Margutti, and R. D. Oviedo. 1997. School yards and nature trails: Ecology education outside the university. *Trends in Ecology and Evolution* 12:115–20.

Forgie, V., P. Horsley, and J. Johnston. 2001. *Facilitating Community-Based Conservation Initiatives*. Department of Conservation, Wellington, New Zealand.

Frazier, J. 2006. Biosphere reserves and the Yucatán syndrome: Another look at the role of NGOs. *Landscape and Urban Planning* 74:313–33.

Guerrero, J., A. Arambiza, L. González, E. Ity, and I. Combés. 2000. *Ayuru, Kerekere, Tüi: Nuestra vida en el Izozog*. Proyecto Kaa-Iya, Santa Cruz de la Sierra, Bolivia.

Holling, C. S. 1978. *Adaptive Environmental Assessment and Management*. John Wiley and Sons, New York.

Legendre, P. 1993. Spatial autocorrelation: trouble or new paradigm? *Ecology* 74:1659–73.

Lombardi, C. M., and S. H. Hurlbert. 2009. Misprescription and misuse of one-tailed tests. *Austral Ecology* 34:447–68.

Margoluis, R., and N. Salafsky. 1998. *Measures of Success: Designing, Managing, and Monitoring Conservation and Development Projects*. Island Press, Washington, DC.

Peters, R. H. 1991. *A Critique for Ecology*. Cambridge University Press, Cambridge.

Selman, P. 2004. Community participation in the planning and management of cultural landscapes. *Journal of Environmental Planning and Management* 47:365–92.

Underwood, A. J. 1990. Experiments in ecology and management: Their logics, functions, and interpretations. *Australian Journal of Ecology* 15:365–89.

Willson, M. F., and J. J. Armesto. 2006. Is natural history really dead? Toward the rebirth of natural history. *Revista Chilena de historia natural* 79:279–83.

Notes

1. The title of a classic song, highly relevant to the ecology of place, written by Julio Numhauser and most famously sung by Mercedes Sosa.

2. Not to be confused with the "inquiry cycle" or "cycle of inquiry" used to describe teaching strategies (Cushman, 1999).

19 *Concluding Remarks*

TAKING ADVANTAGE OF THE POWER OF PLACE

Ian Billick and Mary V. Price

Introduction

As editors we have been surprised by how engaged we became in these personal accounts of how places revealed themselves under the pressure of sustained scientific investigation. One gets few glimpses of the process by which understanding unfolds in the formal scientific literature, where insights are presented in isolated papers that, like the tiles of a mosaic, form a coherent picture only as one steps away to see them in a larger context. Even then the picture is incomplete, for the tiles rarely capture all of the bits and pieces of informal information and reasoning that went into them. Our current system of publication provides only rare opportunities for scientists to fill in the bits and pieces, to tell the story of how they came by their knowledge. This is unfortunate, because it is only after the whole story is told that we see how science really works.

The essays in this volume demonstrate that sustained place-based research is a powerful way to gain both general and local ecological understanding. The narratives also provide insight into why the ecology of place works. In this concluding chapter we recapitulate by reviewing several themes that run through the case studies, pointing out major insights that could not have been made with any other approach. We then suggest ways, beyond the types of investment discussed in the last part, for scientists and society at large to promote the acquisition and use of ecological knowledge by taking advantage of the power of place.

Recapitulation

No one would dispute that the place-based research programs described in this book have had significant conceptual impact on ecology. Findings from a number of them are familiar from textbooks and popular books as well as the primary literature; they have attracted prestigious awards; and they have stimulated major new research areas. There is no way we can do justice to the wealth of knowledge these programs have produced in this concluding essay; all we can do is pick examples to illustrate a few major points.

In chapter 1 and our introductions to the parts of this book, we suggest that place-based research is long-term and field-based, that it involves an iterative process of model building, that it integrates diverse types of information during model development and evaluation, and that it benefits from comparison of well-studied sites as well as from formal or informal interdisciplinary exchange.[1] What insights have been spawned by these elements of the ecology of place?

Insights resulting from long-term study

Long-term study is obviously needed to amass the longitudinal data on long-lived organisms or genetic lineages that are required to answer certain questions. Without it Armitage (chapter 7) could not have discovered that marmot social behavior reflects a strategy that maximizes individual direct fitness, and Grant and Grant (chapter 6) could not have assessed the strength of inbreeding depression or the role of hybridization in species evolution.

A long time series is also absolutely essential to document ecological change and to recognize unusual events. Hubbell (chapter 15) needed data on community dynamics to test alternative theories for tropical tree diversity, and Paine et al. (chapter 11) would not have been able to detect potential effects of global change on intertidal communities if they had not monitored seawater pH over a decade. Long-term demographic data allowed Estes (chapter 8) to identify the potential for shifts from bottom-up to top-down control of sea otter populations, and study of the thistle-herbivore system for some time before invasion of an introduced weevil allowed Louda and Higley (chapter 14) to see the consequences of weevil movement onto native thistles.

A long time spent with a system also presents opportunities to take advantage of serendipity. Grant and Grant (chapter 6) learned that the consequence of episodic selection is contingent on environmental conditions, because they happened to be present during sequential droughts.

Insights resulting from model building

Models are used in place-based research in several different ways. As chapters by Estes (chapter 8), Peckarsky et al. (chapter 9), Ostfeld and Jones (chapter 10), Paine et al. (chapter 11), Curtin (chapter 12), and Hubbell (chapter 15) demonstrate, place-based research often generates a qualitative conceptual model that is used to explain system properties. As new observations suggest additional model components, the model is modified and evaluated for its ability to predict an expanded set of qualitative or quantitative observations. This sort of modeling approach is a powerful way to understand complex systems whose properties are determined by multiple interacting and contingent factors that act on a variety of spatial and temporal scales.

Several case studies in this book also made quantitative use of models. Contributors evaluated general theory by translating it into a model tailored to the particular system. This then allowed the research groups to test general theory by making quantitative predictions for their particular systems. Without modeling alternative diversity-generating processes quantitatively, Hubbell (chapter 15) would not have been able to conclude that although density dependence occurs, it acts on too small a spatial scale to maintain the observed tree diversity; or that niche-assembly processes cannot account for temporal dynamics in species composition of the forest, and that forest dynamics instead are best fit by a "dispersal assembly" model. Similarly, Paine et al. (chapter 11) used multispecies models to evaluate the strengths of direct-versus-indirect interactions, and the role of disturbance and successional dynamics in producing observed patterns of species abundance. Peckarsky et al. (chapter 9) were able to evaluate the relative importance of direct mortality through consumption versus indirect prey behavioral responses for the observed effects of trout on stream insect populations only by incorporating both processes into a prey population model. Ostfeld and Jones (chapter 10) could not account for the temporal dynamics of gypsy moths and their primary mouse predators until they embedded that pairwise interaction in a larger interaction-web model that included temporally pulsed acorn production by oaks as well as feedbacks involving other oak consumers.

Insights resulting from integration of diverse information

Place-based research takes advantage of diverse kinds of information to suggest hypotheses, design and interpret experiments, evaluate models, and ask new questions. Attention to stray natural history anecdotes, data about environmental context or history, or seemingly anomalous results frequently leads to novel insights. Virtually all of the case studies described in this volume suggest that natural history is critical for knowing such things as which species are interacting, the nature of the interaction, and the scale on which the interaction operates. Ostfeld and Jones (chapter 10) relied heavily on knowledge of history to understand the origins and possible future of the forest interaction web under continued human influence; and Estes (chapter 8) relied on knowledge of the history of the fur trade to understand where to seek and how to interpret inter-island variation in kelp forests or terrestrial vegetation. Without information about the ecological context of finch populations over a series of drought episodes, Grant and Grant (chapter 6) would not have realized that evolutionary responses of finch populations to drought depend on characteristics of their seed resource, which in turn reflect long-lasting consequences of previous episodes. Louda and Higley (chapter 14) realized that increased leaf sugar concentrations associated with plant physiological response to stress can mediate levels of herbivory only because they attended to the casual observation that herbivory was greater in the sun than in the shade, because they measured leaf nitrogen and sugar as well as defensive chemicals, and because they followed up on the odd result that herbivory was not related to nitrogen.

Insights resulting from comparison of well-studied sites

A number of insights have arisen from attending to similarities and differences of well-studied sites. Pulliam and Waser (chapter 4) point out the potential that cross-site comparison—of niche expression, in their case—holds for understanding the fundamental processes that govern the distribution and abundance of species. Rundel (chapter 5) credits the emergence of a profitable focus on plant functional groups and their role in structuring communities to the comparison of similarities and differences in climate, disturbance, soils, and phylogenetic history of Mediterranean-climate regions. Finally, Estes (chapter 8) could not have unraveled the role of mammalian carnivores in structuring whole ecosystems without taking advantage of inter-island variation in their abundance that was due to historical accident rather than intrinsic variations in their ecological properties.

Insights resulting from interdisciplinary collaboration

A theme that runs through the narratives in this book is the value of formal or informal interdisciplinary exchange or collaboration, including that between empiricists and modelers. To develop their current understanding of the forest interaction web, Ostfeld and Jones (chapter 10) recruited collaborators with diverse taxonomic expertise and various empirical and modeling skills. Hubbell's forest dynamics project (chapter 15) involves the efforts of taxonomists; demographers; physiological, population, and community ecologists; cartographers; soil scientists; and theoreticians. Louda and Higley's insights into the nature of plant-herbivore interactions (chapter 14) rested on dialogue between physiologists, entomologists, and ecologists.

A more subtle benefit of collaborative place-based research is that it brings a diversity of perspectives as well as skills. Several narratives provide examples in which overcoming the bias of single investigators led to novel

TABLE 19.1. Multiple elements contribute to insights in place-based research. We have chosen one or two examples of insights from case-study chapters, suggesting elements of the ecology of place approach that contributed to them in a significant way.

Author (chapter number)	Insight	Contributing elements				
		Long-term study	Iterative model building[1]	Diverse types and sources of information	Comparison of well-studied sites	Interdisciplinary exchange or collaboration
Rundel (5)	Community convergence reflects the relative success of plant functional groups			x	x	x
Grant and Grant (6)	Hybridization can be important to evolution	x	x			x
Grant and Grant (6)	Outcome of repeated selection episodes is contingent on environment	x		x		x
Armitage (7)	Social behavior maximizes individual direct fitness		x	x		x
Armitage (7)	Social behavior molds population dynamics	x	x	x		x
Estes (8)	Human exploitation of mammalian carnivores has cascading landscape effects	x	x	x	x	x
Ostfeld and Jones (9)	Masting in oaks molds a forest interaction web	x	x	x	x	x
Peckarsky et al. (10)	Nonconsumptive predator effects on prey communities are strong in open systems		x	x		x
Paine et al. (11)	Single, "keystone," species can have profound impacts on community composition		x	x	x	x
Paine et al. (11)	Community dynamics in sessile communities reflects patchy post-disturbance succession		x	x		x
Curtin (12)	Spatial scale of ecological processes matters for management.	x	x	x	x	x
Louda and Higley (13)	Plant–herbivore interactions are mediated by physiological stress, not just defensive chemistry		x	x		x
Louda and Higley (13)	Flower-feeding insect guilds are not saturated		x	x		
Hubbell (14)	"Dispersal assembly," not "niche assembly," accounts for tropical tree diversity	x	x	x	x	x

[1] Sustained but not necessarily long-term study that involves successive rounds of model building and evaluation

insights. Paine et al. (chapter 11) originally viewed environmental monitoring as not worth the effort and did not initiate monitoring until new investigators pushed for it; the resulting water quality data eventually provided unantici-pated insights into some consequences of global change for marine intertidal systems. Similarly, new perspectives were the impetus for Peckarsky et al. (chapter 9) to broaden their initial stonefly-mayfly focus to higher and lower trophic levels, to initiate experimental studies of insect responses to fish chemical cues, and to think on a landscape scale.

The intersection of multiple elements in place-based research

Most place-based research programs incorporate more than one of the ele-ments of ecology of place. We illustrate this point in table 19.1, which extracts examples of insights from chapter narratives and identifies the elements that made them possible. In most cases, we can easily identify multiple contrib-uting elements.

Taking Advantage of the Informative Power of Place

Despite the demonstrable contributions of sustained research in single places to ecological understanding, we do not take full advantage of this power of place either in our management of natural lands or in our science. Aldo Leopold's efforts to integrate science with management (Waller and Flader, chapter 2) notwithstanding, land management programs still too of-ten lack a strong science component, and little use is made of the wealth of natural history observations that managers accumulate during their time in the field (e.g., Curtin, chapter 12). Furthermore, the modern scientific en-terprise makes place-based research programs difficult to sustain. Funding for both basic and applied science generally comes via piecemeal support for projects that are evaluated on the basis of individual merit, not contribution to a larger effort. There are few funding mechanisms to produce the sort of synthetic understanding of places illustrated in this book. There also is little infrastructure to support the acquisition of natural history information that is foundational to understanding a place, but often perceived as tangential or old-fashioned (e.g., Ostfeld and Jones, chapter 10; Paine et al., chapter 11). An emphasis on productivity (especially publication rate) and novelty as criteria for career advancement discourages long synthetic papers, as does the scar-city of publication outlets for such work. And finally, a push toward special-ization makes it difficult to forge the interdisciplinary communication and collaboration that are such important elements of place-based research.

Many of these problems could be alleviated if we gave "place" more cen-tral standing in our science. By this, we do not just mean increasing the sort of investment mentioned in part 5—investment in maintaining relatively undisturbed natural places, field stations, place-based data management systems, and educational programs. Investing in these things certainly increases our capacity for place-based research. But we cannot fully use that capacity to harness the power of place without some fundamental changes in how we assign priorities for land management and ecological research. The following are some ideas for change.

Often natural lands are managed for values such as recreation, cattle production, wildlife, or timber—not for knowledge. Incorporating a greater research component into the management prescription for natural lands would improve our ability to extract ecological understanding from them. As an example, consider wilderness areas. In the United States these are managed so strongly for recreational and aesthetic values that anything other than strictly observational research is well-nigh impossible. This is ironic since Aldo Leopold (1949), who was instrumental in establishment of wilderness areas, recognized that "recreation is not their only, or even their principal, utility," and advocated "wilderness . . . as a laboratory for the study of land-health." Rangelands and managed forests have similar potential to inform us, if we let them. The time their managers spend in the field could contribute much more to natural history knowledge if programs similar to the park guard inquiry described by Feinsinger et al. (chapter 18) were implemented, and the resulting insights captured in local databases. Applied management projects have much potential to test ecological ideas if, as Krebs (chapter 13) notes, they "have a firm foundation in ecological ideas and, if management actions fail, time and money are made available to find the source of the failure." Expanding collaborative science programs of the sort described by Curtin (chapter 12) that bring scientists, managers, and resource users together, or that deploy citizen scientists, would extract more knowledge from our natural places.

Changes in funding priorities also are needed to facilitate ecology of place. Although entire place-based research programs exhibit many attributes that are highly valued by funding agencies, they are rarely evaluated as a whole. Instead, their component projects are considered individually, independent of the broader effort. Under such conditions, questions that are important for building local understanding can easily be seen as globally uninteresting. Supplementing existing programs with ones that place greater weight on a proposed project's contributions to the understanding of a well-studied place would make it easier to build upon previous work, to incrementally improve the conceptual model of a system, to bring in diverse perspectives from visiting scientists, to promote collaboration between modelers and empiricists, or to accumulate natural history information about a site. A cost-effective way of achieving this would be to expand current programs that already provide research experience for undergraduate students to also include projects of graduate students, postdoctoral fellows, and assistant professors that supplement ongoing place-based research. Such programs would not only contribute to local and general understanding, but would also train young scientists in collaborative, place-based science.

Another factor that affects funding for place-based science is the common perception that generalization can only come through extensive replication. This perception devalues projects that attempt to generalize from one or a few intensively studied sites, relative to those that generalize from many, superficially studied sites. Greater investment in programs to promote a synthetic understanding of single places, rather than a macroecological pattern analysis, would be effective in promoting place-based approaches. Such programs might include support for within-site synthesis,[2] cross-ecosystem and cross-discipline working groups, symposia, or comparative studies similar to those described by Rundel (chapter 5) for Mediterranean ecosystems.

While funding can provide strong incentives for ecology of place, deeper changes in the culture of science are also needed to take full advantage of the approach. All too often, the primary criterion used for professional advancement in science is the rate of publication of primary research articles. This presents a disincentive to achieving the unique product of place-based research—synthetic understanding of how a system works—because synthesis is retrospective by nature and does not generate novel discoveries. Emphasis in career evaluations on conceptual contributions, rather than publication number or novelty, would help to combat this trend. But evaluating the conceptual contributions of place-based research can be problematical, because it is multidisciplinary and synthetic. Conceptual contributions are often judged in terms of specific disciplinary areas; understanding how multiple conceptual threads together determine properties of single systems may not be seen as novel or as carrying any one field forward. This difficulty in evaluating contributions made by place-based research can affect what gets published as well as professional advancement.

Although journal outlets do exist for reviews and synthetic papers, they target a different sort of synthesis than what emerges from a focus on place. Most reviews and synthetic pieces take a meta-analytical approach that addresses particular broad conceptual issues or subfields by examining trends across studies. It's not clear that place-based synthesis fits this model, because it integrates studies done at one place to arrive at an overview and explanation for the properties of the system. Virtually all of the place-based syntheses that we are aware of have been published as book-length monographs or edited volumes. Examples include Likens and Bormann (1995),

Grant and Grant (2008), Waller and Rooney (2008), and McDade et al. (1994). Because writing a book is a major undertaking (as we editors can attest!) and books are expensive, insights from place-based research would be more readily available if they could be published as shorter monograph-style syntheses in journals.

One final idea came to us as we contemplated the personal narratives in this book and looked back over our own scientific journeys. For us, as for all too many students even today, learning how to do research was like learning how to swim by being tossed into deep water. We finally figured things out, but only by trial and error after a lot of flailing about. Students who have the good fortune to work at a site with other investigators benefit as apprentices do, from observing how experienced researchers tease out the secrets of place. With today's Internet technology, that benefit could easily be spread more widely by posting accounts like those in this book on the Web sites of professional societies.

Taking Advantage of the Motivational Power of Place

Throughout this volume we have emphasized the power of place to advance useful scientific understanding. This is not, however, the only value of place. Ecological understanding is not an end but rather a means for humans to learn how to live sustainably; and as Aldo Leopold acknowledged in the "world of wounds" passage quoted by Waller and Flader (chapter 3), people will act on knowledge only if they are attuned to and care about the world around them.

The key to Leopold's success in inspiring generations to commit to conservation and restoration was his recognition that place motivates. Leopold used the evocative power of natural places to make people want to walk outdoors with eyes open, to convince us that science complements rather than detracts from nature's aesthetic and spiritual and recreational values, and to motivate scientists and citizens alike to take action to protect what he called "land health." As Waller and Flader note, he set the example of a scientist "willing to roll up his sleeves to work on practical, community-based efforts in conservation and to engage the wider public in conservation issues."

It is clear from the narratives in this book that this motivational power of place is what draws ecologists back into the field year after year. That same power can draw citizens and scientists together in the collaborative effort that is needed to solve environmental problems. The trick, as Leopold recognized in his concept of the land ethic, is to build a sense of place that sees humans as part of the ecological system, not separate from it.

References

Grant, P. R., and B. R. Grant. 2008. *How and Why Species Multiply: The Radiation of Darwin's Finches.* Princeton University Press, Princeton, NJ.

Leopold, A. 1949. *A Sand County Almanac and Sketches Here and There.* Oxford University Press, New York.

Likens, G. E., and F. H. Bormann. 1995. *Biogeochemistry of a Forested Ecocsystem.* Second edition. Springer-Verlag, New York.

McDade, L. A., K. S. Bawa, H. A. Hespenheide, and G. S. Hartshorn, eds. 1994. *La Selva: Ecology and Natural History of a Tropical Rainforest.* University of Chicago Press, Chicago.

Waller, D. M., and T. P. Rooney, eds. 2008. *The Vanishing Present: Wisconsin's Changing Lands, Waters, and Wildlife.* University of Chicago Press, Chicago.

Notes

1. We use the term "interdisciplinary" in a broad sense, to include different approaches within a discipline (e.g., theoretical versus empirical ecology) as well as different fields (e.g., physiology versus behavior, botany versus zoology).

2. An example of such a program is the U.S. National Science Foundation's OPUS (Opportunities for Promoting Understanding through Synthesis) program, which supports synthesis of a body of related research projects conducted by a single individual or group of investigators over an extended period.

CONTRIBUTORS

Paul A. Aigner
Donald and Sylvia McLaughlin Reserve
Lower Lake, California

Aylin Alegre
Biokarst and Instituto de Ecología y
Sistemática
Havana, Cuba

Samara Álvarez
Parque Nacional El Cajas, Ecuador

J. David Allan
School of Natural Resources and
Environment
University of Michigan
Ann Arbor, Michigan

Kenneth B. Armitage
Department of Ecology and Evolutionary
Biology
The University of Kansas
Lawrence, Kansas

Ian Billick
Rocky Mountain Biological Laboratory
Crested Butte, Colorado

Maikel Cañizares
Biokarst and Instituto de Ecología y
Sistemática
Havana, Cuba

Geovana Carreño
Instituto Boliviano de Investigacion Forestal
Facultad de Ciencias Agricolas
Santa Cruz, Bolivia

Charles G. Curtin
MIT-USGS Science Impact Collaborative
Department of Urban Studies and Planning
Massachusetts Institute of Technology
Cambridge, Massachusetts

Félix Daza
Wildlife Conservation Society
Merida, Venezuela

James A. Estes
Department of Ecology and Evolutionary
Biology
University of California
Santa Cruz, California

Peter Feinsinger
Department of Biological Sciences
Northern Arizona University
Flagstaff, Arizona

Mireiza Figuera
Wildlife Conservation Society
Merida, Venezuela

Susan Flader
Department of History
University of Missouri
Columbia, Missouri

Lainet García
Biokarst and Instituto de Ecología y
Sistemática
Havana, Cuba

B. Rosemary Grant
Department of Ecology and Evolutionary
Biology
Princeton University
Princeton, New Jersey

Peter R. Grant
Department of Ecology and Evolutionary
Biology
Princeton University
Princeton, New Jersey

Leon G. Higley
Department of Entomology
University of Nebraska
Lincoln, Nebraska

Stephen P. Hubbell
Department of Ecology and Evolutionary
Biology
University of California
Los Angeles, California

Center for Tropical Forest Science
Smithsonian Tropical Research Institute

Clive G. Jones
Cary Institute of Ecosystem Studies
Millbrook, New York

Sharon E. Kingsland
Department of History of Science
and Technology
Johns Hopkins University
Baltimore, Maryland

Catherine E. Koehler
Donald and Sylvia McLaughlin Reserve
Lower Lake, California

Charles J. Krebs
Department of Zoology
University of British Columbia
Vancouver, British Columbia
Canada

Edith Lanz
Wildlife Conservation Society
Merida, Venezuela

Svaťa M. Louda
School of Biological Sciences and
Department of Entomology
University of Nebraska
Lincoln, Nebraska

Angus R. McIntosh
Biological Sciences
University of Canterbury
Christchurch, New Zealand

Andrew Noss
Wildlife Conservation Society
Santa Cruz, Bolivia

Richard S. Ostfeld
Cary Institute of Ecosystem Studies
Millbrook, New York

Robert T. Paine
Department of Biology
University of Washington
Seattle, Washington

Barbara L. Peckarsky
Department of Zoology and Entomology
University of Wisconsin
Madison, Wisconsin

Catherine A. Pfister
Department of Ecology and Evolution
University of Chicago
Chicago, Illinois

Mary V. Price
Department of Biology
University of California, Riverside
Riverside, California

H. Ronald Pulliam
Odum School of Ecology
University of Georgia
Athens, Georgia

Edmundo Rivera
Rescate en Tus Manos
Santa Cruz, Bolivia

Alejandra Roldán
BIOTA
La Paz, Bolivia

Philip W. Rundel
Department of Ecology and Evolutionary
Biology
University of California
Los Angeles, California

Brad W. Taylor
Biological Sciences
Dartmouth College
Hanover, New Hampshire

Donald M. Waller
Department of Botany
University of Wisconsin
Madison, Wisconsin

Nickolas M. Waser
Department of Biology
University of California, Riverside
Riverside, California

J. Timothy Wootton
Department of Ecology and Evolution
University of Chicago
Chicago, Illinois

INDEX

Numerals in italics indicate tables and figures.